CAMPBELL

CHEMISTRY APPLIED

TO THE MANUFACTURE OF

SOAP AND CANDLES

A THOROUGH EXPOSITION OF THE PRINCIPLES AND PRACTICE OF THE TRADE, IN ALL THEIR MINUTIÆ BASED UPON THE MOST RECENT DISCOVERIES IN SCIENCE AND IMPROVEMENTS IN ART

Elibron Classics
www.elibron.com

Elibron Classics series.

ISBN 1-4021-6406-8 (paperback)
ISBN 1-4021-0349-2 (hardcover)

This Elibron Classics Replica Edition is an unabridged facsimile of the edition published in 1847 by Carey and Hart, Philadelphia.

CHEMISTRY APPLIED

TO THE MANUFACTURE OF

SOAP AND CANDLES.

A THOROUGH EXPOSITION OF THE PRINCIPLES AND PRACTICE OF THE TRADE, IN ALL THEIR MINUTIÆ, BASED UPON THE MOST RECENT DISCOVERIES IN SCIENCE AND IMPROVEMENTS IN ART.

BY CAMPBELL MORFIT,

PRACTICAL AND ANALYTICAL CHEMIST.

ILLUSTRATED WITH

ONE HUNDRED AND SEVENTY ENGRAVINGS ON WOOD.

FEED THE FLAME.

PHILADELPHIA:
PUBLISHED BY CAREY AND HART.
1847.

PHILADELPHIA:
T. K. AND P. G. COLLINS, PRINTERS.

PREFACE.

THE limited sources of information relative to the practical details of the art of soap and candle making, the still more meagre explanations of the chemical changes occurring in the processes pursued in these arts, the scattered and confused state in which this information exists, have all created a desideratum among those who pursue these trades, which the present work is designed to supply.

Excepting a few pages in the works of Parke, Ure and Parnell, and a small book of remote date by an Edinburgh operative, we have nothing, in English, relative to the important manufacture of soap; while, in French, save the treatise by Dumas, there has been but little recently written. The small manuals of Thillaye and Normand, though containing much that is useful, are yet very imperfect, and in acknowledging our indebtedness to both for valuable and frequent assistance, we have freely taken the liberty of so modifying what has been drawn from them as to make it conformable to the improvements which chemical science, in its bearing upon the subject, has lately undergone. Combining observation and practical experience with theoretical knowledge, the author, in presenting older facts worthy of repetition and a greater amount of new precepts, has adopted such arrangement and language in their elucidation as will be clearly intelligible to any workman of ordinary comprehension.

In order to remedy the defect which has continued the savonnier a mere empirical operative, and withheld the means of acquiring accomplished skill in his art, we have prepared a systematized mass of information to which he can refer for relief, in rectification of errors produced through ignorance, and for instruction qualifying him as a finished adept in the principles and practice of his trade. There is an accompanying explanation to every process, and though the work embraces the newest improvements, still nothing of worth has been left unsaid because of its antiquity.

In those instances where the value of one foreign word is equal to two or more in our own language, we have given the preference to the French *technical term;* but, to satisfy any objections of the fastidious to these innovations, as they might be called, we have, at the same time, given their English signification. Thus, we have preferred the more unexceptionable expression *savonnier* to that of *soap-boiler,* which is stigmatized, however undeservedly, as harsh and opprobrious, and rather reproachful to the worthy operatives in this art.

Basing our facts upon the opinions of the most authoritative referees, we have supported them with a larger proportion of original information, and so classified the whole as to make it applicable to the thorough instruction and enlightenment of the soap and candle manufacturer.

CONTENTS.

CHAPTER I.

CHAPTER II.

CHAPTER III.

CHAPTER IV.

CHAPTER V.

CHAPTER VI.

CHAPTER VII.

CHAPTER VIII.

CHAPTER IX.

CHAPTER X.

CHAPTER XI.

CHAPTER XII.

CHAPTER XIII.

CHAPTER XIV.

CHAPTER XV.

CHAPTER XVI.

CHAPTER XVII.

CHAPTER XVIII.

CHAPTER XIX.

CHAPTER XX.

CHAPTER XXI.

CHAPTER XXII.

CHAPTER XXIII.

CHAPTER XXIV.

CHAPTER XXV.

CHAPTER XXVI.

CHAPTER XXVII.

CHAPTER XXVIII.

CHAPTER XXIX.

CHAPTER XXX.

CHAPTER XXXI.

CHAPTER XXXII.

CHAPTER XXXIII.

CHAPTER XLII.

CHAPTER XLIII.

CHAPTER XLIV.

CHAPTER XLV.

CHAPTER XLVI.

CHAPTER XLVII.

CHAPTER XLVIII.

CHAPTER XLIX.

CHAPTER L.

CHAPTER LI.

CHAPTER LII.

CHAPTER LIII.

CHAPTER LIV.

ERRATA.

Page 31, 5th line from bottom, for 59 read 60.

41, 11th " for *herbaceii* read *herbaceous.*

48, 19th " omit *forms.*

" " for *from* read *form.*

50, 3d line from top, for *sold* read *sold for.*

68, 14th " bottom, for *chemist* read *alchemist.*

81, 12th " for *is* read *are.*

83, 19th " for 12° read 12° C.

84, 20th " omit *Pinus abies.*

133, 9th " for IV. read XLI.

142, 20th " for *phocerine* read *phocenin.*

177, 9th " for *make* read *make it.*

181, 7th " for *steam* read *steam conduit*

197, 29th " for *soda* read *sodium.*

222, 31st " for *desired* read *required.*

243, 23d " for *fibrin* read *fibre.*

260, 13th " for *soap* read *soaps.*

265, 21st " for *this* read *this latter.*

265, 24th " for *absorbs* read *absorb.*

300, 11th " for δερμος read θερμος.

349, 3d " for *causes* read *cause.*

395, 8th " for *similar to* read *similarly as*

404, 12th " omit the *comma* after *wood*

426, 12th " for *every fat* read *all fats.*

430, 22d " for *preceding* read *following.*

432, 5th " for *level* read *bevel.*

432, 6th " " "

440, 12th " for *rid* read *ridded.*

452, 19th " for *is* read *are.*

483, at top line, for *which* read *and.*

484, 9th line from top, for *treadle* read *trestle.*

CHEMISTRY APPLIED

TO THE MANUFACTURE OF

SOAP AND CANDLES.

CHAPTER I.

INTRODUCTORY REMARKS.

THERE are perhaps no other two articles bearing so importantly upon the household economy as those of soap and candles: they are truly materials of necessity, and as such are indispensable to the wants of both the wealthier and poorer classes of the community. The consequent extent of its trade should long since have procured a more liberal effort in furthering the improvement of the quality of soap, but as that proper spirit of emulation and enterprize, promotive of such desirable ends, has as a general thing been lacking among the savonniers of this country, the consumer has realized but little from the ingenuity and skill of the native manufacturer, and scarcely more from the adoption of foreign improvements. The fact may be ascribed partially to the apathetic indifference of operatives to a knowledge of the principles upon which their art is founded, or perhaps an inability to apply that knowledge; and in a measure to the meagre sources whence information can be derived. In either case the ignorance is lamentable, and in these enlightened days, when the rapid strides of improvement are overtaking every branch of manufacture, is a sad comment upon the culpable inactivity of those practising an art positively scientific, being truly chemical in its nature, and which, by well-directed observation, some skill and a slight modicum of intelligence, could long since have been made to emerge from the darkness in which it has slumbered for years, and been raised to an eminent state of improvement. But there is not that commendable strife for superiority among the artisans of the United States which animates the exertions of the tradesmen of Europe, else the savonnier, instead of continuing merely a "soap boiler," would become an enlightened man in regard to his occupation; and the trade which he follows would assume, by the superior excellence and elegance of its products, an exalted position, in lieu of the degradation which, through the illiteracy of its followers, has until lately adhered to it. The dawn of a new era in the arts and manufactures of this country, impressively admonishes the intelligent craftsman, of the necessity of a speedy and energetic compliance with those requisites tending to a perfect mastery of his trade in its present state of perfection, and of an ability not only to keep pace with, but to give an impetus to its progressive advance. A select few have caught the inspiration of the times, and by their active intelligence, have already contributed an Herculean assistance in resuscitating their profession, which, so lately as five years since, was, comparatively speaking, in a most deplorable slough of stagnancy. To lend a share of aid in dissolving this cloud of ignorance which still mystifies the majority of a worthy class of manufacturers, is the author's desire; and this book is written with a view of qualifying the reader to become not only a practical operative, but an accomplished artisan, thoroughly initiated in the intricacies of his trade, able to explain the rationale of its processes, and competent to carry them into practical operation.

The first mention of the use of soaps is by Pliny, who attributes its discovery to the ancient Gauls, from whom a knowledge of its manufacture passed to the Romans,* and thence to other nations. In the United States at the present day, the commerce in soap is immense, there being, independent of the very extensive home consumption, an export trade of nearly a half million of dollars. But the great mart for this article is France, where the skilful ingenuity of the educated savonniers has done what is a natural consequence of education, promoted the trade to a dignified position among the arts, and procured for its products a preferred consumption throughout the civilized world. In toilet soaps especially, she is far in advance of any other nation, whilst the commoner species have maintained a superiority at least one grade higher than the same sorts made elsewhere. The reason is obvious. A French workman is not a mere automaton. He applies his mind as well as his hands to his occupation, and having previously fortified himself with a thorough knowledge of its principles, can ply the art with every success, and feels himself not only competent to imitate any improvement of an ambitious competitor, but to return the compliment by an equally valuable evidence of skill. So it is in this country, and to a few such individuals is solely attributable the recent melioration in the quality of eastern made soaps; that of the western factories, through the incompeteney of their directors, still possessing the characteristics of inferior soap.

The professional ability of such men as Hull, of New York, and Hyde, of Baltimore, the evidence of whose

* Miss Starke, in her letters from Italy, records the discovery of a complete soap-making shop in the ruins of Pompeii, destroyed by the eruption of Vesuvius, in the 79th year of the Christian era.

competency is in the excellent quality of their soaps; of I. D. Brown, who contributed his judgment and experience in the arrangement of that elegant nonparcil of soap laboratories, owned by Coffin and Landell, in Philadelphia; of J. R. Graves, the successful competitor for the medals of the Franklin Institute, in 1844, and 1845, and of E. Roussel, and Jules Hauel, the manufacturers preeminent in this country of toilet soap, has not only advanced this art to a high degree of perfection, but enhanced its importance and increased its usefulness; whilst their success serves as an encouraging example to their more tardy and uninformed co-savonniers.

You, then, of the trade, who till now have been dilatory and indifferent, rouse up both your energy and ambition, and fail not to profit by the good fortune with which the enterprising ingenuity of the above-named operatives has been favored. The chances of success are equal for all of you; it is true, a well applied industry does much in the accomplishment of good results, but it must be backed by a thorough intelligence as to the business, and this intelligence is communicated in the forthcoming pages. It remains with you whether its usefulness shall redound to your individual interest as well as to that of the public generally.

All soaps, whether of the soft or hard, toilet or domestic species, owe their cleansing influence to what is termed a *detergency*, or, in other words, a power of rendering soluble, in water, the dirt of the clothes and the skin, for washing which, it is almost exclusively used. Both of the alkalies, potash and soda, are detersive in their action, and the former, used by the Hebrews, Egyptians, and Greeks, hastened the cleansing as well as the destruction of their clothes; whilst the Hindoos, taking the Ganges for a wash-tub, made up for the absence of alkali and soap by an extra exertion, much labor and loss of time. But, as the corrosive action of free alkali injures the fibre of stuffs, an effect experienced even in our day, when soda is thrown into the wash water, this action must be remedied by substituting soap, which, by its slight excess of alkali, rendering soluble in, and miscible with water, all the dirty grease of the clothes and oily exudations from the pores of the skin, is at the same time detersive itself; for though composed of oil and alkali definitely united, it still possesses the influence of the latter without any of its hurtfulness. The slight excess of alkali in the soap uniting with the grease, causes its solubility in water, and also the suspension therein of all the dusty particles which through its agency were adherent to the clothes and skin. The application of fuller's earth and magnesia is based upon a different principle, viz., that of absorption, by which the greasy spots are abstracted and thus rendered more easily removable by mere mechanical rubbing, there being no chemical effect produced through their agency.

The quality of the water, too, is quite important to a perfect cleansing of clothes, for those called "*hard*" waters, by decomposing the soap or rendering it insoluble, impair its efficacy. For these waters the "marine soap" is most appropriate, but much the better way would be to use rain water, which is very pure, or still more preferable, distilled water, which can be caught from the exhausting pipes of neighboring steam-engines.

The quantity of soap consumed by a nation, says Liebig, would be no inaccurate measure whereby to estimate its wealth and civilization. Of two countries with an equal amount of population, the wealthiest and most highly civilized will consume the greatest weight of soap. This consumption does not subserve sensual gratification nor depend upon fashion, but upon the feeling of the beauty, comfort, and welfare attendant upon cleanliness; and a regard to this feeling is coincident with wealth and civilization. The rich in the middle ages concealed a want of cleanliness in their clothes and persons under a profusion of costly scents and essences, while they were more luxurious in eating and drinking, in apparel and horses With us, a want of cleanliness is equivalent to insupportable misery and misfortune.

"Soap is one of those articles few in number, which are entirely consumed by use, leaving no residue of worth, as soap-suds have not as yet been profitably applied to any practical purposes; therefore, as its money value is continually disappearing from circulation, it requires a constant renewal," and hence the importance of the trade which embraces its manufacture, and the necessity of propelling it as fast as possible to perfection, both for the good of mankind and the influence its advance will occasion to the progress of other similar arts.

CHAPTER II.

THE DIGNITY OF THE ART AND ITS RELATIONS TO SCIENCE.

INDUSTRY is the element of success in the physical, as knowledge is power in the intellectual world, and when both are combined, they tend to the consummation of all that can be gained by the honorable enterprise of man. Every pursuit of life which connects chemical arrangement in any degree with either the powers of mechanism, or their results, partakes more or less of the character of an abstract philosophy. The workman, who, without knowing the *rationale* of his process, derives the gaudy colors from the dullest objects, and the Prussian blue from the offal of the shambles; the *Savonnier* who presents to domestic cleanliness, through all the gradations of usefulness up to the most delicate taste in the toilet, the results of his oils and his alkalies, is each a chemist; though in his limited sphere and his uninstructed course he feels himself to be a mere man of handicraft. It is for this reason that we should desire to create an emulative spirit in such vocations, so that art with knowledge as her handmaid, may lead her followers to the highest condition of their calling, and by adding instruction to labor, elevate the workman and dignify his trade.

There is no pleasure so great among those which pertain to the rewards of success as that which comes from the toil of the hands well and honestly engaged; there is none more satisfactory to the mind, than that which unites the knowledge of what to do with the act of well doing; and when these qualities are associated, and are supported by perseverance, they are the sure basis of present respectability and future independence.

In this country, happily, there are no authorized distinctions, and those that casually and necessarily exist in the social condition, proceed either from the advantages of industry or of mind. It is within the compass of every parent from the facilities afforded by our system of common schools, if not from his own means, to qualify his children for the highest stations; and it is within the probable fortune of every son to reach some of those elevations. Labor, properly directed by education, must secure reputation to the individual and benefit to the community. It is not in the learned professions that either fame or fortune is always found; neither is it political success that yields either greatness or content: those who aim at such ends must lay aside the implements of their shops, and resign themselves to the contingencies of party; for, though ultimately gained, they often destroy the happiness of the present and insure no certainty for the future. It is by continuing in the steady tracks of their calling that good men, aided by many helps from science and from study, have advanced themselves to high places among their fellow men, and are enabled thus practically to illustrate the dignity of labor and the value of education.

There is no incongruity between occupations, however manual, and a knowledge of the principles which are interwoven with the work or its products. Artisans are like travelers, though on different roads; but, in every route of life whether operative or scientific, the course is the more facile as the instructed mind is enabled to shed its light upon the pathway.

If trade is followed merely for the sake of its moderate gains, it will always keep its votaries in a dull road of antiquated forms and habitual error; but if men are taught in their earlier days, that education is the best preparative for the work-shop, they will bring with them into their future laboratories the powers of other influences than those of mere ingenuity; they will come with new levers, those of the cultivated intellect; and with the inspirations of a laudable ambition and a forecast of the future, will strike out inventions and improvements to elevate their names and benefit mankind.

The theory which accounts for the union of oils and alkalies in the formation of soaps or the decomposition of fats into their crystaline acids, is as abstruse as that which expounds some of the most refined instances of affinities. The manufacture of soaps and candles is, therefore, strictly scientific, and they who desire to advance in that art must not merely make, but make understandingly; so that each result may be an advancing lesson for their own profit, while it enables them to repay the benefits they have derived from chemistry by furnishing from their experience new lights to their fellow men.

CHAPTER III.

AFFINITY AND CHEMICAL EQUIVALENTS IN REFERENCE TO THE ART

CHEMICAL affinity is that attraction existing between bodies, disposing them to a closer connection, so that by their union the formation of new products, endowed with new properties, may result. This power of affinity forms the basis of chemical science, because it is by its practical effects that the operator is enabled to fathom the causes of phenomena by the decomposition and synthesis of natural and artificial substances. All bodies are either simple or compound;—the term *simple* being synonymous with *elementary*, and is applied to such as have never been decomposed, that is, resolved into constituent parts, and is in opposition to the expression, *compound*, given to such substances as have been, by experiment, proved to consist of two or more elementary components. Thus, for instance, the metals are elementary bodies, and when they are found in nature combined with oxygen in the form of oxides; with sulphur as sulphurets; with carbonic acid as carbonates; they have become so in the course of time by reason of their affinity for these conjunctives which they have abstracted from other bodies. They then take the name of *chemical* compounds in contradistinction to a *mechanical* compound, which is nothing more than an admixture of any several substances—the union being, as it were, a mere physical effect; thus, for example, tallow is a chemical compound, because it consists of fatty acids and a base united together in definite proportions to form salts, whereas, by melting this same body together with wax, the production of a mechanical compound is illustrated. When potassa is added to this tallow, the greater affinity of its natural acids for that body causes its decomposition or separation into its constituent parts; whereas, by division with a knife or other mechanical means, the smallest particles separated still contain the original proportion of acids and base. It will be seen from this, therefore, that in order to a thorough understanding of chemical action generally, there must be some intimacy with the laws by which it is governed, and, as it is our place to supply this information we proceed to the duty, premising that we shall be indebted for much assistance to a very lucid article upon the subject in that valuable work, *Booth* and *Boye's Encyclopædia of Chemistry.*

"Affinity is not always exerted between bodies; thus fluorine does not combine with oxygen, but readily with metals. Carbon does not unite with mercury, but combines with oxygen, iron, &c. It is therefore elective in its character, preferring one body to another."

Thus, for instance, in making stearic acid block, when, in the first step, lime is added to the melted suet, the fatty acids which the latter contains by reason of a greater affinity or preference for the lime, leave their natural base and combine therewith, and thus the original suet consisting of an oleo-stearate of glycerin is transformed into an oleo-stearate of lime, the natural base glycerin, at the moment of its elimination, becoming dissolved in the water which is added to facilitate the action. This is an instance of what is termed *simple elective affinity*, by the influence of which an original combination is broken up and a new one formed; it is, however, applicable but to those circumstances where a third body in its action upon a compound has a stronger affinity for one of its ingredients than they have for each other; for example, the process of rendering soda ash caustic, or, in other words, depriving it of carbonic acid, is due to this power, for, as will be learned hereafter, soda ash consists chiefly of carbonic acid, and soda united as carbonate of soda. Now when lime is added, that being a stronger base, breaks up the original union, and by reason of its greater affinity, abstracts the carbonic acid from the ash, becomes itself carbonated, and in that state is insoluble, whilst the free soda entirely devoid of acid is obtained in solution. The following diagram will illustrate the changes; the water added in the process serving as the solvent of the soda.

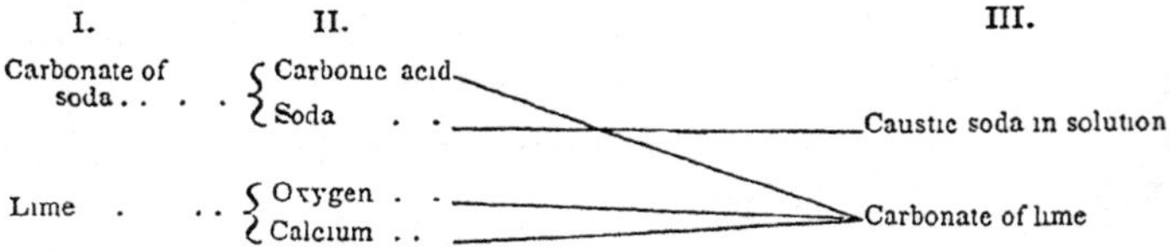

When there are two compounds brought into contact, the number of affinities are increased; thus, for instance, let us take common soft soap which is an oleo-stearate of potassa, and add to it a solution of acetate of lead;—here then are four affinities in action, exhibiting an instance of what is called *double elective affinity*, or double decomposition, as it is sometimes styled, and explicable as follows:

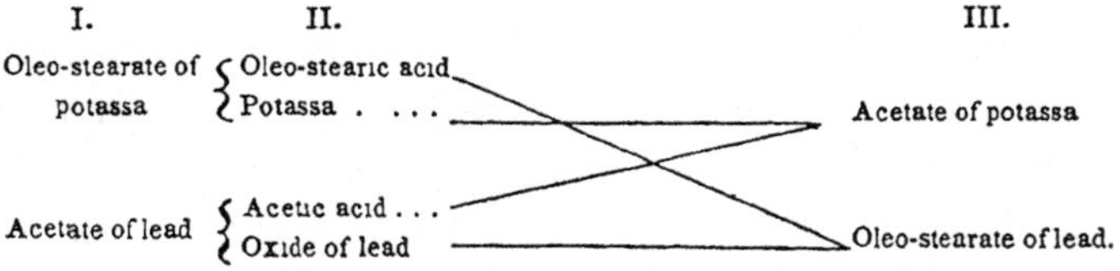

The first column designates the substances in their riginal state; the second, their composition or constituency; and the third, the results and the title of the compound newly formed. To explain further, the oleo-stearic acid of the soap having a preference for the base of the acetate of lead, leaves its own and goes over, and forms with it an oleo-stearate of lead, whilst that base which it abandoned takes up with the acetic acid, similarly situated, and forms acetate of soda, which, unlike the other new compound, remains dissolved in the menstruum. This kind of affinity is of the utmost importance, for it is solely by its action that many compounds are formed. There are certain conditions and circumstances attendant upon combination by affinity; it must be sufficiently powerful to overcome the opposing forces of gravity, cohesion, and elasticity.

"The bodies must be in immediate contact, since the action is at insensible distances. It is rarely sufficient to pulverize two solids, however fine, to induce their chemical union, for their particles do not admit freely of motion, but by constant and rapid trituration many points of surface are brought into contact nearly at one time, and combination may ensue; in this manner sulphur and finely divided copper unite with the evolution of heat. The affinity of bodies is, therefore, promoted by everything which tends to their close approximation; in solids, by their pulverization and intermixture, this attraction residing in the ultimate particles of bodies; in gases, by their spontaneous diffusion through each other, which occasions a more complete intermixture than is attainable by mechanical means; and between liquids, or between a liquid and solid, by the adhesive attraction which liquids possess, which must lead to perfect contact, and also by a disposition of liquid bodies to intermix, of the same physical character as gaseous diffusion. Elevation of temperature has certainly often a specific action in increasing the affinity of two bodies, but it also often acts by producing a perfect contact between them, from the diffusion or vaporization of one or both bodies. Hence, no practice is more general to promote the combination of bodies than to heat them together. Analysis has shown that the composition of bodies is fixed and invariable; thus, 100 parts by weight of water are uniformly composed of 88.9 oxygen, and 11.1 hydrogen, (or 8 to 1,) whether the water be from rains or springs, or formed artificially by the burning together of oxygen and hydrogen in any proportion. If we observe the *measure* of the two gases, that is, the bulk which 88.9 and 11.1 parts respectively occupy, we are struck with the remarkable fact, that there is exactly twice as much hydrogen as oxygen. Nature here points out a very simple proportion where two gases unite; but the greater number of substances are only known in a solid or liquid state, where their union by measurement cannot as well be observed. The former is called combination by *weight*, the latter by *volume;* and, although the composition of a body is usually expressed by weight, yet we are frequently led to the result by an examination of their combination by volume. The numbers expressing these two modes of combination are very different. If we examine several bodies containing one element in common, we find, what may be inferred from the composition of water, that that element unites with very different quantities of other elements. Thus, in the following table the ratio of oxygen to the other constituent is very different:

Composition of Oxides.

Water.		Oxide of Copper.		Oxide of Zinc.		Oxide of Lead.		Oxide of Silver.	
Oxygen	88.9	Oxygen	20.2	Oxygen	19.1	Oxygen	7.2	Oxygen	6.9
Hydrogen	11.1	Copper	79.8	Zinc	80.9	Lead	92.8	Silver	93.1
	100		100		100		100		100

"But the relation between the oxygen and the other constituent in these oxides will be seen more distinctly by stating their composition in such a way as to have the oxygen expressed by the same number in every case, or made equal to 8 parts. Thus,

Water.		Oxide of Copper.		Oxide of Zinc.		Oxide of Lead.		Oxide of Silver.	
Oxygen	8	Oxygen	8	Oxygen	8	Oxygen	8	Oxygen	8
Hydrogen	1	Copper	31.7	Zinc	32.3	Lead	103.7	Silver	108.3
	9		39.7		40.3		111.7		116.3

"It therefore follows that 8 parts of oxygen
Combine with 1 part of hydrogen,
31.7 parts of copper,
32.3 " " zinc,
103.7 " " lead,
108.3 " " silver.

"These numbers seem to belong to the substances to which they are attached; for when the compounds of sulphur with the same bodies are examined, the same numbers are obtained as shown by the following table; the only difference being, that the quantity of sulphur is rather more than twice that of the oxygen.

Composition of Sulphurets.

Of Hydrogen.		Of Copper.		Of Zinc.		Of Lead.		Of Silver.	
Sulphur	16 1	Sulphur	16 1	Sulphur	16.1	Sulphur	16 1	Sulphur	16.1
Hydrogen	1.	Copper	31 7	Zinc	32 3	Lead	103.7	Silver	108 3
	17 1		47 8		48 4		119 8		124 4

"In the same manner 35.4 parts of chlorine combine with 1 hydrogen, 31.7 copper, 32.2 zinc, 103.7 lead, 108.3 silver; and, indeed, if a large number of bodies be examined, the same remarkable relation is observed.—(*Graham.*)

"Now it is evident that we might express the composition of all bodies by per centages, as in the 1st table above given; but the 2d and 3d reveal to us a certain fixed numerical relation between the weights in which bodies combine, and which the memory can retain more easily than the variable numbers of per centage composition. For the sake of uniformity there should be a standard or unit of comparison, which once adopted, fixes a particular number to every substance. Such numbers are called the Combining Proportions, Combining Weights of Bodies, Equivalent Numbers, Equivalents, Atomic Weights. The elements of water are each taken as unity for equivalents or combined weights, just as water itself is assumed as unity for the specific weights of bodies, and hence arise two series of equivalents according as hydrogen or oxygen is employed as the unit of comparison. In the one, hydrogen is assumed as unity; in the other, oxygen is assumed to be 100. The following table of the substances introduced above will show the relation between the two series.

			H = 1		O = 100
Hydrogen	1	equivalent	1	or	12.5
Oxygen	1	"	8	"	100.0
Sulphur	1	"	16.1	"	201.1
Copper	1	"	31.7	"	395.7
Zinc	1	"	32.3	"	403.2
Chlorine	1	"	35.5	"	443.75
Lead	1	"	103.7	"	1294.5
Silver	1	"	108.3	"	1351.6

"These proportional numbers belong to one proportion of the body to which they are attached, and in the present work they are usually termed *equivalents*, and their numbers *combining weights;* thus, 1 eq. oxygen has a combining weight of 8 by the hydrogen scale, or 100 by the oxygen scale, and so of each of the others.

"The hydrogen scale is preferred in the present work, according to which the combining weight of a simple substance is the quantity of it which unites with 8 parts of oxygen to form a protoxide or a strong base.

"There are two laws of combination to which the preceding remarks lead us. The first is that *bodies unite with each other by proportional numbers which become fixed when referred to a standard of unity, and are called equivalents or combining weights.*

"The second law, deducible from the first, is that *where bodies unite in more than one proportion, they combine in simple multiples of their equivalents, and in no intermediate proportions.* There are at least two series of combinations embraced in this law.

"1st Series. A unites with 1, 2, 3, 4, 5, &c., of B.

"2d Series. A unites with 1, 1½, 2, 2½, &c., of B.

"The first series is exemplified by the subjoined compounds.

Water is composed of - -	Hydrogen	1	Oxygen	8	1
Binoxide of hydrogen - -	do.	1	do.	16	2
Carbonic oxide - - -	Carbon	6	do.	8	1
Carbonic acid - - - -	do.	6	do.	16	2
Nitrous oxide - - - -	Nitrogen	14.19	do.	8	1
Nitric oxide - - - -	do.	14.19	do.	16	2
Nitrous acid - - - -	do.	14.19	do.	24	3
Hyponitric acid - - -	do.	14.19	do.	32	4
Nitric acid - - - -	do.	14.19	do.	40	5

"In all these compounds the ratio of the oxygen is expressed by whole numbers. In water the hydrogen is combined with half as much oxygen as in the binoxide of hydrogen, so that the ratio is as 1 to 2. The same relation holds in carbonic oxide and carbonic acid. The oxygen in the compounds of nitrogen and oxygen is in the ratio of 1, 2, 3, 4 and 5.

"The following compounds exemplify the second series:—

Protoxide of iron -	consists of iron	27.18	Oxygen	8	1
Peroxide - - -	do. do.	27.18	do.	12	1½
Protoxide of manganese	Manganese	27.72	do.	8	1
Sesqui-oxide - -	do.	27.72	do.	12	1½
Bin-oxide - - -	do.	27.72	do.	16	2
Sulphurous acid -	Sulphur	16.12	do.	16	2
Hyposulphuric acid -	do.	16.12	do.	20	2½
Sulphuric acid - -	do.	16.12	do.	24	3

"Instead, however, of expressing the combinations of B by 1½, 2½, &c., by doubling the two elements, the 2d B will be in whole numbers; thus the peroxide of iron will be 2 eq. iron $= 27.18 \times 2 = 54.36$ and 3 eq. oxygen $= 8 \times 3 = 24$, and hyposulphuric acid = 2 eq. sulphur with 5 eq. oxygen.

"The number of equivalents which may enter into a compound, is subject to considerable variation, as will appear from the following examples:

One equiv.	oxygen	+	One equiv.	hydrogen	form	water.
Two equivs.	oxygen	+	One "	hydrogen	"	peroxide of hydrogen.
One equiv.	oxygen	+	Two equivs.	copper	"	suboxide of copper.
One "	sulphur	+	Three "	oxygen	"	sulphuric acid.
Two equivs.	sulphur	+	Two "	oxygen	"	hyposulphurous acid.
Two "	iron	+	Three "	oxygen	"	peroxide of iron.
Two "	sulphur	+	Five "	oxygen	"	hyposulphuric acid.
Two "	manganese	+	Seven "	oxygen	"	hypermanganic acid.

(*Booth and Boye.*)

"2. *Combination by volume.* Frequently termed theory of volumes. It was observed at the commencement of the section on combination by weight, that water was differently composed by weight or volume, and that the latter offers a very simple ratio. A closer examination of the subject exhibits a close agreement between the laws governing combination by weight and volume.

"Humboldt and Gay Lussac first observed that hydrogen and oxygen unite to form water in the exact ratio of 2 to 1; and the latter, pursuing his investigations with other gases and with vapors, discovered that the principles deducible from that fact admitted of general application to such bodies. He proved in his essay on the 'combination of gaseous substances with each other,' that they unite by volume in very simple ratios. The following views are such as either were established by Gay Lussac, or the results of subsequent investigations.

"1. Gases and vapors unite in the simple ratio of 1 to 1, 1 to 2, 1 to 3, &c. Thus,

100 vls.	hydrogen	+ 50 vls.	oxygen	=	water.
100 "	ammoniacal gas	+100 "	chlorohydric acid	=	sal ammoniac.
100 "	do.	+100 "	carbonic acid	=	carbonate of ammonia.
100 "	do.	+200 "	do.	=	bicarbonate of ammonia.
600 "	hydrogen	+100 "	vapor of sulphur	=	sulphuretted hydrogen.
600 "	oxygen	+100 "	do.	=	sulphurous acid.
100 "	hydrogen	+100 "	vapor of iodine	=	iodohydric acid.

"The volumes of compound gases and vapors always bear a very simple ratio to the volumes of their elements. This will appear from the following table, in which all the substances are supposed to be in the gaseous state:—

Volumes of Elements.				Volumes of resulting compounds.
100 Nitrogen	+	300 Hydrogen	- yield	200 Ammonia.
50 Oxygen	+	100 Hydrogen	- -	100 Water.
50 Oxygen	+	100 Nitrogen	- -	100 Nitrous oxide.
100 Sulphur	+	600 Hydrogen	- -	600 Hydrosulphuric acid.
100 Sulphur	+	600 Oxygen	- -	600 Sulphurous acid.
100 Chlorine	+	100 Hydrogen	- -	200 Chlorohydric acid.
100 Iodine	+	100 Hydrogen	- -	200 Iodohydric acid.
100 Bromine	+	100 Hydrogen	- -	200 Bromohydric acid.
100 Cyanogen	+	100 Hydrogen	- -	200 Cyanohydric acid.
100 Oxygen	+	100 Nitrogen	- -	200 Nitric oxide.

"The law of multiples (see above) is equally demonstrable by means of combining or eq. volumes as by combining or eq. weights. The annexed tabular view will justify this statement:—

Volumes of Elements						Resulting Compounds
100	Nitrogen	+	50	Oxygen	yield	Nitrous oxide.
100	do.	+	100	do.	- -	Nitric oxide.
100	do.	+	150	do.	- -	Hyponitrous acid.
100	do.	+	200	do.	- -	Nitrous acid.
100	do.	+	250	do.	- -	Nitric acid.
100	Hydrogen	+	50	do.	- -	Water.
100	do.	+	100	do.	- -	Binoxide of hydrogen.
100	Carbon vapor	+	50	do.	- -	Carbonic oxide.
100	do.	+	100	do.	- -	Carbonic acid." (*Turner.*)

The changes resulting from the combination of substances are manifold and various, both as regards color and otherwise, sometimes in density; for instance, the usual expansion of volume with aëriform substances. In the combination of solids or liquids, there is generally a condensation of volume, and the different states engendered by this action are diverse, a solid arising occasionally from the union of two gases, as in the case of the junction of carbonic acid with ammoniacal vapor; from a gas and a solid, as in the case of the oxidation of metals. Liquidity, as the result of the union of two gases, viz. oxygen and hydrogen, is produced by the formation of water. The evolution of heat which is attendant upon chemical combination is spoken of under another head.

In the first part of this chapter, when citing the decomposition of tallow by lime as an instance of simple elective affinity, we referred to its constituency as being of stearic, margaric, and oleic acids united with glycerin. These three ingredients are termed its *proximate constituents*, in contradistinction to its ultimate components, of which the former are made up. Thus, carbon, oxygen, and hydrogen, are the ultimate components of suet and of each of its fatty acids and base, and sulphuric acid and soda, the proximate constituents of Glauber's salt, as the diagram will show:

	Proximate constituents	Ultimate components
Sulphate of soda	Sulphuric acid	Sulphur Oxygen
	Soda	Sodium Oxygen

We have heretofore said, that bodies combine in certain definite ratios and proportions; now in order to a convenience of expression in an abbreviated form, chemical bodies and compounds are represented by certain symbols and groups of symbols, so placed together as to furnish at a glance the actual composition of the substance, that is, its atomic composition, "which means the relative proportions of the equivalents of its constituents as well as the ultimate or per centage composition." In this way a full knowledge is conveyed to the reader, at sight, of the constitution of a body; a very important consideration and a great improvement upon the old-fashioned mode of using the unmeaning term of "copperas" for instance, instead of the more appropriate title of sulphate of iron which, when written symbolically Fe. O. SO_3+7 HO conveys the knowledge, that it is composed of one equivalent of oxide of iron, one equivalent of sulphuric acid, and seven equivalents of water; a fact that could not be communicated by the old method of expression. Below is a table showing the composition of crystallized sulphate of magnesia (Epsom salts) represented in its several constituencies.

	Equivalent or atom	Atomic weight	Per centage composition
Magnesia	1	21	16.1
Sulphuric acid	1	40	32.1
Water	7	63	50.8
Sulphate of magnesia	1	124	100.0

There are 59 substances in the accredited list of elements, which, by union with each other, form all the known compounds in nature. Each of these elementary bodies, by way of ready convenience, is represented by a symbol and a compound of any two or more of them, by grouping together the symbols of their elementary or proximate constituents. The symbols adopted are generally the initial letters of the Latin name of the element, as will be seen in the following list of simple bodies with their symbols and equivalent numbers annexed.

Names of Elements	Symbol	Equivalents		Names of Elements	Symbol	Equivalents	
		Oxygen = 100	Hydrogen = 1			Oxygen = 100	Hydrogen = 1
Aluminum	Al	171.2	13.7	Mercury (Hydrargyrum)	Hg	1265.[illegible]	101.4
Antimony (Stibium)	Sb	1612.9	129	Molybdenum	Mo	59[illegible].5	48
Arsenic	As	940.1	75.3	Nickel	Nk	369.7	29.6
Barium	Ba	856.9	68.7	Niobium	Ni	?	?
Bismuth	Bi	[illegible]	71.1	Nitrogen	N	175.0	14.0
Boron	B	136.2	10	Osmium	Os	1244.5	99.[illegible]
Bromine	Br	978.3	78.4	Oxygen	O	100	8
Cadmium	Cd	696.[illegible]	55.[illegible]	Palladium	Pd	665.[illegible]	53.4
Calcium	Ca	256.0	20.5	Pelopium	Pe	?	?
Carbon	C	75	6	Phosphorus	P	392.3	31.4
Cerium	Ce	574.7	[illegible]	Platinum	Pt	1233.5	98.8
Chlorine	Cl	442.6	35.5	Potassium (Kalium)	K	[illegible]	39.2
Chromium	Cr	[illegible]	[illegible]	Rhodium	R	651.4	52.2
Cobalt	Co	369.0	29	Ruthenium	Ru	?	?
Tantalum / Columbium	Ta	2307.4	184.9	Selenium	S	494	39.6
Copper (Cuprum)	Cu	395.[illegible]	31.7	Silicon	Si	277.[illegible]	22.2
Didymium	Di	?	?	Silver (Argentum)	Ag	1351	10[illegible]
Fluorine	F	233.[illegible]	[illegible]	Sodium (Natrium)	Na	290	23.3
Glucinum	G	331.3	26.5	Strontium	Sr	547.[illegible]	43.[illegible]
Gold (Aurum)	Au	24[illegible]	199.2	Sulphur	S	201.1[illegible]	16.1
Hydrogen	H	12.5	1.0	Tellurium	Te	801.7	64.2
Ilmenium	Il	?	?	Thorium	Th	744	59.8
Iodine	I	1579.5	126.[illegible]	Tin (Stannum)	Sn	[illegible]35.29	[illegible]
Iridium	Ir	1233.5	[illegible]	Titanium	Ti	303.66	24.3
Iron (Ferrum)	Fe	339.2	27.2	Tungsten or Wolfram	W	11[illegible]3	94.[illegible]
Lantanum	La	?	?	Vanadium	V	[illegible]56.9	68.7
Lead (Plumbum)	Pb	1294.5	103.7	Uranium	U	2711.4	217.3
Lithium	Li	80.3	6.4	Yttrium	Y	[illegible]	32.2
Magnesium	Mg	158.3	12.7	Zinc	Zn	403.2	32.31
Manganese	Mn	345.9	27.7	Zirconium	Zr	420.2	33.7

Now, by way of explaining the value of this table, let us admit that it is desired to express the metal *lead* according to a more expeditious method than by writing out its title in full. It is done by marking down the two first letters of its Latin name, Plumbum, thus Pb: but, suppose this lead is in the state of oxide, then, by annexing O, thus PbO, you convey the knowlegde of its composition by mere letters, and, by still further extending the formula as follows, $PbO + SO_3$, you have written in five letters the compound called sulphate of lead, consisting as is readily seen of one equivalent of oxide of lead, and one equivalent of sulphuric acid. These expressions are called the formulæ of compounds, and the equivalent of that above mentioned, viz., sulphate of lead, according to the hydrogen scale, is 151.90, explicable as follows. By reference to the preceding table, it will be seen that the combining equivalent of lead is 103.7 and that of oxygen 8; therefore, as oxide of lead consists of one atom of each, by adding the two together you ascertain its atomic weight = 111.7: now this equivalent of oxide of lead, when converted into a sulphate, unites with one equivalent of sulphuric acid, the equivalent of which is 40, because it is composed of 1 equivalent or 16 parts of sulphur, and 3 equivalents or 24 parts of oxygen. The sum of the atomic weights of its two constituents, oxide of lead and sulphuric acid, viz., $111.7 + 40 = 151.7$ is the atomic weight of the compound or sulphate of lead. The symbols in short represent the chemical equivalents of the elements of a body; for example, stearic acid read from a formula, as $C_{68}H_{66}O_5$, $2HO = \overline{St}_2HO$, is directly known to be composed of 68 atoms of carbon, 66 of hydrogen, 5 of oxygen, and 2 of water. Again, by the connection of the symbols H and O, water is expressed, these letters being definite in their meaning and signifying at the same time but one equivalent of these elements, viz., 1 part of hydrogen and 8 parts of water on the hydrogen (as unity), scale and 12.5 parts of hydrogen and 100 parts of oxygen on the oxygen (as unity) scale. By a repetition of the symbol, or by the prefix of a corresponding figure to it, any number of equivalents can be indicated, thus, chromic acid, which consists of 1 equivalent of chrome and 3 equivalents of oxygen, is expressed by Cr 3O or Cr, OOO, or, still more preferable, by the annexation of a smaller figure, Cr, O_3.

The association of symbols as before said, expresses combination: thus, FeO represents a compound of an equivalent, or one proportion of iron and one of oxygen, or the protoxide of iron: SO_3, a compound of one equivalent of sulphuric acid and three of oxygen; that is, one equivalent of sulphuric acid and sulphate of iron itself, consisting of one equivalent of each of the preceding compounds, may be represented as follows. (*Graham.*)

$FeOSO_3$ or
$FeO+SO_3$ or
FeO, SO_3

The algebraic sign +, for which the comma is frequently substituted, indicates in the above formulæ the distribution of the elements of the salts into its two proximate constituents, oxide of iron and sulphuric acid, in order that it may be more distinct. The propriety occurs frequently of using both the *comma* and + in the same formula, as in writing out the composition of the double sulphate of iron and potassa, $=FeO, SO_3+KO, SO_3$, in which case the comma is employed to indicate combination more intimate in degree, or of a higher order than the +; viz., of the oxide with the acid of each salt, while the combination of the two salts themselves is expressed by the sign +. (*Graham.*)

The small figures in the preceding formulæ on the right hand of a symbol do not apply to any other than that to which it is immediately attached, but the larger figures placed antecedent, and on the same line with the symbols, relate to the whole compound expressed; as in the formula of stearic acid before given, there is so much carbon, oxygen, and hydrogen, and 2HO, that is, two equivalents of water. And to further exemplify: $3NO_5$, $5SO_3$, $7CrO_3$, signify respectively three equivalents of nitric acid, five of sulphuric acid, and seven of chromic acid, and 2PbO, two equivalents of oxide of lead. The interposition, however, of a comma (,) or (+) prevents the influence of the figure extending further (*Graham*). Thus

$$2PbO, CrO_3, \text{ or}$$
$$2PbO + CrO_3$$

is two proportions of oxide of lead, and one of chromic acid or the subchromate of lead. To make the figure apply to symbols separated by the comma (,) or plus (+) sign, it is necessary to enclose all that is to be affected within parenthesis, and place the figure before them: Thus,

$$2(PbO, CrO_3)$$

signifies two proportions of the chromate of lead. (*Graham.*)

The mode adopted in the expression of vegetable and animal acids, so as to distinguish them from those of mineral origin, is by surmounting the first, and in some instances the first two initials of their title with a dash: for example, we write Acetic acid $\overline{A}$ and Citric acid $\overline{C}$. In scientific treatises, however, the symbol is frequently accompanied with a formula in parenthesis expressive of its composition, as instanced in the case of the former acid: thus, $\overline{A}=(HO+C_4H_3O_3)$; and in that of margaric acid as follows: $\overline{M}=(2HO+C_{68}H_{66}O_6)$.

This chapter may at first sight have a formidable aspect to the tyro, but he must not be easily deterred from reading it over carefully, for a very limited study will yield a complete understanding of the whole subject, and afford that facility in comprehending chemical action and its results, without which it will be difficult to make progress in any scientific art.

CHAPTER IV.

ALKALIES.

The general characteristics of alkalies are their ready solubility in water, power of neutralizing the strongest acids, (this also being a test of their purity and strength,) their capability of changing many vegetable blue colors into green, and some yellows to brown, or of restoring the colors of those which have been changed by acids. Among these, the yellow of turmeric and litmus, reddened by an acid, are exceedingly delicate tests of the presence of an alkali. The term was formerly applied to the ashes of plants, but is now generally used to designate potash, soda, and ammonia, called respectively, also, vegetable, mineral and volatile alkali. Even when combined with carbonic acid, for which they have a great affinity, and attract it from the atmosphere, the above three exercise the discoloring power before mentioned, a property not possessed by the alkaline earths, lime and barytes, when in a similar state of combination. These three alkalies have also an acrid and somewhat urinous taste. Ammonia and potash are energetic solvents of animal matter, and the three combine with oils so as to form soaps. They unite with water in every proportion, and also with alcohol.

Potassa, soda, and lithia, are formed by the union of oxygen with their metallic bases. The affinity of these metallic bases for oxygen is so powerful as to enable them to decompose the coldest water, and even ice instantaneously, with the disengagement of hydrogen. To the above might be added baryta, strontia, lime, and magnesia, which have certain properties in common with the alkalies proper, and on this account are sometimes classed with them. These are called alkaline earths, a term originating from the fact that whilst in many respects they are possessed of alkaline properties, they otherwise resemble the earths. The property before spoken of as that possessed by the carbonated alkalies, viz., solubility in water, and power of changing blue and yellow colors, is the chief mode of distinguishing between the alkalies and the alkaline earths—a power not belonging to the carbonates of the latter.

We have rated ammonia among the alkalies, because, though differing from them in its organic constitution, it yet has similar properties. We shall make no further mention of any, save those used in the manufacture of soap.

Lime, or *hydrated oxide of calcium.*—Lime has been known from remote antiquity, and is found in nature under different forms. Combined with sulphuric acid it is called sulphate of lime, gypsum and plaster of Paris, and when in a crystaline form, selenite or alabaster. Combined with carbonic acid it constitutes chalk and limestone, which, when it assumes a crystaline structure, is marble. In these forms lime is more or less scattered in different localities throughout every part of the terrestrial globe, and composes a goodly portion of it. Lime is obtained by exposing oyster shells or any of the carbonates of lime to a red heat in kilns specially constructed for the purpose: in this way all the carbonic acid is driven off and lime remains, more or less pure, according to the quality of the limestone used. The product, thus obtained, is called *quick lime.* When sprinkled with water it becomes very hot and crumbles down into a dry powder termed *slaked lime*, or hydrate of lime.

The heat evolved in this instance is that which served to keep the water in a state of liquidity and now given up because no further needed, the water having, as an acid, combined with the lime to form a salt, viz: the above hydrate of lime. This process takes place also, upon exposure for some weeks to air, the lime then falling into powder by the absorption of moisture and a portion of carbonic acid gas, so that in such case a part of the lime having gradually returned to its original state of carbonate, has lost its causticity. To explain more fully the phenomena produced by the contact of water with lime, let us suppose that some water be added to recently burned lime; it is immediately absorbed with the disengagement of heat, a great quantity of aqueous vapour is evolved, the lime crackles, then swells, and eventually falls to powder. If a fresh quantity of water is thrown upon the remaining unaffected fragments, it is absorbed with an accompanying hissing sound. Three hundred degrees, Centigrade thermometer, is the temperature estimated as produced by this operation, and in working upon larger masses of material, the heat is sufficiently intense to ignite powder. This evolution of great heat by the slaking of lime has been applied to the warming of apartments, and to that end several patents have been obtained for proper and convenient apparatus, but I believe the result has not been satisfactory. Instances have been known of the destruction of lime houses by fire, by reason of water dripping upon the lime contained in them. Lime, in a state of purity, is white, easily pulverizable, has a warm caustic taste, and, assisted by the presence of water, destroys, to a certain extent, the organic tissue by direct contact. Its specific gravity is about 2.3. It becomes highly luminous when heated to a full redness, and the injection of the inflamed oxygen and hydrogen gases combined, against a lump of it, is what makes the well known "Drummond Light." Hydrate of lime is very sparingly soluble in water, and what is remarkable, cold water solves it more readily, as well as in greater quantity, than hot water. Thus, according to Dalton, one grain of lime requires for solution 1270 grains of water at 212° F.; 972 grains of water at 130° F.; and 778 grains of water at 60° F.; consequently, on heating lime water, which has been prepared in the cold, deposition of lime ensues. Philips likewise observed that water at 32° is capable of dissolving twice as much lime as at 212°. One of its peculiar properties is its tendency to absorb carbonic acid and thus become a carbonate, in which state it is insoluble in water, and hence are owing the deposit and incrustation which form in and upon lime waters or solutions and milks of lime. Upon this property depends the preparation of the caustic leys, of which we will speak hereafter. Lime, as was demonstrated by Davy, is a metallic oxide, and consists of

Calcium, (its metallic base)	71.91
Oxygen	28.09

Lime, that is, the oxide of calcium, combines with the acids and forms salts, but as they are inappropriate to this work, we shall make no note of them. We will merely say, however, in passing, that the aqueous solutions of the salts of lime have the property of decomposing the ordinary soaps, and thus forming a lime base soap, which is insoluble; hence the impropriety of using hard, or water containing lime, for washing clothes, or for other detergent operations.

The most delicate test of the presence of lime is oxalate of ammonia, a solution of which, when added to water supposed to contain lime, will, if that body is present, throw down a white almost insoluble precipitate of oxalate of lime. This reagent serves sufficiently well for all technical purposes, though a chemist, in making a professional analysis, would have to carry his examination further, in order to distinguish whether the precipitate may not be oxalate of baryta or strontia, both of which are likewise sparingly soluble.

Potassa, or Protoxide of Potassium.—This alkali is indiscriminately called, in commerce, vegetable alkali, sal tartar, pearlash, potash, hydrated protoxide of potassium, &c. It exists in the ashes of vegetables, combined with carbonic, sulphuric, nitric, hydrochloric, oxalic and tartaric acids, but each and every kind of vegetable ashes are not equally rich in potassa, for some are very meager in that constituent. As a comparative table of the potassa product of different plants will be useful to the soap maker, we insert one below, compiled from the analyses of authoritative chemists.

Names of the Plants	Quantity of Ashes	Proportion of Alkali.	Names of the Plants.	Quantity of Ashes	Proportion of Alkali
100 PARTS.			100 PARTS.		
Willow	2 8	0 285	Fumitory	21.9	7.9
Elm	2 36727	0.39	Hop stems	10.000	3.015
Oak	1.35185	0.15343	Windsor Bean Stalks	10.000	4.129
Poplar	1.23476	0.07481	Common Nettle	10.67186	2.5033
Yoke-elm	1.1283	0.1254	Common Thistle	4.04265	0.53734
Beech	0.58432	0.14572	Fern	5.00781	0.6259
Fir	0.31740	0.7318	Large Rush	3.85395	0.72234
Vine shoots	3.379	0.55	Heath	2.9019	0.84
Corn stalks	8.86	1.75	Horsechestnut (bark)	18.460	4 840
Absinthe or Wormwood	9.744	7.3	Centaury	8.44	2.008

The above table demonstrates: 1st, that the herbs and grasses yield much more ashes than the woods, (with the exception of the bark of the horsechestnut,) and that these same ashes are richer in potash; 2d, that of all the herbaceous plants, the fumitory is that which gives the most alkali. Experience proves that the younger parts of a tree, and especially the leaves, furnish the most potash. The knowledge of this alkali dates from time immemo rial, and though we offer no proof of its antecedence to soda, it may be as well to remark that ashes were calcined by the Egyptians, and used as caustics. It is called potash from having been prepared for commercial purposes, by the evaporation in iron pots of the lixivium of the ashes of wood fuel. Of all the Greek philosophers Aristotle is the first who announced that the ashes of the rush and reed give an abundant ley. Pliny speaks of alkaline leys, but we find the first data upon alkali in the writings of the Arabs; for it was under this name of alkali that both potash and soda were confounded until 1745, when the works of Pott, Duhamel, and Margraaf announced the difference as determined by evident proofs. These two alkalies took their names;—the potash, of vegetable alkali;—the soda, of mineral alkali. The word alkali is compounded of the Arabic name *kali*, which is that of the plant from which it is extracted, and of *al*, which indicates the superiority or greater strength of the salt over the plant whence it is obtained. The title of mineral alkali, however, is as well applicable to potash as to soda, since the former has been found in angite, obsidian and many other minerals.

As a general rule, it may be regarded that the shrubs are richer in potassa than the trees, and that the former succumb in this respect to the *herbaceii*. Mostly, however, the product of the trees being equal to 1, that of the shrub is in the ratio of 3, and that of the plants in the ratio of 5. The trunk of the tree affords less alkali than the branches, these latter less than the fruits, and the fruits again less than the leaves. It is also well attested, that the pith of a tree exceeds in potash constituent, the more solid part, so that the quantity of potash seems to be inversely to their hardness. Saussure says that those plants which transpire the most are those which give the most potash; that the bark yields more than the wood;

in fine, that the evergreens are poorer in alkali than those plants which drop their leaves in winter.

It was for a long time mooted whether the alkalies existed already formed in the plants, or were the product of their combustion, a question now decisively settled in the affirmative; for the ashes of those woods which have remained for a long time in water, for instance of float or drift wood, do not yield alkali. We will add, however, that combustion seems to augment the quantity. Fontenelle has also announced that those plants which have undergone an incipient state of putrefaction, furnish a greater product. But this increase of quantity doubtless must arise from the absorption of a certain quantity of carbonic acid gas by the alkali.

In the crude state called potashes, it is composed of such constituents of burned vegetables as are very soluble in water and fixed in the fire, and hence contains, besides potassa, other salts and impurities. When this crude potash is calcined, it is partially purified by the action of the fire upon some of its foreign salts and admixtures, and becomes an impure carbonate of potassa, called in commerce *pearlash.*

The process for the extraction of potash from the ashes of plants, varies according to the substances from which it is to be drawn. For instance, when plants are used, select those directed in the preceding paragraphs. It is better to gather the ripe, yet not perfectly dry plants, for Chaptal and Lavoisier both say that, though combustion augments the quantity of alkaline product, it must not be too rapid, else the yield is much lessened. The manufacture of potash is largely carried on in many of the states of the Union, and the process followed is partially that used by the soap boiler for leeching his ashes, or similar to it, and is as follows, recollecting, however, that in the latter instance the ashes are already at hand, whereas in the manufacture of alkali from the plants, they must first be reduced to ashes before they can be lixiviated.

The plants should be piled in stacks before they become too dry. Then dig into the ground ditches of three feet depth and eight feet width, well plastered and cemented with compact clay. This method is preferable to that of burning them in pyramids, because the combustion is slower and the wind cannot easily blow away the products.

As soon as the ditches are well dried, the plants are thrown in and burned, a small quantity at a time, and when the pits are full, settle down their contents with sticks of wood. When the ashes are cold, leech them with cold water, adding five per cent. of pure lime, to ensure a greater causticity, and evaporate the lixivium or solvent liquid to two-thirds. These residuary two-thirds are run off into troughs and allowed repose for eight or ten days, that the greater portion of the least soluble constituents, salts and the contained foreign matter, may separate and precipitate to the bottom. After this the clearer supernatant liquor is drawn off with a lead syphon, or through a spigot, (properly adjusted in the vat or trough so that its barrel is just above the sediment,) into cast iron kettles, and evaporated to dryness. This evaporated mass, called *salin*, is finally fused at a red heat into compact masses which are gray on the outside and pink-colored within. By subjecting this potash to the heat of a reverberatory furnace, the carbonaceous matter and the most part of the sulphur are dissipated, and the excess of water evaporated. Then the *salin* acquires a more or less white color, and loses ten to fifteen per cent. of its weight In this state it is called pearlash. Whatever may be the beauty in appearance of potash thus obtained, it is nevertheless far from being pure. By further lixiviating this pearlash, in a cistern having a false perforated bottom covered with straw, and evaporating the clear ley or lixivium to dryness in clean flat iron pans, and constantly stirring it towards the end, you form it in white lumpy granulations, and as thus treated it is called *salt of tartar*. This purification of pearlash is founded upon its greater solubility in water than are the neutral salts which debase it.

Analyses of the ashes of plants.—The ashes of plants are composed of a number of saline constituents, metallic oxides, &c. Of this number are, among the oxides, those of alumina, iron, manganese, and silica; among the salts, the carbonates and hydrochlorates of lime, magnesia, potassa, and soda; the hydriodate of potassa (that is, supposing with Claubry that the iodine exists in *varech* under this form), the nitrates of lime, magnesia and potassa; the subphosphates of lime, magnesia and potassa; the sulphates of potassa and soda; and sulphur and bromine.

All these saline constituents do not exist simultaneously in the same plant; those most frequently met with are the subcarbonates and phosphates, the hydrochlorates and sulphates of potassa, also the subcarbonates and subphosphates of lime, the subphosphate of magnesia, the chloride of sodium (hydrochlorate or muriate of soda), the silicate and oxides of iron and manganese.

The marine plants, such as the salsola, the fucus, &c., in place of the subcarbonate and sulphate of potassa, &c., give carbonate and sulphate of soda. Chaptal and Fontenelle have announced that certain plants, such as the *tamarix gallica, &c.*, cultivated upon the coast or a short distance from the sea, yield salts having soda as their base, whilst those reared ten leagues distant produce potash salts. After what has been said of the constituent principles of ashes, it is easily seen that the potash of commerce must and does contain always more or less of foreign matter, and that such impurities render it unfit in that state, to be used in the facture of soaps. These foreign matters vary its purity and are also a great inconvenience to its use in dyeing operations, for the quality of the potash has a great influence upon the excellency of the color.

We give below Vauquelin's accurate estimate of the amount of alkali in 1152 parts of each of the following commercial potashes:

In 1152	parts of	American potash,	he found		857	of potassa.
"	"	Russian	"	"	772	"
"	"	Dantzig pearlash		"	683	"
"	"	Vosges	"	"	444	"
"	"	Triers	"	"	251	"
"	"	American	"	"	754	"

To obtain the potassa pure, take equal weights of pearlash and quick lime slaked into powder, dissolve the former in from six to ten times its weight of water, and boil both together for thirty minutes in a clean iron pan. The lime abstracts the carbonic acid from the salt of potash and becomes carbonate of lime, which, being insoluble, precipitates. When the potash salt has been entirely deprived of carbonic acid, a little of the clear liquid taken from the pan will fail to effervesce when acid is added thereto. It is a remarkable fact that the decomposition is never complete, if the carbonate of potassa or pearlash has been dissolved in less than the prescribed quantity of water. Liebig says, that a concentrated solution of potash decomposes carbonate of lime, and consequently hydrate of lime could not, under similar circumstances, decompose carbonate of potassa. The pan, being covered by a lid, may be allowed to cool and settle, after which the supernatant clear solution is drawn off through a lead syphon into a clean iron basin and rapidly evaporated until an oily liquid remains. This liquid, poured into moulds, forms the stick potassa of the shops. To obtain a purer article, requires the use of platinum vessels, more care and a further extension of the process. But this is of no import to the art of which we write

This fused potash is a solid, white, crystaline mass of sp. grav. 1.706, fusible at a heat under redness, and, like the commercial potash, very deliquescent.

Davy, in 1807, discovered potassa to be an oxide of a metal to which he gave the name of potassium.

Lussac and Thenard give, as their analysis of pure potassa,

Potassium - - - -	100
Oxygen - - - - - -	20

Potash ley.—The solution of potash, or potash ley, has a slight but peculiar characteristic odor, an acrid taste, and is highly caustic. It absorbs carbonic acid with great avidity from the air, on which account it should be kept in closed vessels. To prepare the ley, use by preference that potash which rates 66 to 68 degrees, and dissolve 30 kilogrammes thereof in a quantum of water sufficient to form a solution which will mark 12° by Baume's areometer. After having heated this solution to 60° of temperature (Centig.) add, in five or six separate portions, fifteen kilogrammes of quick lime, taking care, however, that one parcel is extinct before another is added, in order that you may evade the heat which would result from a too rapid application of the lime and which might cause a spattering of the contents of the vessel. When all the lime has been put in, the mixture is run up to and then kept at boiling heat for an hour and a half, after which it is allowed to repose, and the supernatant liquor drawn off and preserved in closed vessels. With the above indicated proportions, a solution of potash should be obtained, marking ten to eleven degrees. If for certain operations, a greater strength is required, this may be attained by evaporating the solution to the degree requisite. The rationale of the above process is as follows. Commercial potash consists of potassa or oxide of potassium, carbonic acid and the sulphates and hydrochlorates of potassa. The lime (which is the oxide of calcium), by reason of its greater affinity, abstracts the carbonic acid from the potash and falls down as an insoluble precipitate of carbonate of lime, with any excess of lime that may have been added, whilst the decarbonated potassa remains in solution. The other two salts, being less soluble, crystalize out on cooling.

The following table exhibits the quantity of *fused potassa* in 100 parts of *caustic ley* at their respective densities.

Sp grav	Pot in 100	Sp grav	Pot in 100	Sp grav	Pot in 100	Sp grav	Pot in 100	Sp grav	Pot in 100
1 58	53.06	1 46	42.31	1.34	32.14	1.22	23.14	1 10	11.28
1.56	51 58	1.44	40.17	1.32	30.74	1.20	21 25	1 08	9.20
1.54	50.09	1.42	37.97	1.30	29 34	1.18	19 31	1 06	7 02
1 52	48.46	1.40	35.99	1.28	27.86	1.16	17 40	1 04	4.77
1 50	46.45	1 38	34 74	1.26	26 34	1.14	15 38	1 02	2.44
1.48	44.40	1.36	33 46	1.24	24.77	1 12	13 30	1 00	0.00

Soda.—According to Pliny this alkali was first discovered by those driven by the tempest into the mouth of the river Belus in Syria, who, having used kali for fuel in boiling their food, observed that its ashes, becoming mixed with the adjacent sands, formed by fusion a vitreous mass. Hypothesis aside, soda was pointed out in the ninth century by the Arab Gebert, not, however, as an alkali *sui generis*, but as being of the same nature as potash. It is since the works of Bergman and others that the difference has been distinctly established.

The commercial demands for soda are chiefly supplied from two sources;—the combustion of marine vegetables which furnishes the impure alkalies called *kelp* and *ba-*

rilla, and the decomposition of common salt or rather of sulphate of soda obtained by the decomposition of salt by sulphuric acid.

Native Soda.—The presence of soda in mineral substances was announced long ere this, for, though Chaptal did not show its existence in volcanic products, he proved that the soda in glass making could be in a measure replaced by lava, &c. Kennedy made known its partial constituency of pumice stone and some of the basalts; whilst Klaproth found it in the chrysolite of Greenland, in the proportion of 0.36.

Soda from the combustion of marine plants.—There are an infinite number of these which yield soda, among which are the *salsola, fucus*, some of the *chenopodii*, the *tamarix gallica*, &c.

The incinerated ashes of the *salsola* fused into gray, forms masses, from the impure carbonate of soda, known in commerce as *barilla*, and imported generally, by way of England, from Spain and the Levant. The ripened plant is cut down and dried, and burned in rude furnaces at a temperature just sufficient to cause the ashes to enter into a state of semi-fusion, so as to concrete on cooling into moderately compact, cellular masses. The most valuable variety of this article is called *sweet barilla*. It has a grayish-blue color, and when exposed to the air for some time, effloresces at the surface. It is hard and difficult to break, and when applied to the tongue, gives a pungent alkaline taste. Its average proportion of free or alkalimetrical soda, according to Ure, is about 17 per cent., though several varieties contain but 14, whilst a few even reach as high as 20 per cent. This soda is chiefly a carbonate, with a little sulphuret and sulphite, and is mixed with sulphate and muriate of soda, carbonate of lime, carbon, &c.

Another mode of manufacturing crude soda is by burning sea weed and the different species of *fuci* and other similar plants, into *kelp*. Formerly, "very large revenues were derived, by the proprietors of the shores of the Scottish islands and the Highlands, from the incineration of sea weed by their tenants, who usually paid their rents in *kelp*;" but since the tax has been taken off of salt, and the manufacture of a crude soda from it has been generally established, the price of kelp has fallen extremely low. This crude matter varies in the proportions of its alkaline constituent according to the source from whence extracted.

Ure analyzed two specimens, the result of which is as follows, as regards their soluble portion.

Sulphate of soda - - - -	8.0	19.0
Soda, as carbonate and sulphuret -	8.5	5.5
Muriates of soda and potassa - -	36.5	37.5
	53.0	62.0

The insoluble matter consisted of

Carbonate of lime - - -	24.0	10.0
Silica - - - - - - -	8.0	0.0
Alumina tinged with iron oxide -	9.0	10.0
Sulphate of lime - - - -	0.0	9.5
Sulphur and loss - - - -	6.0	8.5
	100.0	100.0

The analyst forgot to mention a very important constituent of kelp. This component is iodine, and kelp the principal source from whence it is obtained; minute portions of bromine being also a concomitant ingredient.

The first of these specimens was from Heisker, the second from Rona, both in the Isle of Skye. From other analyses besides these, the composition of kelp is proven to be too variable and meagre in soda power, to be used as the alkaline material in soap making. The very low price that *soda ash,* which is nothing more than dry crude carbonate of soda, from the decomposition of sea salt, is now sold, has enabled it to entirely supersede the use of kelp.

A solution of caustic soda or soda ley, is obtained by decomposing the *soda ash,* (dissolved in four or five times its weight of water) by means of half its weight of hydrate of lime; the same points being attended to, as in the preparation of potash ley. The solution thus obtained contains foreign salts, as the chloride of sodium and sulphate of soda, &c.; but Mr. W. Blyth says, that by concentrating it considerably, these salts become insoluble and precipitate. We give below Richter's table of the quantity of *caustic soda* contained in *leys* of different densities.

Sp grav	Soda per cent	Sp grav.	Soda per cent	Sp. grav	Soda per cent	Sp grav	Soda per cent
1.00	0.00	1.12	11.10	1.22	20.66	1.34	31.67
1.02	2.07	1.14	12.81	1.24	22.58	1.35	32.40
1.04	4.02	1.16	14.73	1.26	24.47	1.36	33.08
1.06	5.89	1.18	16.73	1.28	26.33	1.38	34.41
1.08	7.69	1.20	18.71	1.30	28.16		
1.10	9.43			1.32	29.96		

Soda is soluble in all proportions of water and alcohol and the *solid fused soda* is obtained by evaporation precisely in the same manner as the corresponding preparation of potash. Soda is distinguished from potassa and other bases by several properties:—1st, all its salts are soluble in water, and it is therefore not precipitated by tartaric acid, chloride of platinum or any other re-agent. 2d, with sulphuric acid it gives a salt which crystalizes in large efflorescent prisms, easily recognized as Glauber's salt. 3d. Its salts communicate a rich yellow tint to flame.

Until 1817, soda was considered as a mineral alkali, but at that time Davy discovered, by decomposing it, that it is an oxide of a metal to which he gave the name of sodium. According to Lussac and Thenard, soda consists of

Sodium - - - - - - - - -	74.42
Oxygen - - - - - - - - -	25.58
	100.00

This combination forms the protoxide of sodium. It is white, very caustic, and, when exposed to air at the ordinary temperature, attracts moisture and deliquesces; if the contact is prolonged it absorbs carbonic acid, becomes efflorescent and passes into the state of a subcarbonate. The oxide of sodium combining with water, forms a hydrate, the composition of which is as follows:

Water - - - - - - - - -	22.34
Oxide of sodium - - - -	77.66
	100.00

Artificial sodas.—This is the title given to the sodas extracted from common sea salt by a process of Leblanc, a French chemist, made known towards the close of the last century. The supply of Spanish barilla being cut off from France during the cessation of the commercial intercourse between France and Spain at the time of the French Revolution, the National Convention made an appeal to the chemists of France, to devise a process in which common salt might be made available as a source of soda. The result was Leblanc's discovery, whose method is that still practiced most generally both in Europe and in this country. The inventor, it is said, neglected by the government of his native country, after having spent all his own means in perfecting his plan, at last died in a hospital. The British government, however, yielded him a handsome reward, and at Greece and Rome, statues of the inventor have been erected.

The process consists in decomposing with sulphuric acid common sea salt, improperly called muriate of soda, it being a combination of two elementary substances, chlorine and sodium, in the proportion of one equivalent (35 parts by weight) of the former and one equivalent (23 parts) of the latter. Its proper designation, therefore, is the chloride of sodium. It contains no water of crystalization. The salt being converted by the sulphuric acid into a sulphate of soda by proper manipulation and arrangements, a reaction takes place during the operation, and may be easily understood by the assistance of the following diagram.

Reaction of strong oil of vitriol (sulphuric acid, containing only one equivalent of water) on common salt (chloride of sodium).

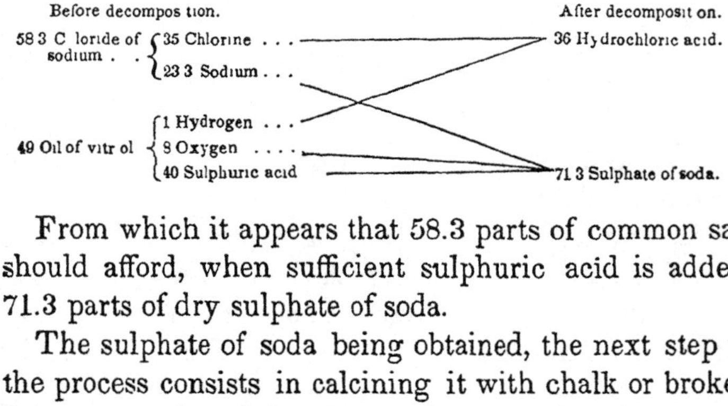

From which it appears that 58.3 parts of common salt should afford, when sufficient sulphuric acid is added, 71.3 parts of dry sulphate of soda.

The sulphate of soda being obtained, the next step in the process consists in calcining it with chalk or broken limestone, and coal dust, in certain proportions, in a reverberatory furnace called the *fluxing* or *black ash* furnace.

The product of this operation, which is known as black ash, ball soda, and British barilla, should contain from 20 to 24 per cent. of available alkali. Its composition varies considerably and with the skill of the furnacer; the principal insoluble constituents are carbonaceous matters, carbonate of lime and a compound of lime and sulphuret of calcium; the soluble ingredients are, carbonate of soda, undecomposed sulphate of soda and common salt, caustic soda and generally some sulphuret or bisulphuret of sodium.

The operation of calcining is the most important part of the soda process. By the fusion of sulphate of soda with coal, the former parts with either the whole or a portion of its oxygen, and is thereby reduced either to the state of sulphuret of sodium only, or else to a mixture of sulphuret of sodium and carbonate of soda.

In effecting this abstraction of oxygen, the hydrogen of the coal plays as important and active a part as the carbon; but in the following diagram, illustrative of this decomposition, the carbon of the coal may be supposed for the sake of simplicity to be the only reducing agent concerned.

Reduction of sulphate of soda by carbon.

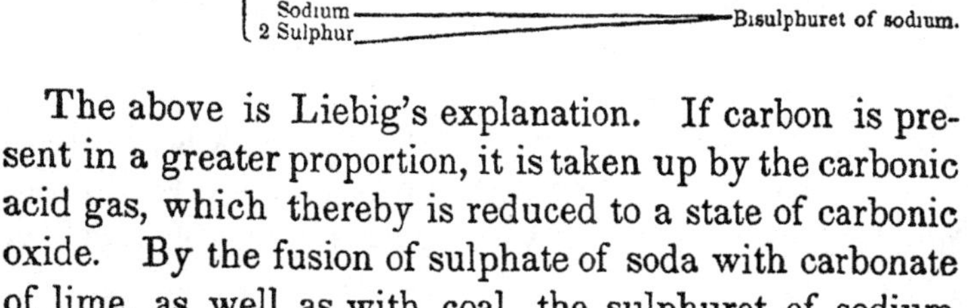

The above is Liebig's explanation. If carbon is present in a greater proportion, it is taken up by the carbonic acid gas, which thereby is reduced to a state of carbonic oxide. By the fusion of sulphate of soda with carbonate of lime, as well as with coal, the sulphuret of sodium produced through the latter is converted into carbonate of soda. The calcium of the carbonate of lime and the sodium of the sulphuret exchange places, and the resulting substances are sulphuret of calcium and carbonate of soda. The calcination of sulphate of soda with both coal and carbonate of lime, is by no means essential to the process.

Three hundred weight of material constitute a charge, and the product of this charge, the black ash, before spoken of, is now subjected to another operation by which is effected the separation of all the soluble constituents from the carbonaceous and other insoluble matters. This process consists in lixiviation, evaporation and calcination, and the residue when ground under edge stones, is the *white ash* or *soda ash* of commerce. The article thus prepared is sufficiently pure for ordinary manufacturing purposes, but Descroizilles says, that some factured sodas bear the just reproach of being too highly charged with hydrosulphurets, the presence of which renders the ash unfit for making white soap and of also possessing at the moment of their fabrication, a degree of alkalimetrical strength, which they very soon lose. Some sodas are so highly charged, that they attract the humidity of the air, and emit a characteristic odor of rotten eggs. Now-a-days, however, the processes are so perfect that both the French and English sodas are uniformly of a beautiful gray ash color, without odor or much variation in specific gravity, and mostly of uniform properties. The sulphuretted sodas, nevertheless, are more suitable than the purer, for making marbled soaps.

To obtain the crystalized carbonate of soda, a further purification and solution of the ash and crystalization must be resorted to. Crystalized carbonate of soda consists of

Carbonic acid - - - - - - -	15.37
Soda - - - - - - - - - - -	21.83
Water - - - - - - - - - - -	62.80
	100.00

Its crystals are colorless, transparent rhomboids, which

readily effloresce in the air and melt in their own water of crystalization.

Ammonia—We now speak of this body in its turn after potash and soda, and though but little applicable to the art of which we treat, it yet has analogous properties to the above, and is besides the basis of that compound known as ammoniacal soap. In its pure state it is a gas at all temperatures, and hence its name *volatile* alkali, in contradistinction to the remaining three (potassa, soda and lithia) which are called *fixed* alkalies, they not being at all volatile at red heat.

Ammonia is a gas of which the *aqua ammoniæ* is a solution, and, in its caustic state, is thus met with in commerce. It has a strong pungent odor, and is recognized by this and its transient alkaline effect upon vegetable colors, and the production of white fumes when it is approached by hydrochloric acid. Ammonia as a gas is a compound of nitrogen and hydrogen, and consists as follows:

Nitrogen	1 atom	=	14	82.35
Hydrogen	3 "	=	3	17.65
			17	100.00

Pure ammonia is obtained in the form of gas by heating a mixture of quicklime and muriate of ammonia. Water absorbs 780 times its volume of this gas, and by this absorption, the volume of the water is increased in the ratio of six to ten. Its specific gravity in the gaseous state is 0.590. The specific gravity of water saturated with ammoniacal gas is 0.900 and contains in weights, water 74.63, ammonia 25.37=100.00.

The title of ammonia originated from the fact of the camel's dung, its original source, collected for the purpose in Egypt, having been burned about the temple of Jupiter Ammon. Davy and Ure have prepared tables which exhibit the per centage of ammoniacal gas contained in ammoniacal solutions of different specific gravities. We compile therefrom as follows:

Table of the strength of caustic ammonia.

Davy.		Ure	
Specific gravity.	Per cent of Ammonia	Specific gravity.	Per cent of Ammonia.
0.8750	32.30	0.8914	27.940
.8857	29.25	.8937	29.633
		.8969	29.038
		.8983	26 751
.9000	26 00	.9000	26.500
.9054	25.37	.9045	25.175
		.9090	23.850
.9166	22.07	.9133	22.525
		.9177	21.200
.9255	19.54	.9227	19.875
		.9275	18.550
.9326	17.52	.9320	17.225
.9385	15 88	.9363	15.900
.9435	14.53	.9410	14.575
.9476	13.46	.9455	13.250
.9513	12.40	.9510	11.925
.9545	11.56		
.9573	10.82	.9564	10.600
.9597	10.17		
.9616	9.60	.9614	9.275
.9692	9.50	.9662	7.950
		.9716	6.625
		.9828	5.500
		.9887	3.975
			2.650

CHAPTER V.

ALKALIMETRY.

ALKALIMETRY is that branch of chemical analysis which is used for determining the degree of richness, in available alkali, of soda ash, potash, wood ash, and alkaline matters generally. The solutions or leys of the above named bodies, as found in commerce, contain besides the alkalies and their carbonates, a variable quantity of foreign matter, and as it is only the alkalies and their carbonates that can be used in the fabrication of soap, it is requisite to have a mode of determining the value in alkali of these materials, respectively. Several processes have been contrived, by which results of sufficient accuracy, for all ordinary purposes, may be obtained with ease after a very little experience.

Descroizilles was the first person who contrived an instrument to define the alkaline strength of potash and soda. This instrument, Fig. 1, which he titled *polymëtre*

Fig. 1.

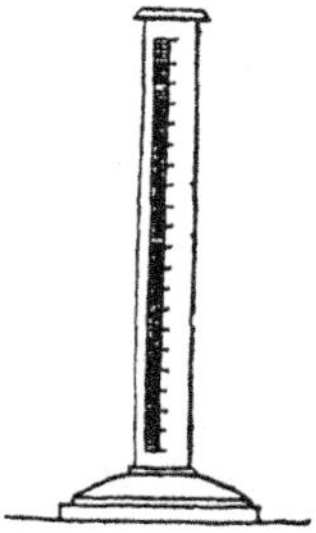

chemique, by reasons of its applicability to a diversity of operations, consists of a proof glass eight or nine inches tall, and about seven or eight lines in diameter, closed at one end but open at the other, and surrounded by a projecting ledge. This glass should contain 70 to 80 grammes of water, divided off into 100 parts, by an accurately proportioned alkalimetrical scale, each representing a half gramme or a half thousandth of a litre of water. This scale, and its figures, and inscriptions, are engraved upon the glass. To render this same instrument applicable to chlorometry and acidemetry, Descroizilles arranged upon the glass other appropriate scales. The alkalimetrical proof liquor is prepared as follows:—Take a new bottle and wash it perfectly clean with distilled water, and weigh in it eight hectogrammes of pure water; mark upon the neck of this bottle with a glazier's diamond, the point of elevation of the liquid, of which then draw out one-half. On the other hand, in an equally well cleansed and dry glass, weigh eight decigrammes, (or a tenth of the weight of water) of sulphuric acid of 66°; pour this acid, little by little, into the water in the other bottle, and shake well at each addition, in order that the heat which is disengaged shall not break the bottle. Wash the glass which contained the acid, with that half of the water which was first taken out and put aside, and unite the wash waters with that in the bottle containing the acid and water together, to which still add more water, (stirring or shaking well all the time,) until the liquid rises even with the mark upon the neck of the bottle.

To be completely appointed for the essays, it is requisite to have the following subsidiary articles, viz., syrup of violets, litmus paper, a small pair of scales, with weights up to ten grammes, a half decilitre tin measure, some experimental glasses, glass rods and funnels, and finally, a glass mortar.

All of these instruments can be had of Stephen Heintz,

glass factor, Kensington, or, together with the materials, of Smith and Hodgson, or Potts, Linn and Harris, druggists, in Philadelphia.

In this method, the experiment consists, in finding what quantity of an acid is required to destroy the alkaline reaction of a known weight of the specimen to be tested. The quantity of acid required for the neutralization of a soda ash, being of course exactly equivalent to the quantity of soda present in the free state and as carbonate.

Here is the modus operandi:

To ascertain the strength, in alkali, of a potash, you take an average of the samples and weigh thereof exactly a decigramme, which, put into a glass, and pour thereon four-fifths of a half decilitre of water, stirring with a glass rod to aid the solution of the potash, which, when perfect, is poured into a tin measure of a demidecilitre capacity, then to receive water to fullness; the contents are then returned anew into the glass and another decilitre of water added. Stir this new mixture several times and leave it to repose, and when it has become settled, slope 'the glass cautiously, and decant the small tin measure full of the clear liquor, and empty this last into another large glass. This done, now arrange upon a plate many drops of syrup of violets, and fill the alkalimetrical glass or tube with test liquor to the point O. Take it now with the left hand, inclining it upon the glass which contains the moiety of the clean alkaline solution; the acid liquor will fall into it by hasty drops, or, in a very thin thread, which can be moderated at pleasure by retarding the entrance of the air; at the same time, with a small rod, assist the mixture and facilitate the development of the carbonic acid, which is manifested by effervescence. When you have emptied the alkalimeter to about the line 40, try if the saturation approaches, by drawing your small rod from the mixture and resting it upon the drops of syrup of violets, which should become green if the potash is not of a very inferior quality, for this color proves that the alkali is not saturated. If, on the contrary, the violet color is not altered, or what would be worse, if it be changed into red, there would be in the first case, an indication of saturation, and in the second, a proof of supersaturation. You therefore stop adding the proof liquor to the alkaline liquor as soon as a feeble reddish tint appears; and observe at what scale the acid liquor has descended, and reckon one degree less, in order to compensate the excess of saturation. The mean term of commercial potashes is 55°, that is to say, they require for their saturation 55 hundredths of their weight of sulphuric acid, or, in other words, absorb and neutralize 55 hundredths of their weight of sulphuric acid in passing to the state of sulphate. Beyond this point they are strong, if they mark less they are weak. In this operation, each time that the proof liquor is poured into the alkaline liquor, it produces an effervescence which is due to the disengagement of carbonic acid from the subcarbonate of potassa, the base of which leaves its acid and unites with the sulphuric acid. It is an established chemical fact, that acids for the most part redden the vegetable blues, and that the alkalies green them; now, every time that the syrup of violets touches with the alkaline liquor and greens it, the proof is that there is free and uncombined potash in the liquor; when, on the contrary, the syrup of violets commences to redden it, we have a certain index that all the alkali is saturated, and that free acid exists in the liquor. To know that the saturation of the alkaline liquor by the acid, or proof liquor, is exact, it must produce no change in the violet's syrup. Litmus paper can be substituted for syrup of violets, but it must be exceedingly delicate, and a distinction must be made between the purplish red tint which carbonic acid produces, and the distinct reddening caused by sulphuric acid. The slight reddening which carbonic acid makes may be discharged by the application of gentle heat.

Sodas.—When about to test the commercial soda of any kind, take in lieu of a decigramme of potash, ten and a half decigrammes of that alkaline material, and proceed as above detailed.

Potashes.—Annexed is a table of results.

American pearlash	1st sort	- - - - - - -	60 to 63	hundredths
" "	2d "	- - - - - - -	50 to 55	"
" Potash,	1st sort in reddish lumps		60 to 63	"
" "	2d " " grayish lumps		50 to 55	"
Russian "		white - - -	52 to 58	"
Dantzic "		" - - -	45 to 52	"
" "		blue - - -	45 to 52	"
Ashes of new firewood	- - - - - - - - - -		$8\frac{1}{5}$	"
" " float or drift-wood	- - - - - - -		$4\frac{2}{5}$	"

Descroizelle found some American potashes to rate at 66 hundredths, and some of the first sort even as high as 72 hundredths.

2d Sodas.

Artificial or manufactured sodas - - - - - -	30 to 35	hundredths
Sal soda crystalized and well dried - - - - -	36	"
Alicant soda - - - - - - - - - - - - - -	20 to 33	"

If the only impurities contained in the soda ash to be tested, are chloride of sodium, sulphate of soda, and insoluble matters, the process just described may afford results of sufficient accuracy; but other bodies are frequently present which interfere with this process by neutralizing a portion of the test acid, such as sulphuret of sodium, sulphite, and hyposulphite of soda, and which are liable to occasion a considerable error by taking up the sulphuric acid. The best method of avoiding error from this source of fallacy, consists in mixing the solution of the ash with chlorate of potassa, evaporating to dryness and heating the residue to redness; the sulphuret, sulphite, and hyposulphite are thereby converted into the sulphate of soda, which does not interfere with the accuracy of the process.

Remember, in applying this test acid to the examination of the carbonates of potassa, that their equivalent is higher than that of soda, and consequently, the number of measures poured from the alkalimeter cannot represent the per centage of potassa as it does that of soda. A simple calculation, however, reduces the numbers for soda to those of potash. For instance, their respective equivalents being very nearly as 4 to 6, the number of measures necessary to neutralize 100 grains of the specimen multipled by 6, and then divided by 4, gives the per centage of potash required.

Mr. Edward Pesier has lately announced a new method of estimating the value of pearlashes. His plan is founded on the increase in density, which the sulphate of soda gives rise to in a pure saturated solution of sulphate of potassa; and the mode of appreciation depends on the employment of a peculiar areometer which he styles a *natrometer*. This principle is deduced from the following facts, founded on experiment:—1st, that a saturated solution of neutral sulphate of potassa always possesses a constant density when it is made at the same temperature; 2d, that the sulphate of soda increases the density of this solution in proportion to its quantity. This is the more perceptible as the solubility of the sulphate of potassa is greatly increased when the two salts are present.

Below is his process, which, he says, is more than sufficient for the purposes of the merchant and manufacturer.

About 200 grammes of water are agitated with 50 grammes of pearlash in a flask of capacity sufficient to hold 600 grammes of water, the solution carefully saturated with sulphuric acid, heated to expel all carbonic acid, allowed to cool, filtered, and the residuary salt washed with a concentrated solution of pure sulphate of potassa until 300 cubic centimetres' solution are obtained; and into this the natrometer is immersed. If the pearlash is pure, the degree of temperature at which the experiment is made, will be found on the red scale of the instrument beneath the spot where the liquid ascends around the tube; if, on the contrary, it contains soda, some degrees more will be indicated, the number of which, when compared with the adjacent soda-scale, becomes converted into so many parts per cent. of soda; for instance, the experiment made at 12° Centigrade, indicates 25° on the natrometer; the 13° in excess are consequently due to the foreign alkali; and on looking for these 13° on the soda-scale, we find that the pearlash contains four per cent. of soda.

The above method is likewise applicable for determining in a short time, the amount of caustic soda in soapmaker's leys. For this purpose, a certain quantity of the ley is saturated with sulphuric acid, an excess of pulverized sulphate of potassa added to it and the whole well shaken; upon which 300 cubic centimetres of the solution are measured off and the natrometer immersed into it. This process, however, is objectionable, where, instead of a carbonate, the soda present is also chloride or sulphate. It, however, is of no consequence where the object is to ascertain the amount of all soda salts in the pearlash.

The *natrometer* consists of two contiguous scales, one of a rose color with the degrees of temperature; it indicates at each degree of the centigrade thermometer, the point of density of a saturated solution of fine sulphate of potassa. The other represents hundredths of soda; the zeros of the two scales coincide. If we experiment at 0° Centigrade, the soda will be directly determined; but if we work at 25° Centigrade, it is known that the point to which the instrument would sink in the solution of pure sulphate of potassa, saturated at this degree, corresponds to eight per cent. of soda. It is, however, there that the zero of the soda scale should be placed; it is easily found by a subtraction.

As experiment has shown that the degrees of soda cannot be equal, and that they are smaller in proportion as they indicate more alkali, the observation should be made on the rose colored scale, which is considered as a measure of equal divisions. From the number found, the figure of the temperature at which the saturation has been made should be subtracted, and the remainder gives accurately the per centage of soda. Thus, in an operation made at 20°, if the instrument sinks in up to 59, we see on subtracting 20 from 59 that the potash contains $39 \div 3 = 13$ per cent of soda.

It is scarcely necessary to take into account the fractions of a degree, as three degrees of temperature occupy only a space equal to one per cent. of soda.

The foregoing are the methods used for determining the value of pots, pearls and sodas, when a stock of either is about being purchased The soap boiler regulates his operations as regards the strength of his ley, in degrees, by a little glass graduated instrument, similar to a hydrometer. It is only approximately correct, but serves the rough purpose of its intention as well, perhaps, as is necessary. They are called alkalimeters, and are made better than the imported, by Stephen Heintz, Glass-blower, Kensington, Philadelphia, who prepares to order all the tubes and glass apparatus appertaining to experimental and analytic chemical operations. A minute description of the hydrometer, and the mode of using it, is given under an appropriate heading.

CHAPTER VI.

ACIDS.

THESE, composing a large class of bodies in the organic and mineral kingdoms, are found sometimes free, but usually combined. In the earlier days of the modern science of Chemistry, it was usual to apply this term to such oxidized bodies as possessed a sour taste, reddened certain vegetable blue colors and neutralized alkalies. It soon became evident that the definition of the term acid must be extended; for, 1st, the hydracids, such as the hydrochloric, contain no oxygen, (which in years past was considered as the sole acidifying principle), excepting when in an aqueous solution; 2d, some bodies analogous to acids and presenting all other characters of acids, are insoluble in water, and therefore can neither redden, blue, nor taste sour, such as the silicic acid (flint) which closely resembles the boracic; and lastly, some acids cannot conceal or neutralize the alkaline nature of potassa or soda, excepting when in excess: thus, the carbonate of potassa and soda, and the sulphate of soda exhibit alkaline reaction. The subsequent development of sulphur salts by Berzelius, led to the farther extension of the term acid to certain sulphurets, such as sulphuretted hydrogen, &c. It is difficult to give a definition of acids which may at the same time be precise and chemically correct. We may say that *an acid is the electro-negative compound of a salt consisting of more than two elements.—Booth and Boye's Encyclopædia of Chemistry*, p. 38.

The acids, a knowledge of some of which is important

to the soap boiler, are arranged into two classes, inorganic and organic. The latter will be studied when we speak of fats and oily matters; those of the former division, connected with our art, will be mentioned directly. Practically speaking, the general properties of acids are a sour taste, and the power of changing certain vegetable blue colors, such as litmus, to red. Their capability of forming salts by saturating or neutralizing earthy or metallic oxides, is a more general characteristic than any other. The insoluble silicic acid and such like, do not of course taste sour, and the boracic, although soluble in water, has scarcely an acid taste, but both neutralize bases—the former by fusion, and the latter, either in that way or in solution.

Carbonic acid (CO_2).—This acid, also sometimes called fixed air and mephitic air, was first found by Dr. Black to exist in limestone and the mild alkalies, and to be expelled from them by heat and the action of acids. It occurs also in the atmosphere, forming about 0.05 of one per cent. in volume; is ejected from volcanoes, caves and cellars; exists in mineral springs, and is thrown off from the lungs in respiration. It is also a product of the vinous fermentation, and the staleness of beer, ale and other fermented liquors upon exposure to air, is occasioned by the loss of this acid, to which is due their agreeable pungency. It is also largely produced by the combustion of carbonaceous fuel, of the diamond, and organic substances, in air or oxygen.

Carbonic acid at ordinary pressures is a transparent, colorless gas, of a slightly piquant odor and sourish taste. It feebly reddens the vegetable blues, extinguishes flame, and when inspired produces asphyxia. Water under the pressure of the atmosphere and at 20° Centig., can dissolve its volume. It is formed of one volume of the vapor of carbon and two volumes of oxygen, these two being condensed into one volume in which the weights are as follows:

Carbon	27.27
Oxygen	72.73
	100.00

The specific gravity of the gas is 1.5277, the air being reckoned as unit. This acid combines with most of the salifiable bases and forms carbonates which are decomposable by all the mineral acids, save the hydrosulphuric and hydroselenic acids, which upon some carbonates are without action. The same decomposing power is possessed by many of the mineral acids also; and the effervescence which ensues, upon the application of an acid to a carbonate, is due to the escape of the gaseous acid, which is set free by the union of the base, to which it was connected, with the acid added, and for which latter the affinity is much stronger.

When carbonic acid has accumulated in wells, cellars, &c., its presence can be ascertained by letting down a lighted taper, which if it be not extinguished, indicates that the air may be safely breathed for a short time. Carbonic acid, being so much denser than common air, may be drawn out of cellars or fermenting tubs by a pump, furnished with a leather hose, which reaches to the bottom. Quick lime, mixed with water on account of its affinity for and power of absorbing it, can be used to purify the air of this aërial acid. In the arts it is used in the preparation of aerated waters; and the "mineral water" of our day, and much of the so called champagne wine, is nothing more than—the first, an aqueous solution of carbonic acid; the latter, purified cider, surcharged with carbonic acid gas, pumped in by means of forcing pistons. Farraday liquefied this gas: when allowed to escape from the receiver by a small jet, it is partially frozen by its own evaporation, and forms a white soft mass like snow, which may be handled, and does not evaporate very rapidly, owing to its low conducting power, although its temperature cannot be more than 148° F. Dr. J. K. Mitchell exhibited and explained these phenomena with improved apparatus, some two or three years since, at the Franklin Institute of this city

Sulphuric acid.—The extensive application of oil of vitriol, has rendered it by far the most important of all acids used in the arts. It is less expensive and in energy surpasses all other acids, possessing by its superior affinity for the base, a power, liberative of most of them from their combinations with bases.

Sulphuric acid, between which and oil of vitriol the difference is in the latter being a hydrate, is a combination of sulphur with oxygen, in the proportion of three equivalents of the latter to one equivalent of the former, and represented by symbols, as follows:

$$SO_3$$

The earliest mention of sulphuric acid is by Basil Valentine, a celebrated chemist, who wrote in the early part of the fifteenth century. The original process by which it was prepared consisted in the distillation of dried copperas, which is the sulphate of the protoxide of iron. The highly concentrated liquor which passed over into the receiver, having an oily appearance and consistence, gave rise to the name of oil of vitriol;—this latter, with the addition of the word green, being the title then applied to the copperas or protosulphate of iron which furnished it. This process is still adhered to at Nordhausen in Saxony, but in most countries has been abandoned for a much better method. The acid of commerce is a transparent colorless liquid, and should mark 66° Baume. At 12° below zero it crystalizes, and further down it can solidify. Submitted to distillation, it volatilizes and passes over without being decomposed, and it is upon this property that its purification is founded; a process necessary to be resorted to, when it is to be used for analytic purposes, as the commercial acid contains more or less foreign matters.

The following table by Ure, shows the quantity of concentrated and dry sulphuric acid in 100 parts of dilute at different densities.

Liquid	Specific Gravity	Dry	Liquid	Specific Gravity	Dry	Liquid	Specific Gravity	Dry
100	1.8460	81 54	66	1.5503	53.82	32	1.2334	26.09
99	1 8438	80.72	65	1.5390	53.00	31	1.2260	25.28
98	1 8415	79.90	64	1 5280	52.18	30	1.2184	24.46
97	1.8391	79.09	63	1.5170	51.37	29	1.2108	23.65
96	1.8366	78.28	62	1 5066	50.55	28	1.2032	22.83
95	1.8340	77.46	61	1.4960	49.74	27	1.1956	22.01
94	1.8288	76.65	60	1.4860	48.92	26	1.1876	21.20
93	1.8235	75 83	59	1.4760	48.11	25	1.1792	20 38
92	1.8181	75.02	58	1.4660	47.29	24	1.1706	19 57
91	1.8026	74.20	57	1 4560	46.48	23	1.1626	18.75
90	1.8070	73.39	56	1.4460	45.66	22	1.1549	17.94
89	1.7986	72.57	55	1.4360	44.85	21	1.1480	17.12
88	1.7901	71.75	54	1.4265	44.03	20	1.1410	16 31
87	1.7815	70.94	53	1.4170	43.22	19	1.1330	15 49
86	1.7728	70 12	52	1.4073	42.40	18	1.1246	14 68
85	1.7640	69 31	51	1.3977	41.58	17	1.1165	13 86
84	1.7540	68 49	50	1.3884	40.77	16	1.1090	13 05
83	1.7425	67 68	49	1.3788	39.95	15	1·1019	12.23
82	1.7315	66 86	48	1.3697	39.14	14	1.0953	11 41
81	1.7200	66 05	47	1.3612	38.32	13	1.0887	10.60
80	1.7080	65.23	46	1.3530	37.51	12	1.0809	9.78
79	1.6972	64.42	45	1.3440	36 69	11	1 0743	8.97
78	1.6860	63.60	44	1.3345	35 88	10	1.0682	8.15
77	1.6744	62.78	43	1.3255	35 06	9	1.0614	7.34
76	1.6624	61 97	42	1.3165	34.25	8	1.0544	6.52
75	1.6500	61.15	41	1.3080	33.43	7	1.0477	5.71
74	1.6415	60.34	40	1.2999	32 61	6	1.0405	4.89
73	1.6321	59.52	39	1.2913	31.80	5	1.0336	4.08
72	1.6204	58 71	38	1.2826	30.98	4	1.0268	3.26
71	1.6090	57.89	37	1 2740	30.17	3	1.0206	2.446
70	1.5975	57.08	36	1.2654	29.35	2	1.0140	1.63
69	1.5868	56.26	35	1.2572	28.54	1	1.0074	0.8154
68	1.5760	55.45	34	1.2490	27.72			
67	1.5648	54.63	33	1.2409	26.91			

Hydrochloric acid.—Muriatic acid, known vulgarly as marine acid, and spirits of salt, is gaseous at the ordinary temperature, but is readily condensed by water. It is in aqueous solution that we meet with it in commerce, and as such has usually a yellow tinge, but when chemically pure, is perfectly colorless. This tinge is owing to the presence of iron, which initiates itself during the course of preparation; the process being conducted in iron retorts, and by the action of sulphuric acid upon common salt, with the assistance of heat. Under the pressure of 40 atmospheres at 50°, the gas is condensed into a liquid of specific gravity, 1.27, and, according to Dr. Thomson, one cubic inch of water absorbs 418 cubic inches of gas at 69°, and becomes 1.34 cubic inch. The following table of his construction gives the specific gravity of hydrochloric acid of determinate strengths, adjoining which we also affix the useful table of Mr. Davy.

Hydrochloric acid.

Atoms of water to one of acid	Real acid in 100 of the liquid	Specific Gravity	Atoms of water to one of acid	Real acid in 100 of the liquid	Specific Gravity
6	40.66	1.203	14	22.70	1.1060
7	37.00	1.179	15	21.512	1.1008
8	33.95	1.162	16	20.442	1.0960
9	31.35	1.149	17	19.474	1.0902
10	29.13	1.139	18	18.590	1.0860
11	27.21	1.1285	19	17.790	1.0820
12	25.52	1.1197	20	17.051	1.0780
13	24.03	1.1127			

Hydrochloric acid.

Specific Gravity	Quantity of acid per cent	Specific Gravity	Quantity of acid per cent
1.21	42.43	1.10	20.20
1.20	40.80	1.09	18.18
1.19	38.38	1.08	16.16
1.18	36.36	1.07	14.14
1.17	34.34	1.06	12.12
1.16	32.32	1.05	10.10
1.15	30.30	1.04	8.08
1.14	28.28	1.03	6.06
1.13	26.26	1.02	4.04
1.12	24.24	1.01	2.02
1.11	22.22		

Its composition is one equivalent of chlorine, and one equivalent of hydrogen, and it has for its symbol HCl.

The liquid muriatic acid has a very sour corrosive taste, a pungent suffocating smell, and acts very powerfully upon a very large number of mineral, vegetable, and animal substances, and is of specific gravity 1.2474. A mixture of two measures of muriatic, and one of nitric acid, forms the *aqua regia*, so called from its property of dissolving gold and the less oxidable metals, platinum, &c. The power of nitric acid to decompose hydrochloric acid (an evolution of chlorine taking place by the oxidation of the hydrogen of the acid) imparts to it this property. For analytic purposes, the commercial acid should be diluted with water and a solution of chloride of barium, (to throw down the sulphuric acid contents,) and then distilled.

Nitric acid.—This acid, called also *aquafortis*, is never found insulated. It was known as far back as the thirteenth century, and is now obtained by distillation, with heat, of a mixture of cubic nitre (nitrate of soda) and sulphuric acid. Commercial aquafortis is generally contaminated with foreign salts, which may be gotten rid of by a re-distillation at a gentle heat, after having been mixed with a solution of nitrate of silver, and then decanted from the insoluble chloride of silver which has abstracted all the chlorine and precipitated. In its dry state, this acid consists of 26.15 parts by weight of nitrogen, and 73.85 of oxygen, or of one equivalent of the first gas, and five equivalents of the second, and is expressed by NO_5.

The pure liquid acid is colorless, odorant, and corrosive; that of commerce has a slight straw color tint. The specific gravity at its maximum of concentration is 1.5543, according to Kirwan, and 1.513 after Lussac. Nitric acid has a great affinity for water, which is necessary to its existence. As, in a number of operations, it is requisite to know the quantity of real acid contained in 100 parts of a mixture of known density, we have added here the following table by Ure.

Specific Gravity	Liquid Acid in 100	Dry Acid in 100	Specific Gravity	Liquid Acid in 100	Dry Acid in 100	Specific Gravity	Liquid Acid in 100	Dry Acid in 100
1.5000	100	79.700	1.3783	66	52.602	1.1895	33	26.301
1.4980	99	78.903	1.3732	65	51.805	1.1833	32	25.504
1.4960	98	78.106	1.3681	64	51.068	1.1770	31	24.707
1.4940	97	77.309	1.3630	63	50.211	1.1709	30	23.900
1.4910	96	76.512	1.3579	62	49.414	1.1648	29	23.113
1.4880	95	75.715	1.3529	61	48.617	1.1587	28	22.316
1.4850	94	74.918	1.3477	60	47.820	1.1526	27	21.519
1.4820	93	74.121	1.3427	59	47.023	1.1465	26	20.722
1.4790	92	73.324	1.3376	58	46.226	1.1403	25	19.925
1.4760	91	72.527	1.3323	57	45.429	1.1345	24	19.128
1.4730	90	71.730	1.3270	56	44.632	1.1286	23	18.331
1.4700	89	70.933	1.3216	55	43.835	1.1227	22	17.534
1.4670	88	70.136	1.3163	54	43.038	1.1168	21	16.737
1.4640	87	69.339	1.3110	53	42.241	1.1109	20	15.940
1.4600	86	68.542	1.3056	52	41.444	1.1051	19	15.143
1.4570	85	67.745	1.3001	51	40.647	1.0993	18	14.346
1.4530	84	66.948	1.2947	50	39.850	1.0935	17	13.549
1.4500	83	66.155	1.2887	49	39.053	1.0878	16	12.752
1.4460	82	65.354	1.2826	48	38.256	1.0821	15	11.955
1.4424	81	64.557	1.2765	47	37.459	1.0764	14	11.158
1.4385	80	63.760	1.2705	46	36.662	1.0708	13	10.361
1.4346	79	62.963	1.2644	45	35.865	1.0651	12	9.564
1.4306	78	62.166	1.2583	44	35.068	1.0595	11	8.767
1.4269	77	61.369	1.2523	43	34.271	1.0540	10	7.970
1.4228	76	60.572	1.2462	42	33.474	1.0485	9	7.173
1.4189	75	59.775	1.2402	41	32.677	1.0430	8	6.376
1.4147	74	58.978	1.2341	40	31.880	1.0375	7	5.579
1.4107	73	58.181	1.2277	39	31.083	1.0320	6	4.782
1.4065	72	57.384	1.2212	38	30.286	1.0267	5	3.985
1.4023	71	56.587	1.2148	37	29.489	1.0212	4	3.188
1.3978	70	55.790	1.2084	36	28.692	1.0159	3	2.391
1.3945	69	54.993	1.2019	35	27.895	1.0106	2	1.594
1.3882	68	54.196	1.1958	34	27.098	1.0053	1	0.797
1.3833	67	53.399						

Aquafortis, strictly speaking, is a weaker nitric acid, whilst nitrous or hyponitric acid, a title which is sometimes improperly given thereto, is a separate and distinct definite compound of oxygen and nitrogen (NO_4). In acting upon the less oxidable metals, such as copper and mercury, nitric acid is itself decomposed, and nitric oxide gas (NO_2) comes off with effervescence and in red fumes.

Boracic acid.—This acid is a compound of oxygen with an elementary substance to which the name of *boron* is applied. It is prepared by mixing with a nearly saturated solution of borate of soda (borax), at its boiling point, a quantity of oil of vitriol, equal to one-fourth the weight of the borax. After filtering the solution, thin crystaline plates are deposited during its cooling. Its application to art will be known when we write upon the facture of candles.

Boracic acid is inodorous and has a weak scarcely acid taste, and affects blue litmus paper like carbonic acid, imparting to it only a light red tint, and not that deep red which the stronger acids produce. At 60° F. it requires 25.66 times of its weight of water to solve it, but only 2.97 times at 212°. It is soluble in alcohol, and the solution burns with a green flame.

Acidimetry.—This has for its object the determination of the strength of acids, without resort to analysis. The most convenient for liquids is the specific gravity, but it cannot always be employed, since the same acid of different strengths may have the same density, and foreign ingredients may affect the density and not the acid. However, the little glass instrument, or acidimeter, made by Heintz, answers all practical purposes. There are more scientific ways of estimating the value of acids, but either of them would be out of place here as involving manipulations requiring the skill of an experienced chemist. We therefore direct attention to the acidimeter, or hydrometer, described in Chap. 36, as sufficiently accurate for most practical purposes.

CHAPTER VII.

ORIGIN AND COMPOSITION OF FATTY MATTERS.

FATS, or fatty matters, as generally considered, are those bodies greasy to the touch, inflammable, and capable of being converted into candle or soap material. In consequence of the different consistencies of organic fats, regardless of their source, whether from the animal or vegetable kingdom, they have been classified under several distinct heads, viz., as *oils*, *butters*, and *suets*; the former are the most fluid, the latter the most solid of the three, whilst the second class holds a medium position. All these fats were originally, and for some time considered as simple products, but the researches of Chevreul, and Braconnot, have corrected this erroneous idea by developing experimentally but accurately, the actual constituency of most fats, showing them to be compound bodies formed partially of liquid, and partially of solid ingredients; these components, however, varying proportionally in the different fats as is shown by the following table.

100 parts yield.	Fluid.	Solid.
Fresh butter in summer	60	40
" " " winter	37	63
Hog's lard	62	38
Ox marrow	21	76
Goose fat	68	32
Duck fat	72	28
Ox tallow	25	75
Mutton suet	26	74
Olive oil	72	28
Oil of sweet almonds	76	24
Palm oil	69	31

The fluid constituent here expressed, is olein, an unerring verification of the presence of which, in fats, is the formation of sebacic acid by their distillation; but the solid portion of the animal greases generally consists of two distinct principles, margarin and stearin, though in some instances one or the other is wanting. For example, the former comprises entirely the more consistent part of human fat, whereas, in most vegetable greases, the solid per centage is so exclusively of margarin, that this principle is somewhat characteristic of them, stearin being seldom found associated with it under such circumstances. But olein, margarin, and stearin, are only the *proximate* constituents of fats, for their *ultimate* components, scarcely varying in any of them, are carbon, hydrogen, and oxygen, which, being solidified, as it were, and confined by certain chemical forces, are united in the proper ratio in the form of fat, and so remain until the influence of some stronger power effects its decomposition. The proximate components above named, are not universally the same in all fats, for among these latter there are a few of peculiar physical and chemical properties, the solid principle of which slightly differs from that of others; thus, for instance, in palm oil, crystaline palmitin occupies the place of stearin in tallow; so also cetine in spermaceti. All these components are themselves compound bodies, being positive and definite chemical unions of their several natural fatty acids with a natural base. This base, in most cases, is glycerin, though in some instances, spermaceti for one, ethal takes its place, and the presence of the former is always proved when the tested fat on being boiled yields *acroleine*, a pungent gaseous body. These fatty acids, by changing the termination of its title into *ic*, take the name of the compound which yields them; thus the acid united with glycerin in olein, is called oleic acid. They differ in their properties, and are distinguishable by certain characteristics peculiar to each. None of these acids, when absolutely pure, is alterable in the air, nor indeed are any of the natural compounds which yield them; the rancidity, therefore, assumed by fats on exposure is owing to absorption of oxygen from the atmosphere, which, promoting a species of fermentation in the impurity of the fat, engenders a corrosive smell; or, it may be, that acting upon a portion of the fat itself, decomposes it into carbonic acid, and some new compounds, one or more of which may possess sufficient volatility to render their odor disagreeable. In some fats, however, for example, that of the goat, volatile acids exist already formed, and it is to these that the peculiar natural scent of such greases is owing, the ordinary temperature of the atmosphere rendering them so diffusive as to be readily perceptible. Pure olein, margarin, and the like, are perfectly inodorous and unalterable in the air, because all their volatile and other taint has been dissipated and removed during their purification.

Vegetable fats exist mostly in the pulpy envelop of the seeds whilst the animal fats are found secreted in the *fat cells* in the cavities of the cellular tissue. Intervening the muscular fibres, between the skin and the flesh, around the kidneys and within the abdominal cavity of the omentum, it is most abundant. In the species of whale, the bony cavity of the head is the receptacle of the secreted spermaceti, and in the case of butter, it is mixed in a liquid state in milk. The molecular formation of fat was first announced by Braconnot and Raspail, who have proved that animal fats consist of small microscopic particles enclosed in sacks, a number of which conglomerated together form a mass. Crumbling the whole in cold water between the fingers, separates these granules which are readily removable, free of membranous matter and water, by means of a sieve. This much explains what fats are, and their composition; the next step in course is a knowledge of their origin and of the means by which they are stored in those receptacles from which we extract them. In the first place, then, the living plant is as much indebted to food for the support of life as is the living animal, and the presence of fat in either instance is an evidence of the conversion or assimilation of a part of the elements of their nourishment into those constituents of which it is naturally formed. Assimilation, or the process of formation or growth, says Liebig, goes on in the same way both in animals and vegetables, and the formation of fat in either is owing to the same cause, and that is, the separation of the oxygen from the elements of the food. Plants feed mainly upon the atmosphere from which they absorb their carbon in the form of carbonic acid, the oxygen of which latter is returned by respiration so soon as the organism of the plant has assimilated the carbon and formed it into fat, and those parts of the structure in which it is an ingredient; the other constituents of the fat being similarly procured from the same source. This transmutation of the atmospheric nourishment is effected by the vital force of the vegetable under the influence of light; the digestion of the food being performed by the leaves of the plant, which act the part of the stomach in animals, in absorbing carbonic acid (CO_2), ammonia (NH_3) and water, (HO), the food requisite for their sustenance, from the surrounding media, which afford not only the elements of fat but of all the various secretions upon which the support of life is dependent; for it is now known that plants breathe, feed and digest, and these functions with them are explicable upon as reasonable bases as those which elucidate similar phenomena in animals; though there is this difference, that "animal life depends upon the absorption of oxygen, which is evacuated again under the form of carbonic acid and water, whilst the plants fix both the carbonic acid and water, and eliminate the oxygen." To sum up, therefore, the formation of vegetable fat, differing nothing in its ultimate constituency from animal fat, and as before said composed of carbon, hydrogen and oxygen, is then owing to absorption and transformation of food, containing these gaseous elements, which, by the vital force and the peculiar organic structure of the plant, is secreted and arranged into entirely new products. Among these products is fat, and the difference between vegetables and animals in the assimilation of their food is a dependence with the latter upon certain influences which, not affecting the former, are yet therein otherwise substituted.

As regards the formation of animal fat, there are two theories; one declaring its production by the conversion of the non-azotized ingredients of food, such as starch, sugar, and the like, by the processes of digestion and assimilation; the other averring that it is not generated in the animal frame, but exists already formed in the manifold products of the vegetable kingdom, serving as food. This latter supposition is advanced and supported by Boussingault, Dumas and Payen, the former by Liebig, and as his reasoning is better founded than that of the other chemists, we regulate our remarks accordingly. It is true that fat taken into the body contributes to the formation of fat, but it has other sources also, for the carnivorous animals, which, as their name designates, are flesh eating, have more muscle but less fat than the herbivorous; and for the reason, that the organic principles of the food of these animals are assimilated by their digestive and other apparatus without being organized. The herbivorous animals, on the contrary obtain the constituents of their bodies from plants, the comparatively limited azotized portion of which goes to form muscle and the other nitrogenized parts of the structure, and the non-azotized, or sugar and starchy components, to the production of fat, &c The absence of these latter ingredients from the food of animals creates a leanness of flesh.

The "first and main conditions of animal life are nutritious matter and oxygen," the supply of which comes from the surrounding atmosphere, and is inhaled at every inspiration, the constant sensation of warmth in the body known as "animal heat, having its source in the mutual action between the elements of the food and the oxygen, conveyed by the circulation of the blood to every part of the body."

The non-azotized constituents of the body are water and common fat; and hence, the latter can be increased by augmenting the quantity of non-nitrogenous food, as is evidenced in the case of a stall-fed pig or cow, but with those animals which naturally require flesh diet, there is no more fat produced than exists in the meat which it consumes, for the composition of the food in no wise differing from that of the body, serves entirely for its reproduction, or rather, the reproduction of those parts wasted or destroyed by the process of respiration.

When the analyses of fats are compared with those of the different starchy matters of such plants, grains, &c., as serve as food for certain animals, it will be seen that their composition in carbon and hydrogen is very analogous, the proportion of one to the other being nearly the same; the only difference is in the oxygen ratio; hence it follows, that starch and the like, by the mere separation of a part of their oxygen in the process of assimilation, may pass into fat; and that the *solidification*, as it were, of these gases eliminates or is the cause of the animal heat before mentioned. Fat, therefore, is a product consequent upon a deficient supply of oxygen to the body, and is derived from the non-azotized elements of their food when these are in excess and oxygen is deficient, this latter being absolutely indispensable to the dissipation of the former; otherwise it becomes assimilated into fat, "the deficit of oxygen being supplied at the expense of the starch, sugar and the like, which, by losing oxygen, gives rise to fat; for the proportions of carbon and hydrogen in sugar, &c., and in fat are exactly the same, that of oxygen alone being different." From this it therefore follows, that farinaceous food and abstinence from exercise are promotive of the increase of fat, rest being a requisite condition, for by locomotion sufficient oxygen is taken into the body to consume all its starch and non-azotized food, and hence there is no fat produced, a muscular leanness ensuing, as in the case of the fleet bounding stag which roves at liberty, yet is nourished by food of the same nature as that consumed by cattle. The penned pig, (want of exercise being equivalent to a deficiency of oxygen,) on the contrary, which eats and reposes merely for digestion, consumes more non-azotized food than is needed for the support of respiration and maintenance of animal heat, and the excess going to the production of fat is deposited in the cellular tissue as oil and suet.

It is an established rule that an animal fed on nitrogenized food becomes lean and muscular, whereas those that subsist upon potatoes, grain, and vegetable diet generally, gain of fat. The excess of carbon taken into their bodies, as a component of their food, when eliminated during the process of digestion being unable to form any part of an organ, is transformed into fat; and here is the reason for the practice of stalling animals in order to fatten them before killing time, though the rationale of the process is scarcely known to any one of the whole fraternity of victuallers who practice it.

From this exposition it is readily seen that there is a

mutual dependence between the animal and vegetable kingdoms; the practical effect of which admirably illustrates a very important instance of the economy of nature. The plants deriving their food from the earth and surrounding media in their turn serve for the nutrition of animals, and these, after the process of decay has extinguished their vital force, again serve in an elementary form to reproduce themselves through the aid of that vegetable matter which is generated from their remains. The plant feeds upon the atmosphere, and having thus nourished itself becomes animal food. The animal perishing restores to the vegetable creation that, which while living it took from it for the support of life, and thus is the process of reproduction constantly maintained; the animal and vegetable food, consisting, independent of certain fixed salts, of carbon, hydrogen, oxygen, and nitrogen being assimilated by appropriate organism and assorted out, as it were, and in mixtures of proper ratio directed to the formation of such part of the structure as its constituents fit it for; and thus is produced simultaneously all the wasted portions of the body, both nitrogenous and non-azotized.

In this most beautiful arrangement there is evidenced a wonderful realization of the truth of that figurative expression which declares, that our bodies are of dust and to dust must return, for the same earth which yields the plant, thereby sustains animal existence, and hence both man and worm alike are recipients of and contributors to its productive power; generated from its elements, dying they are again resolved into the same, and become a source of nutriment for new creations of life.

CHAPTER VIII.

SAPONIFIABLE FATS.

THE fatty matters which are saponifiable, are of both vegetable and animal origin. Their characteristics are, a voluminous white or reddish flame when burned, a greater or lesser solubility in alcohol and insolubility in water. As arranged in groups, the first class rests upon their degree of fusibility; thus, under this head might be placed those which are liquid at 10° to 15° Centigrade, and to which is given the name of *oils*. In the second class are ranged those fats which are of a soft consistence at 18° Centigrade, and which are styled *butters*. The third class comprises the greases which are yet more solid than those of the preceding. The fourth embraces the suets, the fusing point of which is about 40° Centigrade; and in the fifth, are contained the waxes which melt at 45° to 65° Centigrade. Finally, the sixth class consists of the resins, the characteristics of which are their solidity, high point of fusion, affinity for greasy bodies, friability and odor.

It is readily perceived that there is much difficulty in determining a correct classification of the fatty bodies. To obviate these inconveniences most authors divide these same bodies into two classes only;—the most simple and convenient arrangement.

The first class comprises those extracts of plants more generally known by the name of oils;—wax, and the resins, by a kind of analogy, also ranging under this head. And the second, consists of the animal fats known as suets, tallow, lard, &c.

Oils.—The discovery of the oils dates back to a remote period, for Abraham (Gen. xv. 17) served it in the lamps, and from his history we read that the Cecrops brought olive oil from Sais in lower Egypt, there acquired the art of extracting the oil, and imparted it to the Athenians. The use of oil was thus carried into Europe, (Herod. book ii, 59.) It seems certain, nevertheless, that though the Greeks at the time of the siege of Troy, knew of the oils, they were ignorant of their mode of application to illumination, for no note is made by Homer, but on the contrary, the heroes used torches of wood for this purpose.

Under the name of oil is designated those liquids having an unctuous consistence and a liquidity, which are retained to a temperature less than 10° to 12°. They are generally lighter than water, and when in contact with paper, produce a spot which renders that part of the paper touched by it semitransparent. They are more or less viscid, of a feeble but sometimes disagreeable taste. If oil is added to and stirred with water, on repose, by reason of its lesser specific gravity, and immiscibility, it rises to the surface of the water and forms a separate and distinct layer; the water beneath and forming the lower stratum having solved out and retaining a portion of the foreign matters originally connected with the oil.

There is great difference frequently in the same kind of oil, varying according to the locality where the plant was cultivated, and also the mode of extracting the sample of oil therefrom. The specific gravity of oil increases with its age, but we will here remark in passing, that the degree of purity of an oil cannot be accurately deduced from its specific gravity, for this itself varies at different times. It is, however, a pretty fair criterion.

List of plants which yield the ordinary oils of commerce.

Plants.	Oils.	Sp. gravity.
Linum usitatissimum	Linseed oil	0.9347
Corylus avellana / Juglans regia	Nut oil	0.9260
Papaver somniferum	Poppy oil	0.9243
Cannabis sativa	Hemp oil	0.9276
Olea Europæa	Olive oil	0.9176
Amygdalus communis	Almond oil	0.9180
Guilandina moringa	Oil of ben	
Fagus sylvatica	Beech oil	0.9225
Sinapis nigra	Oil of mustard	0.9160
Helianthus	Oil of sunflower	0.9262
Brassica napus	Rapeseed oil	0.9136
Ricinus communis	Castor oil	0.9611
Prunus domestica	Plum kernel oil	0.9127
Thœbroma cacao	Butter of cacao	0.892
Cocos nucifera	Cocoa nut oil	
Cocos butyracea	Palm oil	0.968
Gossypium barbadense	Cotton seed oil	
Brassica campestris oleifera	Colza oil	0.9136
Prunus cerasus	Cherry stone oil	0.9239
Æsculus hippocastanum	Horse chestnut oil	0.927
Pinus abies	Pine top oil	0.285

Among the different species of oil, there are those which can be raised to a high temperature without volatilizing, but which, when brought near to the point of ebullition, begin to be decomposed, become blackened, and emit thick whitish fumes. Others, on the contrary, are distillable, without fear of alteration, provided always, however, that the heat does not exceed their point of ebullition. These latter are designated as volatile oils, and will be spoken of hereafter. The first, called fixed oils, are insoluble in water, and but slightly soluble in alcohol, castor oil excepted.

Here is Planche's table of the solubility of certain oils in 1000 drops of alcohol, at 40° Baume.

Oil of ricini (Castor oil) - - - - -	all proportions.
" poppy (year old) - - - - -	8 drops.
" linseed - - - - - - -	6 "
" poppy (fresh) - - - - - -	4 "
Nut oil (juglans) - - - - - -	6 "
Beech oil - - - - - - - - -	4 "

Olive oil	- - - - - - -	3 drops.
Oil of sweet almonds	- - - - -	3 "
Nut oil (corylus avellana)	- - - -	3 "

The fixed oils exist in the seeds of plants, and sometimes, but rarely, in the fleshy pulp of fruits. The oleaginous grains contain at the same time fecula, and a kind of mucilage, which renders them miscible with water, giving with this liquid a white liquor, known by the name of *emulsion*, or milk of almonds, when it is with that fruit prepared; and it is by reason of this property that some seeds are called emulsive. The f t oils are contained in that part of the seed which gives birth to the cotyledons, but not in the plumula and radicle. Of all the families of plants the cruciform is the richest in oleiferous seeds. One root, alone, contains a fat oil, viz., the *cyperus esculenta*. The fixed oils are, at the atmospheric temperature, nearly all liquid, some, however, as the palm, butters of galam and cocoa, &c., are more or less consistent; they are greasy, some colorless, but generally of an amber shade and occasionally greenish, the color probably being due to a peculiar principle held in solution. The fixed oils are mostly soluble in sulphuric ether, but the exact solvent power of this liquid has not been determined. It, however, is an excellent solvent, and is used in analysis for extracting oils from other bodies, after which it is separated by evaporation. Acetic ether has a less solvent power, and its ability in that respect is represented by the following table of Planche.

Quantity of acetic ether.	Names of the oils.	Quantities dissolved.
1 part	Castor oil	8 pts. and over
2 parts	Nut oil (juglans)	1 part
2 "	Linseed oil	1 "
2½ "	Beech.oil	1 "
3 "	Poppy oil	1 "
4 "	Sweet almond oil	1 "
5 "	Olive oil	1 "
7 "	Nut oil (corylus)	1 "

Several fat oils, mixed with one or two per cent. of sulphuric acid, instantly assume a dark green, or brown hue, and when allowed to stand quietly, after some time, deposit a coloring matter. It consists in a chemical combination of the sulphuric acid with a body, thus separated from the oil, which becomes in consequence more limpid, and burns with a brighter flame, especially after it has been washed with steam and clarified by repose or filtration. Any remaining moisture may be expelled by the heat of a water bath.

In close vessels, oils may be preserved fresh for a long time, but when exposed to air or oxygen gas they undergo a gradual alteration, losing their liquidity, becoming thick and viscid, and some even attaining to solidity, especially when they are in thin strata. In becoming solid they acquire more color and solidity. These last take the name of drying or siccative oils, a property which procures their use in varnishes and mixed paints. Of this number are the oils of linseed, nuts, poppy, grapeseed, &c. Others, again, do not dry up, yet thicken, become less combustible and contract a disagreeable smell. They are then called rancid and *non-siccative* oils, in contradistinction to the first mentioned drying oils. Among them are the oils of olive, colza, sweet almonds, of nuts and castor. Saussure has announced the phenomena occurring during exposure, as follows. A layer of nut oil one-quarter of an inch thick, and three inches diameter, enclosed with oxygen gas over quicksilver, in the shade, absorbed only three times its bulk of gas during eight months, between December and August, but in the ten days following the first of the latter month, it had absorbed 60 times its volume. At the end of October, the time at which the diminution of the volume of the gas became nearly insensible, this oil had taken up 145 times its volume of oxygen gas, and given out 21 times its volume of carbonic acid gas, and without any production of water. The oil thus treated, formed a kind of transparent jelly, which no longer spotted paper. To a similar absorption may be attributed the elevation of temperature which happens when wool or hemp, besmeared with olive or rapeseed oil, is left in a heap, circumstances under which it has frequently taken fire, and caused the destruction of cloth mills and store houses. For instance, by way of illustration, if paper, linen, wool, tow, cotton, mats, straw, moss or soot, be slightly imbued with linseed or hempseed oil, and placed in contact with the sun and air, especially when in heaps or piles, they become, very soon, spontaneously hot, emit smoke and finally burst into flames. If linseed oil and ground manganese be triturated together, the soft lump so formed will speedily become firm, and in a short time take fire. This spontaneous combustion is owing to a rapid abstraction of oxygen from the atmosphere ; the hydrogen of the oils being inflamed by the caloric eliminated from the gas at the moment of its absorption by the oil, and until then, existing as latent heat, and serving to retain the oxygen in an aeriform state.

Oils possess the same ultimate constituents, carbon, hydrogen and generally oxygen, and sometimes a little nitrogen. The fat oils exposed in a retort to a temperature sufficiently high to effect distillation, are in part decomposed; carburetted hydrogen gas (illuminating gas) being disengaged, and a brownish-yellow oil of a very strong piquant odor, passing over into the receiver. The residue is a small quantity of carbonaceous matter, hence the acrid and irritant taste of a ragout, when the cook allows the metallic vessel in which it is cooked, to become too strongly heated. It is upon this property that is founded their conversion into illuminating gas, a simple apparatus for which purpose was invented by Messrs.

J. & P. Taylor, England, as far back as 1815. Mr. Crutchett, of Cincinnati, Ohio, of late years, has put in his claim for an invention applicable to this purpose, and with success, I believe, for I have myself been in several steamboats and buildings in the western country, lit up with oil gas produced by his apparatus. The oils dissolve both sulphur and phosphorus by the aid of heat. Their action upon oxides, or that of the oxides upon them will be explained further on, when speaking of the theory of saponification. Chlorine and iodine act upon oils even in the cold, and from the decomposition of the latter, result the hydrochloric and hydriodic acids and other combinations, which have not as yet been examined.

Relative to the elementary principles of oils, De Saussure concludes that the less fusible fats contain more carbon and less oxygen, and that oils are more soluble in alcohol proportional to their greater content of oxygen.

Olive oil.—The *olea europea,* a tree regarded by the Greeks as the emblem of peace, is one of the prettiest in nature, and occupies so distinguished a place in agriculture, that Varron called it the "first of all trees." Its origin it is supposed was in Greece or Asia Minor. It is impossible to say exactly when the discovery was made of the oil, and its application to the wants of man. It seems to have been known from antiquity, for its mode of facture is mentioned in Holy Writ, and what is remarkable, it has but little varied ever since. In the different countries of Spain, as Catalonia, Valencia, &c., whence comes the most delicious oil, ignorance and prejudice, favorable to the old routine, have prevented all amelioration. The olive oil, well prepared, is of a bright yellow, sometimes inclined to greenish, as is the case with that from Roussillon. Its taste is mild and agreeable; it is unctuous to the touch, transparent when very pure, insoluble in water, slightly soluble in alcohol and ether. Heated up to 315° F., it leaves a spot upon paper not removable by heat, a property distinguishing it from the volatile oils. Exposed to high heat it is partially decomposed, one portion volatilizing and running over in an altered state; more colored, of a strong odor, lighter, and of greater fluidity. It clarifies after some months, and deposits a blackish viscid substance called "*oil refuse*," with which is made a black soap. There are three kinds of olive oil in the market. The best, called virgin salad oil, is obtained by gentle pressure of the fruit, in the cold; the common sort is procured by stronger pressure, aided by the heat of boiling water; and thirdly, an inferior kind, by boiling the residuum or *marc* of the olives (after these two pressures) with water, whereby a good deal of mucilaginous oil rises to the surface, and floats on it. The latter serves chiefly for making soaps. A still more inferior oil is obtained, by allowing the mass of bruised olives to ferment before subjecting it to pressure. A few degrees above 32° F. it begins to deposit some white granules of margarin, especially if the oil has been originally expressed with heat. At 22° it deposits twenty-eight per cent. of its weight of solid constituent, which is fusible again at 68°, and affords seventy-two per cent. of olein.

The pure salad oil is used for table and domestic purposes, the inferior kinds in the manufacture of soaps. Its superiority over other sorts of oil, and higher price render it liable to adulteration, more especially with lard oil, which sometimes forms as high as sixty or seventy per cent. of "*surfin Lucca huile*." To guard against such impositions, we give below a means of detecting them. The lard oil always contains more or less of stearin, which it drops at a much higher temperature than that at which the margarin falls from the olive oil. This enables an approximately correct judgment. To be critically accurate, the solid deposit must be separated by filtration, saponified, the resulting soap decomposed, and the eliminated acid carefully dried. The point of fusion then carefully ascertained will determine whether it is stearic or margaric acid; if the former, then it is proper to infer the presence of lard oil.

Oil of poppies, too, is not an unfrequent falsification of olive oil; it is detected by the addition of hyponitric (nitrous) acid to the latter, which, if pure, becomes solid or, more properly speaking, transformed into elaidine, whilst if there is but $\frac{1}{10}$th of poppy oil present, the concretion is prevented, a greasy paste only being produced.* The composition of olive oil is oleic and margaric acids in union with glycerin; the more solid constituent, or margarate of glycerin, being in the proportion of about twenty to twenty-eight per cent.

Oil of sweet almonds.—This oil is obtained by pressing the bruised fruit of the *amygdalus communis*, (previously cleansed of the adherent furz and dust, by being well shaken and sieved,) in strong bags, between iron plates, slightly and uniformly heated by hot water or steam. The oil thus obtained, is filtered and kept in carefully stoppered bottles, as it is prone to become rancid. It is of a bright color, with a feeble taste of almonds, and after a time, deposits a mucilaginous substance which must be

* M. Diêsel has recently announced that olive oil is colored green by ordinary nitric acid. On the other hand, one drachm of olive oil, sophisticated with rape oil, and mixed with 12 drops of nitric acid, exhibited a strong yellowish-gray color, and this reaction is perceptible even when the adulteration forms but $\frac{1}{10}$th. When poppy oil is the adulteration, the mixture is rendered yellowish-white by nitric acid, and not brownish, as in the previous case. Pure nitric acid exhibits the reaction in a far less degree. Dr. Bley adds, that the samples must be judged of in the course of 12 hours after the addition of the acid, as pure olive oil is likewise colored more or less brown after a longer interval.—*Archiv. du Pharm.*, xlvi. p. 287.

immediately separated by filtration, as it engenders rancidity. The tree which furnishes the nut is native to Africa, but cultivated in Spain, Italy, and France, is of several varieties, and their fruits are usually distinguished as sweet and bitter almonds. Sweet almonds were analyzed by Boullay and others; the bitter by Vogel.

Sweet Almonds by Boullay.		*Bitter Almonds by Vogel.*	
Fixed oil	54.0	Fixed oil by pressure	28.0
Emulsin	24.0	Emulsin	30.0
Uncrystalized sugar	6.0	Uncrystalized sugar	6.5
Gum	3.0	Gum	3.0
Woody fibre	4.0	Woody fibre	5.0
Seed pellicles	5.0	Seed pellicles	8.5
Acetic acid	0.5	Loss in water } Fixed and volatile oil, &c. }	19.0
Water	3.5		
	100.0		100.0

The important constituents are fixed oil, and emulsin in both varieties, and amygdalin and volatile oil in bitter almond. The fixed oil, as above said, is obtained by expressing both sweet and bitter almonds, but usually the former. It generally congeals at a lower temperature than olive oil. It is soluble in ether, and in six parts of boiling and twenty-five of cold alcohol.

Emulsin is an albuminous substance.

Amygdalin is a crystalizable constituent of the bitter almond, and its source is the almond cake, or *marc*, which remains after the expulsion of the fat oil by pressure between hot iron plates. It is very soluble in water, soluble in boiling alcohol of ninety-four per cent., but scarcely so in cold anhydrous alcohol. Its formula is $C_{40}H_{27}NO_{22}$. It was discovered, subsequent to the foregoing analyses, by Roubiquet, and Boutron-Charlard. The volatile oil of bitter almonds, it is now conceded, does not exist ready formed in the fruit, but, according to Liebig and Wohler, is produced from amygdalin by the presence of emulsin and water. Amygdalin and emulsin being in distinct cells in the cotyledon of the bitter almond, together with the fixed oil, cannot act on each other, and are still further prevented when the water is expelled, by drying. The fixed oil is separated by pressure, and upon boiling the residual cake with alcohol, the emulsin is coagulated, and amygdalin extracted. When the cake is moistened with cold or lukewarm water, the odor and taste of prussic acid, (which, together with formic acid, sugar, &c., are also cotemporaneous results,) and of the volatile oil, are at once perceived, but if the quantity of water be too small, a portion of amygdalin is undecomposed. The decomposition commences instantly, but requires some time for a complete transformation, so that but little oil would be obtained by hastily distilling the mixture, as the emulsin would be coagulated and farther action cease. The same result takes place when the pressed cake is thrown into boiling water, the emulsin being instantly coagulated and rendered inert. To obtain the volatile oil, the almond cake, after the removal of the fixed oil, is made into a thin paste with lukewarm water, left for twenty-four hours in a closed carboy, and distilled in a water or chloride of calcium bath. One hundred parts of amygdalin yield forty-seven parts of crude oil; these forty-seven parts contain 5.9 parts of free prussic acid. This oil is highly poisonous; has a golden yellow color, an agreeable odor, an acrid bitter taste, burns with a white flame, is soluble in ether and alcohol, and has a specific gravity of 1.0436. Sulphuric acid forms with it a thick crimson red liquor, which by dilution with water, gives a yellow emulsion. The volatile oil is used for flavoring soaps, &c.

The fixed oils of almonds mostly used in our art, are not very frequently met with in commerce pure and unadulterated, their greater portion being either olive or lard oil. Indeed, I know of instances of pure olive oil merely flavored with that of almonds, being sold as the genuine article, having been present in a large drug house when the oil was being "made up." To detect such frauds resort must be had to tests, to be convinced of the purity of the article before purchasing it, and for these refer to Chap. 9. The composition of fixed almond oil is oleic, margaric, and perhaps a little stearic acid in combination with glycerin, and it belongs to the non-drying class of oils

Beech oil.—This name is applied to the fruit of the *fagus sylvatica.* The oil is prepared by pressure in the cold, in a manner similar to that by which the preceding oils are made. It has an amber color, a mild savor, and is inodorous. Although inferior to olive oil, it is sometimes substituted therefor as an aliment, and frequently forms a part, as an adulteration, of that which is sold for a pure article.

Rapeseed oil.—This oil is prepared from the seeds of the navette, or *napus brassica,* by heating them with hot water and pressing them. The oil is yellow and of a peculiar smell, and very viscous. This viscidity and the coloring matter retained by the oil, lessen its combustibility and occasion much smoke during its burning. It is purified by adding to it two-hundredths of its weight of sulphuric acid, and agitating well the mixture with double its volume of water. After eight or ten days, decant the oil which swims to the surface of the water and forms the upper stratum, and filter it through a displacing funnel, the holes in the diaphragm of which are furnished with cotton wicks of a decimetre in length.

In this operation the sulphuric acid precipitates the coloring matter under the form of greenish flocculæ; the water uniting with the acid, drops these flocculæ from the solution which it formed of them, and being insoluble in the oil they fall to the bottom. The oil is racked off clear by filtration, as above directed. A mild heat assists this operation, which is equally applicable to the purification of other oils. At 25° F. it becomes a yellow mass, which, by pressure, yields forty-six parts of solid constituent, fusing at 50°, and fifty-four parts of olein or liquid component.

Oil of mustard. This oil is obtained by pressing the bruised seeds of the *sinapis alba* and *nigra.* Thus is procured from twenty to twenty-five per cent. of a very mild yellow oil, which concretes when cooled a little below 32° F. It is a remarkable fact, says Fontinelle, that the mustard seed, after the expression of the mild oil therefrom, becomes more medicinally active by reason of the volatile oil which is then in a state of incipiency. This volatile oil is not contained in the seed, but results from the reciprocal action of water and an albuminous substance in the seed, called *myrozine* by Bussy, upon another crystalizable principle in it, *myronic acid,* which is soluble in water, and apparently is an acid, though this is not positive, as but little is yet definitely known respecting it. This oil is the cause of the acidity of black mustard, a property worthy the consideration of manufacturers of mustard. As in the case of volatile oil of almonds, the application of boiling water (to the mustard), of alcohol, of acids, or of alkalies, which coagulate the albuminous body, prevents the formation of the volatile oil. This volatile oil is not used by the soap factor, therefore we shall make no further note of it.

Oil of grape seeds.—This is an oil which deserves some attention. It is prepared from the seeds of the *vitis vinifera* by pressure. It is yellow, slightly odorous, and of a mild taste when it is very pure and fresh. By age it becomes viscous, thick, and rancid. It should be purified as is directed for the oil of navette.

Oil of poppy.—This name is given to that oil yielded on pressure by the bruised seeds of the *papaver somniferum.* It is whitish yellow, inodorous, of an almond taste, and when pure is less viscous than most oils, and remains liquid even to 0°. It has none of the narcotic properties of poppy juice. It is soluble in ether in every proportion, and is used as a means of detecting the presence of spirits of turpentine in essential oil; full particulars as to the process will be given under the proper head. This oil is purified as those of the other oleaginous grains. It belongs to the class of drying oils.

Nut oil.—This comes from the seeds of the *juglans regia* by either cold or hot pressure. That prepared by hot pressure has a peculiar taste and odor, the latter being recognizable during combustion, a general attendant of which is much smoke. This oil is siccative, and the nuts which yield it must be freed of their ligneous envelop before being expressed. The cold pressed oil is almost colorless, of an agreeable odor and taste, and syrupy consistence. Exposed to the air, it rapidly rancidifies, and becomes as clear as water, particularly when it is put into large shallow vessels with water at the bottom. This oil, thus altered, is used for fine colors, in painting, it having a more drying quality than linseed oil, and like it, congealing into a white mass at the same temperature. The hot pressed oil is very much colored, of a strong odor, and full of mucilage.

The oil of noisette or *fructus avelanæ,* also sometimes called nut oil, is similarly obtained. This also is a drying oil, and in its properties not unlike the above oil. The composition of both this and the preceding oil, is in all probability olein and margarin, or properly speaking, oleate and margarate of glycerin.

Linseed oil.—This oil is obtained in its greatest purity by cold pressure, but by a steam heat of about 200° F. a very good oil may be extracted. The usual method is to torrefy the seeds (*linum usitatissimum*) in order to destroy the mucilage, to bruise, and then submit them to pressure. The oil is of a greenish-yellow color, and has an odor peculiar to itself. It is siccative especially when boiled with litharge, becoming then reddish and clear after repose. It dissolves in five parts of boiling and forty of cold alcohol, and in 1.6 parts of ether. When kept for some time in a partially open cask, and in a cool place, it deposits a portion of its solid constituent along with a brown powder. This solid deposit is difficultly saponifiable.

The composition of linseed oil is oleic and margaric acid in union with oxide of glyceryl, or in other words, olein and margarin.

Castor oil.—This oil is drawn by pressure from the seeds of the *palma christi* or *ricinus communis*, previously deprived of their envelop, bruised and boiled in five times their weight of water. When very pure, this oil is bright yellow, inodorous, insipid, thick, liquid, many degrees below zero, siccative, and inspissates on exposure to air without losing its transparency. The oil as we get it, has generally, however, an unpleasant smell and nauseous taste. It is heavier than most other fixed oils, and is soluble in all proportions in cold absolute alcohol; weaker alcohol of specific gravity 0.8425, takes up about three fifths of its weight: adulterations with other fixed oils may thus be detected, as the latter are but slightly soluble in this fluid. It is used in medicine as a cathartic and vermifuge, and in the arts for making a very beautiful but rather expensive soap.

According to Lecanu, and Bussy, it gives by distillation a colorless volatile oil, very odorous, pungent, and crystalizable by cold. It is separated by washing and by distillation, with water. It is limpid and colorless, has a peculiar odor, acrid taste, and specific gravity of 0.815; is soluble in alcohol and ether, but insoluble in a solution of potassa.

A soap of castor oil, when decomposed by a mineral acid, gives two solid and one liquid oily acid. The two former are 1st, the stearicinic, margaritic, or ricino-stearic acid, which crystalizes in pearly scales, and fuses at 266° F. Its high melting point distinguishes it from its concomitant acids, the ricinic and ricinoleic. It is further detected from them by its partial decomposition when subjected to distillation, and by the insolubility of its salt of magnesia in alcohol. 2d. *Ricinic acid*, a crystaline body, solid at ordinary temperatures and fusible at 72° F. 3d. *Ricinoleic or elaïodic acid*, a yellowish-colored liquid at 32° F., but becoming crystaline at many degrees below it.

Castor oil, therefore, is a compound of stearicinate, ricinate, and ricinoleate of the oxide of glyceryl, or glycerin, which, according to the old idea of the construction of fats, would be called respectively, as uncombined bodies, *stearicinin*, *ricinin*, and *ricinolein*.

Castor oil, solidified with nitrate of suboxide of mercury, or with nitrous acid, forms a yellow waxy transparent mass which, washed with water and dissolved in boiling alcohol, furnishes *palmine* in confused crystaline grains This new fat, soluble in alcohol and ether, is easily saponified, like elaïdine, by alkalies, glycerin being disengaged and palmate of soda or potassa being formed. When this soap thus produced is solved in water, and hydrochloric acid added to the solution, it is decomposed, and the *palmic* acid separates, on cooling, in a crystaline mass. When pure, this acid strongly reddens litmus paper, decomposes the carbonates, fuses at 122° F., and crystalizes in white silky needles. Alcohol and ether solve it in all proportions. According to Playfair, palmic acid is represented by $C_{34}H_{32}O_5$, but when obtained by saponification with potassa, has an additional equivalent of water, making it $C_{34}H_{33}O_6$. The rancidity is owing to the presence of a peculiar volatile acid called the œnanthylic acid.

Palm oil.—The appearance of this oil in commerce dates back a long time, though it was not until late years that its present extensive employment commenced. It is procured by expressing, as also by boiling the fruit of the *elais avoira*, a tree growing in Guinea and Guiana. The fruit is of the size of a pigeon's egg, and yellow colored. In the envelop or covering, called the *caire*, is a yellow unctuous substance which is given as food to the cows, monkeys, and other animals. From the nut contained in the envelop there is also extracted a species of butter called *quio quio* or *thio thio*.

Palm oil has a buttery consistence, and when fresh, is of a violet odor, a mild taste, and an orange color, which bleaches by age. These characteristics distinguish it from the factitious article, which is nothing more than tallow and lard melted together, colored with turmeric, and aromatized with powdered orris root. Palm oil is lighter than water, melts at 84 F., becomes rancid and pale by exposure to air, and dissolves in boiling alcohol, but separates on cooling. Ether dissolves it in all proportions.

Palm oil is chiefly used for admixture with fats, to form yellow soaps, and also for toilet soaps, a delightful one of which, Graves, of this city, has the skill in making both free of color, and retentive of its violet odor.

Palm oil, when saponified, reveals its composition, which is the oleate and palmitate of the oxide of glyceryl, or, according to old notions, *olein* and *palmitine*. The soap of this oil yields, when decomposed by a mineral acid, a mixture of *palmitic* and oleic acids, which mixture dissolved in boiling alcohol, gives crystals of palmitic acid on cooling. Pure *palmitic acid* is in the form of brilliant plates, completely similar in appearance to margaric acid and of the same point of fusion, 140° F. It is insoluble in water, but forms a transparent emulsion with alkaline carbonates. *Palmitine*, or palmitate of glycerin (oxide of glyceryl), when pure, is crystaline and of brilliant whiteness, is very slightly soluble in boiling alcohol, but dissolves in boiling ether in all proportions, and is deposited on cooling in extremely small crystals.

The *elais avoira* or *guineensis*, grows also in Brazil, but the demand is supplied from the West Coast of Africa, where it is extensively cultivated as an article of trade. It has of late years so increased in commercial importance, that, according to Mr. Jamison, in his pamphlet on the importance of the trade between England and West Coast of Africa, the trade in palm oil has been the principal instrument in the reduction of the traffic in slaves.

"The peculiar and characteristic deep orange red color of the oil as it reaches us, renders it unfit to be used in its natural state for making either palm or white soap. The substance which is the source of color, is not, however, proper to the pure oil, the constituents of which, when in a state of chemical purity, are colorless; it is a product of the spontaneous decomposition of a peculiar principle derived from the palm, and is capable of being decomposed and decolorized at a high temperature by the usual bleaching agents." These must therefore be resorted to for the removal of this coloring matter, which is effected by either of the methods (all new and valuable) about to be named.

Mr. Watt not long since secured a patent in England, by which he proposed to use the nascent oxygen produced by the decomposition of chromic acid which consists of one equivalent of chromium, and three equivalents of oxygen $= CrO_3$. This method, though very successful, is not economical, and therefore, not so available for practical purposes. The old plan, too, of employing nitric acid, though not objectionable on the score of cost, has yet the disadvantage of impairing the peculiar violet odor of the oil, in which consists much of its value, whilst at the same time it only destroys the coloring matter without perfectly bleaching the oil—the soap into which it is made being of a grayish-yellow hue. The oil operated upon by the first mentioned chromic acid process, produced a soap of most beautiful whiteness, and rich with the characteristic odor of the oil,—facts which I have verified by my own experiments. The expense attendant upon this process per ton of oil, labor included, reaches ten dollars, and I strongly recommend it, regardless of its cost, where a soap is desired of excellent whiteness and flavor.

You first heat the oil by steam in a common boiler for thirty minutes or more, then allow it repose until having cooled to 130° F., when it must be drawn off from the sediment and water, formed by the condensation of the steam, into wooden vessels, each capable of containing about half a ton. To each vessel containing the half ton of oil, so far deprived of foreign matter, make the following addition. Twelve and one-half pounds of bichromate of potassa dissolved in water, so as to make a saturated solution, four pounds of concentrated sulphuric acid, and about twenty-five pounds of strong hydrochloric acid, well mixed together. These are Mr. Watts' proportions; some bleachers take ten pounds of bichromate of potassa, and forty pounds of commercial hydrochloric acid. After well agitating this mixture with the oil thoroughly for a few minutes, the oil changes in color, becoming first black, then dark green, and soon afterwards light green, when a thick froth appears on the surface. This appearance of a light green color and a froth, is an indication of the completion of the process. If a sample of the oil, when taken out and allowed to settle, does not appear sufficiently decolorized, an additional portion of the bleaching mixture may be added. The process is complete in from ten to fifteen minutes. The whole is now allowed a repose for a half hour or more, so that the aqueous solution of chloride of chromium, sulphate of potassa, and other foreign matters may subside; after which, the oil is drawn off clear, from this sediment, into a wooden cask, mixed with a little water, and heated for a short time by the introduction of steam. After one more settling it is ready to be racked off for use.

The rationale of this process is as follows:—the coloring matters of the oil are oxidized, and thus rendered soluble, by the oxygen of the chromic acid of the bichromate of potassa, decomposed by the muriatic and sulphuric acids. The chromic acid having yielded its oxygen to the coloring matter, becomes reduced to the state of a green oxide of chromium, which, by the action of hydrochloric (muriatic) acid, is converted into chloride of chrome, and the sulphuric acid takes up with the liberated potassa. This, and the coloring matters, when the oil has settled, are found in solution with the water at the bottom of the cask.

Mr. Watson's patent for bleaching palm and other oils, is founded as the above upon the blanching power of oxygen, but, in this instance, the patentee substitutes for bichromate of potassa, the manganate of potassa or chameleon mineral, a body rich in oxygen, which he decomposes by acids, whilst it is in contact with the melted fat. You take of oil and place it in a tightly bound wooden cask, and add thereto one-twentieth of its weight of the manganate of potassa dissolved in water, and acidulated to the taste. The whole is to be well mixed, and gradually heated from 150° up to 212° F., and maintained at that temperature for an hour. On account of the tendency of the dissolved manganate to spontaneous decomposition, it should be added to the dilute acid mixed with the fat, previously melted at the lowest temperature consistent with its fluidity.

"Chlorine is not so well adapted for bleaching palm oil as chromic acid, for though chlorine readily decomposes the coloring matter, it exerts at the same time an injurious action on the oil itself, causing it to become of a brown color, and injuring its consistence. This bleaching agent has been used, however, for decolorizing palm oil on the large scale; the materials for generating the chlorine, viz., black oxide of manganese, sulphuric acid, and muriatic acid, being introduced directly into the melted oil. Considerable inconvenience was formerly experienced in this process, from the slowness with which the minutely divided oxide of manganese subsided from the oil, but this is now completely surmounted by enclosing the manganese in a cloth bag."

"In a method of bleaching palm oil introduced into the soap manufactories of England about nine years since, the discharge of the coloring matter was effected by exposing the oil, contained in a strong cast iron boiler, built over a furnace, to the temperature of 450° F., which completely decomposes the coloring matter without being injurious to the oil. By the time the oil near the surface became heated to 450°, that near the bottom of the boiler frequently acquired a temperature higher than 600°, when it became carbonized and partially converted into gas. A considerable portion of the oil, hence, frequently acquired a black color from intermixture with the carbonized oil at the bottom, and another inconvenience in this plan consisted in the explosions which sometimes resulted from the disengagement of inflammable gas. For these and some other reasons this method has been very generally abandoned.

"If freely exposed to the air, the palm oil may be completely decolorized at a temperature far below 430° F. and even below 212° F. The process requires a considerably longer time, however, with these comparatively low temperatures."

Mr. Cameron's late process for bleaching palm oil is thus recorded in the *Transactions of the Society of Arts* of England. "The oil is introduced into a cast iron kettle, of three or four tons' capacity, and fixed over a small furnace, and furnished with two or more pipes for the circulation of steam and a horizontal revolving fan of sheet iron for the purpose of agitating the oil. By means of a fire in the furnace underneath, the oil is heated to the temperature of 230° F.; the fire is then withdrawn, but the temperature of the oil is maintained at that point by introducing high pressure steam from a boiler loaded with fifteen pounds on the square inch of the valve, until the oil is sufficiently deprived of its color. During the whole process the oil requires to be agitated for the purpose of bringing it into more complete contact with the air, the oxygen of which is the principal agent concerned in the decoloration. The agitating fan may be revolved at the speed of about six revolutions per minute. Four tons of oil may be thus bleached in ten hours, at one operation, and at an expense of half a ton of small coal. All that is essential in this mode of bleaching, is obviously the exposure of the oil to the combined influence of air and heat, many modes of accomplishing which might be devised. In a process for bleaching oil and fatty matters of this nature, patented by Mr. Arthur Dunn in 1843, the air is forced below the surface of the oil through pipes, by means of a blowing apparatus, and then allowed to rise through the liquid in numerous small streams; the oil being maintained at a temperature between 170° and 230° F., by steam pipes or other means. A hood communicating with a chimney is placed over the vessel containing the oil, for the purpose of conducting away the unpleasant vapors which are disengaged.

"The oxidizing action of the air on the impurities in palm oil, seems to be accelerated by combining the influence of light with that of heat. With the view of accomplishing this, the oil is sometimes placed in thin strata, in a large shallow uncovered vat. Several vats, each about one foot in depth, may be heated by means of steam passed through serpentine leaden pipes proceeding from a common boiler: one extremity of each tube may terminate in a receiver for the condensed water which may be returned to the boiler. Each vat is first two-thirds filled with water; when this is become hot, a sufficient quantity of palm oil is introduced, to form, when fluid, a stratum of about two inches in depth, which should be maintained during the whole process at a temperature as near 212° F. as possible. To preserve an equal temperature at different parts of the vat, the steam may be admitted at opposite ends and circulated in opposite directions by two distinct serpentine tubes. Both air and light having access, the time required for the decoloration of the oil at a temperature near 212°, is from ten to fifteen hours; the thinner the stratum of oil the more rapid is the process. M. Payen has ascertained, in operating on the small scale, that the rapidity of the bleaching action is not sensibly lessened by loosely covering the containing vessel with a plate of glass, provided the renewal of the air over the surface of the oil is not much interfered with, because if exposed to a temperature of 212° in a *closed* glass vessel, so that light and heat have access but not air, the oil does not become blanched. Hence, Payen suggests to manufacturers the propriety of adopting such an expedient on the large scale as will, while not excluding the air, partially prevent the great loss of heat which occurs from the extensive surface of oil presented to the air in the open vessels commonly used in this mode of bleaching the oil. While fluid, the blanched oil retains a fawn-colored tint, but when cold and solid is grayish-white."

As before said, nitric acid is often used for bleaching palm oil, which by this means loses its violet odor and is liable to become colored by time. Two gallons of acid, of specific gravity 1.400, are taken for and mixed with five tons of the oil.

Additional to my own experience the opinion of Mr. Parnell, a distinguished English chemist, favors the preference of that method aforementioned, devised by Mr. Watt, and in which bichromate of potassa is the decomponent material of the coloring matter. "The oil bleached by this method may be preserved without re-acquiring color, and the soap made from it does not become tinged like that made from oil bleached by the air and heat, or by nitric acid. When the oil is to be used for making yellow soap and not white soap, the most convenient mode of bleaching, having regard to economy, is by the action of the air, the oil being *moderately* heated. It seems that the color does not re-appear to so great an extent when the oil has been moderately heated, say to 180° F., as when it has been maintained at a temperature above 212°; it should, however, be heated to 212°, at the commencement of the process, and then be allowed to cool a little. The oil does not become sensibly bleached when only moderately heated, so long as water is present; but when all moisture is dissipated, the blanching commences immediately"

"When rancid palm oil is mixed with a ley containing sulphuret of sodium, the free acid which it always contains when in that state, creates the disengagement of a considerable quantity of sulphuretted hydrogen gas; owing to the decomposition of the alkaline sulphuret by the acid. In mixing the materials for yellow soap, sufficient sulphuretted hydrogen is sometimes emitted to stupefy the workmen."

In conclusion, it must be remarked that in the process of bleaching the palm oil by heat, the temperature should not exceed the given point, else the result will be a decomposition of the oil, and its resolution into entirely new bodies. The operation of bleaching in this way is founded upon the destructibility of the coloring matter at a lower temperature than that at which the oil becomes altered.

Nutmeg butter.—This concrete oil is extracted from the fruit of the *myristica moschata*, and is met with in commerce in solid square cakes, yellow colored, and of an aromatic odor. It is perfectly soluble in four parts of boiling alcohol, two liquid fats, independent of the volatile oil, being retained whilst a solid constituent deposits on the cooling of the solution. The first two have not been examined or designated, but the latter has been styled by Playfair, *myristicine* or *sericine*, from the silky appearance of its crystals. This salt fuses at 88° Cent., is saponifiable by the alkalies, and the resulting soap, when decomposed by a strong acid, presents its components in the form of sericic acid and glycerin. The sericic acid is soluble in alcohol, and cools from its solution therein as white brilliant scales, which melt at 49° Cent., and congeal into a distinctly crystaline mass.

Bay laurel butter.—This is a ceraceous fat which is deposited in yellow flocculæ, from a cold decoction of bay-berries in boiling alcohol. By saponification, glycerin is eliminated, and a new fatty acid, to which Marrson gave the name of *lauro-stearic acid*. Its point of fusion is between 42° and 43° Centigrade. Besides the *lauro-stearine*, this butter contains also olein.

Cocoa butter.—This fat is extracted by pressure from

the steamed nuts of the *cocos nucifera, cocos butyracea*. It is a white unctuous grease, melting at 22°, and congealing at 18° Cent.; becomes rancid readily, and forms with soda a solid soap of a disagreeable odor. It has the consistence of lard, and when liquefied, is perfectly clear and transparent. Its great solubility in alcohol serves to distinguish it from other fats and oils. It consists of oleine, a solid constituent, perhaps stearin (but said to be by Pelouze and Boudet identical with elaidine), and a crystalizable acid, the coco-stearic or cocinic acid. The soap formed by its union with alkalies, when decomposed by a strong acid, gives up the fatty acids in a free state; and the coco-stearic acid, when purified by repeated solutions in alcohol and crystalizations, becomes an inodorous and brilliantly white crystaline porcelain-looking mass, with a fusing point at 35° Cent., and is unalterable by distillation. Cocoa butter is, however, difficultly saponifiable on a large scale, and requires for its separation from alkaline leys, more saline matter than the tallow and palm soap.

Oil or butter of galam.—This is a buttery fat, the product of the Phulwana tree, or *bassia butyracea*, which, by reason of its similarity to palm oil, is often confounded with it. It comes from Africa, and has a taste and smell not unlike those of cocoa-butter; becomes rancid by exposure, melts at 22° Cent., is insoluble in water, and but slightly so in alcohol. What its constitution is, I have been unable to ascertain; but from its similarity in other respects, I infer its analogy, in its constitution, to the butter above described, with the exception of the cocinic acid, for there is every probability that this is replaced by a corresponding acid peculiar to the galam oil itself.

Athamantine, from the root of *athamanta creoselinum*, is a crystaline fat-like body, containing *valerianic acid*, united to a body, *oreoselone*, which supplies the place of glycerine.

There are several other trees producing fruits or kernels, which by pressure yield a vegetable concrete fat. Among these is the "butter tree," a remarkable plant in the interior of Africa, from which is obtained a firm white grease, keeping well even in that climate for a year, without salt. The Illupie tree, *bassia langifolia*, and the Madhuca tree, *bassia latifolia* of Bengal, yield each a fatty butter. We have merely alluded to these instances, so that no fact bearing upon our art may be unmentioned. A further notice of them is unnecessary, as they seldom are found among the raw materials used by the savonnier.

NOTE.—The first five pages of this chapter are mostly from Thillaye.

CHAPTER IX.

THE ADULTERATION OF OILS.

M. Heidenreich has found, by the application of a few drops of sulphuric acid to a film of oil, upon a glass plate, a means of ascertaining its purity. The glass plate should be laid upon a sheet of white paper, and a drop of the acid let fall on the middle of ten drops of the oil to be tried. This information will be particularly valuable, for in the purchase of stock, the savonnier will find himself necessitated to resort to it frequently for protection against fraud, as the practice of sophisticating oils has of late become extensively prevalent.

With the *oil of rapeseed* and *turnip seed*, a greenish blue ring is gradually formed at a certain distance from the acid, and some yellowish-brown bands proceed from the centre.

With *oil of black mustard*, in double the above quantity, also a bluish-green color.

With *whale* and *cod oil* a peculiar centrifugal motion, then a red color increasing gradually in intensity, and after some time it becomes violet on the edges.

Olive oil, pale yellow into yellowish green.

Oil of *poppies* and *sweet almonds*, canary yellow, passing into an opaque yellow.

Oil of *linseed*, a brown magma, becoming black.

Of *tallow* or *oleine*, a brown color.

In testing oils, a sample of the oil imagined to be present should be placed alongside of the actual oil, and both should be compared in their reactions with the acid.* A good way of approximating to the knowledge of an oil is by heating it, when its peculiar odor becomes more sensible.

Lamot, a chemist in Paris, finds that colza oil (analogous to rapeseed oil) may be tésted, relative to its sophistication with cheaper vegetable oils, by the increase of density which they impart to it; this becomes very evident when the several oils are heated to the same pitch. The instrument which he calls an *oleometer* is merely an hydrometer with a very slender stem. He plunges it into a tin cylinder filled with the oil, and sets this cylinder in another containing boiling water. His oleometer is so graduated, as to sink to zero in pure colza oil heated as above, and he finds that it stops at 210° in linseed oil, at 124° in poppy seed oil, at 83° in fish oil, and at 136° in hempseed oil—all of the same temperature. By the increase of density, therefore, or the ascent of the stem of the hydrometer in any kind of colza oil, he can infer its degree of adulteration.

* If the drop of acid which has been added is agitated with a glass rod, the reactions are different. The rape and turnip seed oils then assume a uniform brown color, without mixture of red; with from five to six drops of acid, the color is of an opaque brownish red, and the edges of the mixture become green.

The oil of black mustard produces the same reaction with from five to six drops of acid, but the color is still more opaque. Whale and cod oil directly become of a brownish red and then pass to a violet color free from any tinge of green. If from five to six drops of acid are added, the color is brighter and the violet shade is sooner apparent.

The oils of olive, poppy, and sweet almonds, become more or less grayish yellow. Linseed oil changes into a brownish-black resinous mass. Olein assumes a dirty brown color. (*Journ. für Prakt. Chem.* 1842.) In these tests it is important to observe what colors are developed at the commencement of the reaction, and then compare them with those produced by a similar operation upon an oil of standard purity.

The presence of a fish oil in a vegetable oil, is readily ascertained by agitation with a little chlorine gas, which blackens the fish oil, but has little or no effect upon the vegetable oil.

Lard, and lard oil, Ure says, are not at all darkened by chlorine.

A specific gravity bottle or globe, having a capillary tube-stopper, would make an excellent oleometer upon the above principle. The vessel should be filled with the oil and exposed to the heat of boiling water, until it acquires a temperature of 212° F., and then weighed. The vessel with the pure colza oil, will weigh several grains less than with the other oils similarly treated. Such an instrument would serve to detect the smallest adulterations of sperm oil. Its specific gravity at 60° F., when pure, is only 0.875; that of southern whale oil is 0.922 or 0.925; and hence their mixture will give a specific gravity intermediate, according to the proportion in the mixture. Here is a key to the detection of admixture of oils, especially when they differ in their specific weights. Ure says he was enabled by it to discover sperm oil in pretended lard oil.

It is a matter of regret, that the cupidity of dealers should render necessary any means for the detection of frauds, for they are not only a cheat upon the confiding purchaser, but a reproach to the morals of trade. The deception is a lamentable one, and particularly so because of its prevalence, for it is a truth, in regard to most of the essential oils, that it is difficult to procure an unadulterated article, except from the manufacturer or importer. The stench of turpentine is apparent in almost every oil that will bear its presence, and so great at times is its proportion in the essences of lavender and cinnamon, that when these latter are used as perfumes for soap, the predominant odor emitted, instead of being the legitimate scent of the aromatic oil, is that of its adulteration. Almond oil, too, as sold in the shops, is frequently nothing more than lard or olive oil, with its characteristic smell disguised by a slight addition of the genuine oil, whilst the essences of bergamot, lemon, neroli, and sundry others are made to soak up as much alcohol as they will bear. We have made these remarks in view of exposing the impositions practised in this way, and we present the tests given in a preceding page, as an excellent means of detecting the most common adulterations of fixed oils, reserving for a future chapter some similar directions in relation to the essential oils.

CHAPTER X.

ACTION OF ACIDS UPON OILS

Experiment has proved that nearly all the strong acids are capable of uniting with many oils, and producing with them unctuous and pasty compounds. This action is much stronger and more evident when assisted by heat, and the compounds formed dissolve in water and lather like the common soap, but are not permanent, and present no advantage in their use. Those acids, whose radicals are nitrogen, chlorine, and bromine, are easily decomposed by the oils Boudet, however, tells us, that the drying oils treated with two or three hundredths of their weight of hyponitric acid do not solidify, whilst the fat oils become firm in a shorter or longer time. The following table indicates the time required at the temperature of 17° Cent., for the solidification of 100 grains of oil mixed with three grains of nitric acid of 38°, and three grains of pure hyponitric acid (nitrous acid).

Oils	Colors that they immediately take	Number of minutes before their solidification.
Olive	Bluish green	73′
Sweet almond	Dirty white	160′
Bitter "	Deep green	160′
Nuts	Bluish green	103′
Castor	Bright yellow	603′
Colza	Brown yellow	2400′

The changes produced in oils by the action of acids vary with the kind of acid employed. When applied in sufficient quantities, the result is the conversion of the natural acids of the fat into various modifications. The purification of oils by small portions of sulphuric acid is due to its agency in separating the mucilage and other parenchymatous matters, either by rendering them insoluble or partially carbonizing them.

When sulphuric acid is added in larger quantities to fat oils, its action is limited to the abstraction of their glycerin, with which it combines and forms sulpho-glyceric acid, whilst the oily acids are set at liberty. Such is the reaction that occurs when tallow or hog's lard is mixed with half its weight of sulphuric acid. But when the proportion of sulphuric acid is increased, particular compounds of that acid with the oily acids, are produced, which dissolve in water; but of these we will speak hereafter under the head of "*savons acides*," or soapy acids.

The action of nitric acid varies with the oil upon which it operates. Those instances in which its influence is peculiar, are recorded under the head of the individual oil or fat to which they relate. Upon the olein of a non-drying oil, its effect is such as to convert it into a solid body, to which has been given the name of elaïdine, and which, by analysis, is found to consist of elaïdic acid in union with glycerin. This action is similar in many other of the fat oils, but the drying oils are not at all influenced by it, except with regard to a change of color.

The elaïdine is solid at ordinary temperatures, and, when treated with alkali, becomes saponified. The resultant soap, when decomposed, yields elaïdic acid and glycerin, though there is a doubt as to the presence of the latter—the discrepancy in composition between the elaïdine and elaïdic acid having given rise to the opinion that it is one of those fats in which the base is *acroleine*. The stearin of fats does not undergo any transformation into elaïdine, nor, in every probability, does the margarin, though the contrary is averred by some authorities. The elaïdine of olive oil is a brilliant white solid, melting at 36° Centig., dissolving in ether, and in 200 parts boiling alcohol. Its specific gravity is 0.8978, and by distillation it yields acroleine, elaïdic acid, and like oleic acid, gives also sebacic acid. Elaïdic acid is a pearly crystalized solid, dropping from its solution in hot alcohol, in the form of silvery scales resembling benzoic acid. The congealing point of pure elaïdic acid is 44° Centigrade. The action of the acids upon oils has been a subject of critical inquiry and examination of late years, and though there are other changes produced besides those we have mentioned, yet, being irrelevant to our purpose, we make no mention of them. It is a very difficult matter to render these changes intelligible to those who have no knowledge of the chemical constitution and reaction of bodies, and so we have confined our remarks to those points upon which it was deemed necessary to write, in order to prevent any confusion to the reader.

CHAPTER XI.

VOLATILE OILS—THEIR PROPERTIES, AND APPLICABILITY IN THE MANUFACTURE OF SOAPS.

THE volatile or essential oils are the product of every part of the plant, but not in all parts of the same plant. In some they are found only in the leaves; in others only in the flowers; in others again they are found in the rind, or in the wood and fruits; and sometimes in the envelops of seeds, but not in the cotyledons. Some few plants, however, as the thyme and scented *labiatæ*, contain volatile oil in nearly all their parts. It occasionally occurs that different parts of the same plant contain different oils; for instance, the orange furnishes one oil from its flowers, another from the leaves, and a third from the rind or epidermis of the fruit, each varying from the other. In most plants the oil is contained in little sacs or vesicles, so well confined that the oil is retained during the drying of the plant, and even for some time after: whilst in other species, again, especially in the flowers, the oil is constantly produced at the surface, and escapes at the moment of its formation. A few oils are obtained by expression, such as those of the oranges and lemons, where the oil resides in the epidermis of the fruit; others, again, which are not contained in vessels, such as those of violet, jasmine, &c., are procured by maceration of the flowers in oil of ben, an inodorous fixed oil, and are used in this state in perfumery. Volatile oils are distinguished from the fixed oils by their volatility, and as they are usually obtained by distillation, they have been called also essential oils. They are considered as the source of the aroma which is exhaled by plants. These oils are of an acid, burning taste; of a very mild, agreeable, or of a pungent and unpleasant odor. Some are colorless; others are yellow, red or brown; others, again, green, and a few blue; and with the exception of those of cinnamon, cloves, mustard and sassafras, are lighter than water. They are not greasy to the touch, and their specific gravity ranges between 0.759 and 1.094; the first number denoting the density of oil of coriander, and the second, that of oil of sassafras. Most of them congeal at different temperatures; some acquire viscidity at the ordinary temperature of the atmosphere, and become solid; for instance, the oils of fennel, anise, &c. They burn with a brilliant flame and much smoke. Although they are styled volatile oils, the tension of their vapor, as well as its specific heat, is much less than that of water, and though volatile at the ordinary temperatures, their boiling point usually is not less than 316° to 320° F. In contact with air or oxygen, they acquire thickness, and are eventually converted into a resin by the absorption of that gas. This absorption varies with different oils, and gives rise to carbonic acid gas, but no water is formed. Light contributes powerfully to this action. A volume of concrete oil of anise absorbed 159 times its volume of oxygen gas in two years, and emitted at the same time 56 volumes of carbonic acid gas. A volume of oil of lavender, during four of the colder months, absorbed 52 volumes of oxygen and gave out two of carbonic acid gas, without the production of water, and without being even yet completely saturated with oxygen. These facts are stated upon the authority and results of Saussure's experiments. "It has been observed that the odor of oils is closely related with this chemical change. Those which oxidate most rapidly have the strongest smell, and the characteristic odor of no oil can be perceived immediately after its distillation in an atmosphere of carbonic acid gas." Essences, therefore, should be preserved in well-stopped bottles.

Omitting any remarks upon the action of the acids and other reagents upon these oils, we shall only consider here, those which operate upon the salifiable bases, mentioning, however, in passing, that they are more or less soluble in alcohol and ether. It is evident from experiment that the essential oils have no great affinity for the salifiable bases, that is to say, unlike the fixed oils, they have but little disposition to form soaps with the alkalies. With much time and difficulty a sort of feeble combination of the oils of cloves, cinnamon and cedar wood, with caustic alkali, is effected; the products of which are styled *savonules*, which are, by way of imparting a definition, *those compounds of resin acid and base endowed with saponaceous properties*. That soap called Starkey's may be classed under the head of "*Savonules*." It consists of soda and essence of turpentine. It is prepared by triturating recently fused caustic soda in a mortar with a little oil of turpentine, added dropwise, until the mixture has acquired the consistence of soap. The compound is to be dissolved in spirits of wine, filtered and distilled. What remains, after the spirit is drawn off, consists of soda combined with a resin formed in the oil during the act of trituration. Here, therefore, it is seen that there is no combination between the actual constituents of the oil and the soda, but merely a union of the soda with the resin developed from the oil by reason of its partial oxidation, whilst being triturated—this resin, like that used for common soaps, being capable, by its acid properties, of combining with alkalies, and solving in alkaline solutions.

CHAPTER XII.

ORIGIN AND COMPOSITION OF VOLATILE OILS.

THE essential oils are the odorous principle of plants. In most plants they exist already formed, secreted by their proper organs, whilst others again are generated only by distillation, from substances not natural to the vegetable, and take their rise during a kind of fermentation which it undergoes. As examples of these latter, we treat hereafter of the oils of mustard seed and bitter almonds. The essential oils are divided into three classes, those containing only carbon and hydrogen, as oil of turpentine; those containing also oxygen, as oil of cloves; and those containing sulphur, as oil of garlic. Another characteristic of essential oils is, that some, by a direct absorption of oxygen, become acidified, and others again, by a more indirect application of the same gas, become resinous. Experience has demonstrated that most of the essential oils consist of two oils, each of a different degree of fusibility, but these constituents have no analogy whatever to the olein and stearin of the fixed oils. Bizio called these principles *sèreusine* and *igrusine*, or, in more modern language, *elaopten* and *stearopten;* the latter of which is separable from the former by pressure, when the oil is congealed at a sufficiently low temperature. If the analyses of the essential oils are compared with those of the fixed oils, the difference in their constitution will appear evident, for, whilst the former are richer in carbon, many of them are much more deficient in oxygen, some even entirely destitute of it, and a few, as said above, contain a little sulphur, and nitrogen also, according to some chemists. The proportion of hydrogen varies in different oils, but as a general rule they are more hydrogenated than the fixed fats. The source whence is obtained the gaseous constituency of these oils, and, as a consequence, the nourishment of the plants which yield them, is the atmosphere, for the carbon and nitrogen; water, for the hydrogen, and sulphuric acid for the sulphur; all of which are withdrawn from the surrounding media containing them, by the vegetable organism, and assimilated as above. The stearopten, or crystalizable portion of those oils, thus constituted, is commonly called the *camphor* of the plant, and not unfrequently forms artificially, when the oils are in contact with water. According to Gerhardt's and Cahour's experiments, the action of fused hydrate of potassa decomposes all oxygen containing essential oils, into an acid and a non-azotized oil.

As regards the camphors of the oils they vary in properties. Some contain only carbon and hydrogen, while others contain both ingredients, besides oxygen; the stearopten of the otto of rose is an example of the former.

Among the essential oils containing oxygen are those of *bitter almonds, spiræa, cinnamon, cloves, cumin, aniseed, valerian, dill, fennel, parsley, caraway, coriander, pimpernel, peppermint, marjoram, lavender, rosemary, basil, thyme, rue, cascarilla, chamomile, wormwood, tea, cardamom, nutmeg, cajeput, rhodium, rose, bergamot, saffron, sassafras, sweet bay, hyssop, cedar.*

Among those destitute of oxygen are the oils of *turpentine, juniper, savine, lemons, cedrat, oranges, copaiva, pepper, cubebs, storax, elemi.*

Those essential oils containing sulphur, are as follows:—Oils of *mustard seed, horseradish, garlic, onions, assafœtida, water pepper and hops. Mustard seed,* contains, also nitrogen.

On the next page is a table of the most important essential oils, with the characteristic color and specific gravity of each annexed, both of which are varied accordingly, as the oil may be more or less adulterated with foreign admixture.

TABLE OF THE PRINCIPAL VOLATILE OILS.

Names.	Botanic name of plant from which extracted.	Color.	The part of the plant which yields the oil.	Sp. grav.
Oil of absinthe	Artemisia absinthium	green	leaves	0.897
" dill	Anethum graveolens	yellow	seeds	0.881
" anise	Pimpinella anisum	"	seeds	0.9858
" ache or parsley	Apium petroselinum	"	roots	
" mugwort	Artemisia vulgaris	"	leaves	
" elecampane	Inula helenium	white	roots	
" badiane	Illicium anisatum	yellow	seeds and fruit	0.987
" angelica	Angelica archangelica	yellow	root, seeds, &c.	
" Portugal	Citrus aurantium	"	rind of the fruit	0.835
" cinnamon*	Laurus cinnamomum	"	bark	1.035
" chamomile	Matricaria chamomilla	blue	flowers	0.924
" cajeput	Melaleuca leucodendra	green	leaves	0.948 at 48° F.
" cascarilla	Croton eleutheria	yellow	bark	0.938
" caraway	Carum carui	"	seeds	0.960
" chervil	Scandix cerefolium	lemon yellow	leaves	
" lemon	Citrus medica	yellowish	rind of the fruit	0.847
" cochlearia*	Cochlearis officinalis	"	root	1.01
" coriander	Coriandrum sativum	white	seeds	0.759
" cubebs	Piper cubeba	yellow	"	0.929
" cumin	Cuminum cyminum	"	"	0.860
" dittany	Origanum creticum	brown	flowers	0.946
" fennel	Anethum fœniculum	white	seeds	0.997
" galangal	Maranta galanga	yellow	root	
" genista	Genista canariensis	"	"	
" juniper	Juniperis communis	green	seeds	0.911
" ginger	Amomum zinziber	yellow	root	
" cloves*	Caryophyllus aromaticus	"	dry flower buds	1.034
" hyssop	Hyssopus officinalis	"	leaves	
" lavender	Lavendula spica	"	flowers and leaves	0.877
" cherry laurel*	Prunus lauroceratus	"	leaves	
" crisp mint	Mentha crispa	white	flowers and leaves	0.969
" peppermint	Mentha piperita	"	"	0.920
" balm mint	Melissa officinalis	"	"	0.975
" motherwort	Matricaria parthenium	blue	"	
" yarrow	Achillea millefolium	blue and green	"	0.92
" marjoram	Origanum marjorana	yellow	"	
" mustard*	Sinapis alba and nigra	deep brown	seeds	1.015
" nutmeg	Myristica moschata	yellow	"	0.92
" neroli	Citrus aurantium	orange	flowers	0.930
" pennyroyal	Mentha pulegium	yellow	"	0.927
" rosemary	Rosmarinus officinalis	white	plants and flowers	0.911 and 0.8886
" sage	Salvia officinalis	green	flowers and leaves	0.905
" saffron*	Crocus sativus	yellow	pistils	
" sassafras*	Laurus sassafras	"	root	1.094
" turpentine	Pinus sylvestris and abies	white	wood and resin	0.872 at 50° F.
" thyme	Thymus vulgaris	yellow or pale green	flowers and leaves	0.905
" rose	Rosa centifolia	white	petals	0.832
" valerian	Valeriana officinalis	green	root	0.9438
" pimento*	Myrtus pimenta	slightly yellow	rind of the fruit	1.021
" rhodium	Convolvulus scoparius	yellow	wood	
" savin	Juniperus sabina	limpid	leaves	0.915
" tansy	Tanacetum vulgare	yellow	plant	0.931
" rue	Ruta graveolens	yellow-green	"	0.837 at 64° F.
" bergamot	Citrus limetta bergamium	yellow	rind	0.878
" serpolet	Thymus serpyllum	light brown		

* Those with a * attached, are heavier than water.

CHAPTER XIII.

THE MODE OF DETECTING THE ADULTERATIONS OF ESSENTIAL OILS, AND THEIR USE IN THE MANUFACTURE OF SOAPS.

ESSENTIAL oils in soap making, are useful solely for their perfume, but as it is requisite to have them pure, we proceed to give some methods for detecting the adulterations to which they are in commerce generally exposed. The most prevalent adulterations are fat or fixed oils, resins, balsam copaiva, and spirits of turpentine. Either of the first three may easily be detected, by putting a drop of the oil on paper, and heating it, for if pure, it will evaporate without leaving any stain or residuum, whilst an oil mixed with any of the above substances, leaves a translucent spot upon the paper. The presence of alcohol,* which is also sometimes used to sophisticate essential oils, is recognized by their being less odorous, and more fluid. The water, too, with which it may be shaken, becomes milky, by reason of the small portion of oil which, by the presence of the alcohol, it dissolves. It is, however, very difficult to detect alcohol when it is contained in small quantities. But the chief difficulty is to recognize the presence of a cheap essential oil in a dear one, to which it is similar. Here taste and

* Borsarelli proposes chloride of calcium as a test for the presence of alcohol. Perfectly dry chloride of calcium dissolves in oils containing alcohol, and forms a liquid stratum at the bottom of the vessel. When but little alcohol has been mixed with them, the pieces change, at least with regard to their form, but in pure oils the chloride of calcium remains unaltered.—(*Archiv. de Pharm.*, xxiv.)

smell are not sufficiently exact tests. By a careful examination of their respective densities the essential oils may, in most instances, be distinguished; but the wholesale way of "*making them up*" now-a-days, renders necessary some sure method for exposing the fraud; for of oil of lavender, so largely used by the savonnier, I myself have examined samples which contained as high as 40 per cent. of spirits of turpentine. The essential oils, especially of marjoram, lavender, spikenard, sage, thyme, rosemary, wormwood, and peppermint, are the most subject to this adulteration. A few years since, Mero discovered a process for detecting the presence of spirits of turpentine, founded on the circumstance that this oil dissolves the fixed oils with great facility; while the essential oils above mentioned, do not. He considers, therefore, that it will serve to indicate the presence of oil of turpentine, mixed with pure essential oils whose powerful smell conceals that of the turpentine. Experience gives the preference to the oil of poppies, for this purpose, because of its uniform consistence, regardless of temperature. About three grammes of oil of poppies are poured into a graduated tube, and an equal quantity of the essential oil to be tested, added; the mixture is then shaken, and should become of a milky white, if the essential oil is pure, whilst it remains transparent, if it contain any oil of turpentine.

The value of this method may readily be ascertained, by first testing a pure essential oil, and then some essential oil of turpentine; if the essential oil is then mingled with the oil of turpentine, even in proportions so small that no advantage could accrue to traders by the admixture, it is found to act like the essential oil of turpentine itself, that is to say, the mixture is not rendered turbid.

To make the experiment successful, the two essential oils should be very intimately blended. The method employed in commerce for this purpose is as follows:—

The pure essential oil and the quantity of oil of turpentine which is to be added to it, are placed in a hot water bath basin, and this is heated until the mixture, which is at first turbid, becomes transparent. The medley which is obtained by adding oil of turpentine in the process of distilling the plants, is detected in the same manner. Mero received a medal for this process from the *Société d'Encouragement*, before the committee of which he proved, by some experiments, that he could determine at once the mixtures which contained five per cent. of oil of turpentine, and was, moreover, able to tell very nearly the proportions of the mixtures. This process, though available for the detection of adulteration in several of the essential oils most in use, is, however, not of general application, for it furnishes no proof of the presence of essence of turpentine in the oils of thyme and rosemary.

M. Voget considers concentrated sulphuric acid as the best reagent for detecting adulterations of essential oils, with oil of turpentine. The peculiar color which the former assumes with sulphuric acid is much altered by the intense reddish-brown color which the oil of turpentine yields; and, moreover, the heat evolved with the oil of turpentine, is greater than that with other oils. In testing, the oils are best dropped upon a glass plate, beneath which is placed a piece of white paper; five drops of the oil are then added to one drop of fuming (Nordhausen) sulphuric acid, and the two are mixed with a glass rod.

Violet and Guenaud, perfumers in Paris, announced, some years since, an instrument for estimating the value of the essential oils by their specific gravity. It is merely a very sensitive areometer, which, by means of a small weight, was rendered proper for testing those oils heavier than water, and without the weight, for such as are lighter than water. Bussy and Chevalier, who reported favorably upon this instrument, verified its exactness by experiments.

CHAPTER XIV.

WAX.

WHETHER wax is of animal or vegetable origin, and what is its nature, were for some time mooted points. According to Huber's researches and experiments, bees, though fed with honey or sugar alone, produced, nevertheless, a very considerable quantity of wax; thus proving that bees are not mere collectors of this substance from the vegetable kingdom. Accordingly, it has been found that the elementary composition of beeswax and vegetable wax is slightly different.

Beeswax is obtained from the honeycomb by pressing out its fluid contents, and washing and melting the residual cakes. It is yellow, has a peculiar honey-like odor, a specific gravity of .960, its fusing point at 150° F., and at a high temperature volatilizes, and burns with a bright white flame. It is solid and insipid. with a granular fracture, insoluble in water, ether and cold alcohol, but solves by 20 parts boiling alcohol. The wax, reduced to thin ribands and exposed to moisture and light, bleaches perfectly white. The same result is effected by the action of dilute nitric acid and chlorine; but in the latter cases the quality of the wax and its fitness for candles, are impaired. White wax is tasteless and inodorous, and in thin segments, translucent.

As a vegetable product, wax forms a part of the green fecula, and the varnish which coats the leaves of certain plants; it is an ingredient of the pollen of flowers, and is found upon some berries, as those of the *myrica angustifolia*, as also the *cerifera*.

The *chamarops humilis*, growing in Cuba, also yields wax abundantly. Mr. Teschemacher has examined it, and describes it as analogous to beeswax in its composition.

Cerosine is the name given to a waxy substance occasionally found on the surface of the sugar cane. It is not saponifiable.

Wax exhibits a partial analogy to fixed oils, in being composed of two immediate principles, called by their discoverer De John, *cerine* and *myricine*. These constituents appear to be different kinds of wax, for cerine exists also in the pollen of many flowers, and serves as an outer covering to the leaves of cabbage, poppy, &c., and as a varnish to various fruits, as the grape, orange, lemon, figs, &c. It is white, fusible at 124° F., and forms 70 to 80 per cent. of beeswax, from which it is separated by digesting that body for some time in boiling alcohol. The myricine remains undissolved, whilst the cerine is taken up in solution, and recovered from it by evaporation. *Cerine* occupies the position in wax that olein does in fats and oils. *Myricine* is a grayish-white solid, vaporizable almost without alteration, and forms the residuum from the solvent action of alcohol upon wax.

According to the researches of Boudet, Boissenot, and Ettling, *cerine* itself is a compound body, consisting of margaric acid and a peculiar waxy substance. *Ceraïne* is obtained as ethal is from spermaceti, and is separated by treating the cerine with caustic potassa, which forms a soap with the margaric acid without acting at all upon the *ceraïne*. Ettling, who critically examined the matter, has announced cerine and myricine, the two constituents of wax, to be isomeric, and in this assertion he is supported by the result of Lewy's experiments. According to their joint authority, these isomeric bodies are composed in 100 parts, of carbon 80.31, hydrogen 13.38, oxygen 6.30.

Berzelius, in the third edition of his *Manual of Chemistry*, states, that wax is converted into a kind of soap by caustic alkalies, but that the combination formed is of difficult solution in water, and separates in a cream-like form on the surface of the liquid; he also states that this cream may be melted into a very hard soap, but that acids separate the *wax* with nearly unaltered properties. Messrs. Warrington and Francis who tested this statement by thorough experiments, obtained results agreeing closely with those described both by Berzelius and Ettling. The substance they obtained by the saponification of wax with caustic potassa, and the decomposition of the compound thus formed, was exactly isomeric, that is, of the same composition as ceraine, if not identical with it, as assigned by Ettling. To this body they have given the name of *pseudo-ceraine*, until, as they say, the slight discrepancies which at present appear to exist between its characters and those ascribed by Ettling to ceraine, shall have been removed by further experiments.

When beeswax is distilled, it yields neither acroleine nor sebacic acid; it would, therefore, appear to contain neither oleic acid nor glycerin.

The yellow coloring matter of wax has not yet been examined.

Gerhard (*Ann. de Chim. et de Phys.*, xv. p. 236) records the results of his examination of beeswax, but as they are at variance with received opinions, I neither make note of them nor comment upon them.

Mineral wax.—Ure, in his *Dictionary of Arts*, describes a mineral wax, or *ozocerite*, as a various shaded brown, translucent, solid, fusible body like beeswax, slightly bituminous to the smell, of a foliated texture, conchoidal fracture, but wanting tenacity to such a degree that it is pulverizable. Its specific gravity varies from 0.900 to 0.953. In Moldavia, Turkey, candles which give a tolerable light are made of it. It occurs at the foot of the Carpathian mountains near Slanik, beneath a bed of bituminous slate clay, in masses of from 80 to 100 pounds weight. Ozocerite consists of different hydrocarburetted compounds associated together, the whole being composed ultimately of, hydrogen 14, carbon 86, very nearly. It is sometimes classed among the bitumens by mineralogists.

CHAPTER XV.

RESINS

The resins are proximate principles, found in most vegetables and in almost every part of them. Their ultimate components are carbon, oxygen and hydrogen, and hence their assimilation to the oils in that respect. There are many varieties, of which the most are solid, though some are liquid. They are obtained by spontaneous exudation from the trees, or by incision. The solid resins are more or less friable, sometimes colorless, but most generally yellow or brown, and of a warm, sharp, bitter taste. They are mostly heavier than water, their specific gravity varying from 1.0 to 1.2. Their general characteristics are fusibility, inflammability, and solubility in ether and alcohol, and insolubility in water. Many resins, when dissolved, redden litmus paper, combine with bases, and possess all the characters of acids; others, again, are neutral, and do not unite with bases. The compounds which they form with the alkalies are called resinous soaps.

Every resin, says Unverdorben, is a mixture of several resins, which are separable from each other by their unequal solubility in hot or cold alcohol, in ether, potassa, and carbonate of potassa, or by the different solubilities of their compounds with metallic oxides in these and other menstrua. This chemist separated from some, five and more resins, all quite distinct substances. The resins become negatively electric when rubbed. "There is every reason to suppose a close relation in composition between the oil and its associated resin, the last being often obviously the product of the oxidation of the former, many of which, by the action of the air, are converted into resins. In this change, the oxidation may be occasioned by a loss to the oil of one part of its hydrogen, which unites with the oxygen absorbed, to form water, and the replacement of the hydrogen lost by oxygen, in equivalent proportions; or it may occur by the combination of the entire oil, as a radical, with oxygen. The point is not decided by the analytic information at present possessed, but Liebig adopts the following as the composition, for instance, of oil and resin of turpentine."

Oil of turpentine $C_{40}\ H_{32}$ or $C_{20}\ H_{16}$
Resin of turpentine $C_{40}\ H_{30}\ O_4$ or $C_{20}\ H_{15}\ O_2$.

The fragrant resins owe their odor to the presence of volatile oils. With the exception of colophony or common resin, none others are used by the soap maker, so we do not particularize, but pass on to the composition of that body.

Colophony or *common resin.*—This substance, found in commerce under the name of resin, and which forms an ingredient of yellow soap, and almost *the* ingredient of "Yankee soap," comes to us from North Carolina, where it is obtained as the residuum, after distilling off the volatile oil from the different species of turpentine. Yellow resin contains some water,—black resin does not. That grade branded "soapmaker's resin," is what is generally used for our purposes.

Though this substance forms an important ingredient in the composition of yellow soap, and is a partial substitute for fixed oil or fat, it is not analogous to it in chemical constitution, nor can it, like those bodies, form with an alkali a proper soap by itself. Resin contains no glycerin, nor any equivalent for that substance. Unverdorben divided it into two different resins, which he named *sylvic* and *pinic acids.* The two are separable from each other by means of cold alcohol, of specific gravity 0.867, which dissolves the alpha resin (pinic acid) so called, and leaves behind the beta resin, or sylvic acid. The salts formed by the union of potassa, soda or ammonia with pinic acid, the title of which is the pinate of each respectively, dissolve in water, but are precipitated by an excess of alkali, or the addition of any alkaline salt. The pinates of other bases are insoluble. Pinic acid has for its formula $C_{40}\ H_{30}\ O_4$, and when exposed in alcoholic solution for some time to the air, absorbs oxygen, and becomes oxypinic acid, and possessed of stronger acid properties, and takes as its formula $C_{40}\ H_{30}\ O_8$.

The *beta resin* or *sylvic acid* is insoluble in water, but easily solves in alcohol, ether, or the fixed and volatile oils. The sylvates of potassa, soda, and ammonia are soluble in water; those of other bases are not solved by water, but most of them by alcohol. Its composition is, according to Fromsdorff, $C_{40}\ H_{30}\ O_4$, which makes it isomeric with pinic acid.

The formation of yellow soap from resin, depends on the direct combination of the resin acids with soda. No glycerin is eliminated, there being no proper saponification. The compounds formed, however, by the union of soda with resin, are not separable from their aqueous solutions by common salt, like true soda soaps, nor do their concentrated solutions become mucilaginous or gelatinous on cooling; nevertheless they produce a lather when agitated.

By distilling or heating pinic acid, a new resin is formed, *colopholic acid,* with stronger affinities for bases. There is doubtless a small portion present in common resin generated during distillation of the turpentine from it. Pinic and sylvic acids being colorless when pure, the brown shade of resin is therefore due to action of the fire.

The white resin, *galipot*, derived from the *pinus maritima*, consists almost entirely of a colorless crystalizable resin named *pimaric acid*, $C_{40} H_{30} O_4$, of the same composition as the preceding resins, but differing from them in properties. When the crystalized resin is dissolved in alcohol, it soon separates as an amorphous powder, which is much less soluble without being altered in composition.—(*Laurent.*)

CHAPTER XVI.

OF ANIMAL FATS AND OILS.

THE fat of animals is secreted near the kidney, in the folds of the omentum, under the skin in the cellular membrane, at the base of the heart, among many of the muscles, and also upon the surface of the intestines. The consistence, color and smell of fats vary with their source, and the animals which yield them. Fat is of a mild, insipid taste; more or less fusible; lighter than water, when fresh; without action upon litmus paper; becomes altered in the air; acquires rancidity by exposure; is insoluble in water, and only partially so in alcohol. "In the *cetaceii* the fats are generally fluid, in the carnivorous animals soft and rank flavored, and nearly scentless in the ruminants; usually white and copious in well-fed young animals; yellowish and more scanty in the old. As taken out by the butcher, it is not pure, for being of a vesicular structure, it is always enclosed in membranes, mixed with blood, blood-vessels, lymphatics, &c." For modes of separating these impurities and rectifying the fats, turn to Chapter IV. Fat unites with the alkalies and forms soaps; the phenomena occurring during the combination are fully explained in the chapter on the theory of saponification. We proceed to describe, separately, each of the fats having connection with our art.

Lard.—This is generally prepared from the adipose matter of the omentum and mesentery of the hog by freeing it with the hand from the membranous matter connected with it, washing with water until colorless, and melting, with moderate heat, until the dissipation of all moisture, which is known, by the absence of crepitation when small portions are thrown on burning coals, and by the transparency of the melted substance. The western lard is rendered differently and by steam. The whole hogs, excepting the hams and shoulders, after being scraped and cleansed, are steamed in a properly constructed cistern, which, when full, is allowed to settle, and the melted lard is then run off into barrels, clear and free from the lower stratum of water and sediment. Of all arrangements for rendering lard by steam, none is more advantageous, I think, than that of Wilson, of Ohio, a patent for which was granted a year or two since. A description of this patent, accompanied by drawings, will be given in another part of this work. Western lard has a granular appearance, owing to the manner in which it is prepared, and it is this characteristic that renders it so eligible for the purpose of making lard oil, as it can be pressed for oil without further granulation. Corn fed lard has the most consistency; mast fed is next in quality, whilst that obtained of hogs fed on distillery refuse, is thin, flabby, and deficient in body. Pure, properly prepared lard is white, rather soft at the ordinary temperatures, inodorous, fusible at 81° F., and has a specific gravity of 0.938 at 50° F. Treated with 100 parts anhydrous boiling alcohol, 2.80 parts of olein are taken up. Exposed for some time to air, it becomes yellow and rancid for the same reasons that other fats are similarly affected, and which have been before mentioned. When saponified with an alkaline solution, glycerin is eliminated; and when the resulting soap is decomposed by an acid, it is converted into the oleic, margaric and stearic acids; 100 parts by saponification yielding 9.0 of glycerin and 94.65 of acids, (oleic, margaric and stearic.) Lard, then, consists of olein, margarin, and stearin, or, to speak more correctly, of the oleate, margarate, and stearate of the oxide of glyceryl.

Mutton suet.—When carefully prepared, this fat is of a beautiful white color and peculiar odor It is very firm and brittle, completely insoluble in water, and but partially so in alcohol, 100 parts boiling alcohol at 0.821, dissolving but 2.272 parts. This fat, which is taken from the sheep, is richer in stearin than the others. It is saponifiable with the alkalies. Its constituents are olein, (oleate of glycerin,) stearin, (stearate of glycerin,) and perhaps a small quantity of hircin, (hircate of glycerin,) and some margarin, (margarate of glycerin.) The melting point of suet is from 98° to 104° F.

Beef tallow, or suet of the Ox.—Tallow and mutton suet are similarly constituted, except that the latter contains no hircin, and that its proportion of stearin is somewhat less. Its general characters are well known; it is firm, brittle, not so white as mutton suet, has a yellowish tint, and is fusible between 98° and 104° F., insoluble in water, and soluble in 40 parts boiling alcohol. It is in commerce, found most generally mixed with mutton suet.

Beef marrow.—This fat is of a bluish-white color, of a weak taste and odor, and is fusible at 113° F.

Bone fat, soap grease and oil lees are, each, an inferior kind of fat, available for making common yellow soap. The first is procured by boiling bones in water; the second, by the expression of heated pork or beef cracklings. The last is the odds and ends, saved in the different processes for making and purifying fish oils. The two former serve for hard soaps—the latter is used for a semi-concreted soft soap known as "oil soap."

Kitchen stuff.—This admixture of all sorts of fats, the refuse of kitchens, is gathered together by persons who make a business of it. After being melted by steam and drawn off clear, it furnishes a fat sufficiently good for yellow soap. Being a heterogeneous compound, it is difficult to determine its constitution in olein, margarin, and stearin, quantitatively.

Human fat.—This fat is more or less fluid, according to the proportion of its solid constituent; it is yellowish, of a feeble taste, somewhat odorate, and forms about the twentieth part of the human body. One hundred parts of cold alcohol dissolve 2.48 of the fat. With the alkalies it gives a firm soap, and Chevreul says, that a soap, made with fat taken from the breast of a woman, when dissolved in water, emitted a cheese-like odor whilst this characteristic was not incident to the soap prepared with fat from the leg.

This fat varies according to the part of the body whence it is taken; it is most firm near the kidneys and under the skin, whilst between the muscular fibres, it is only of the consistence of a semi-congealed oil, and near the movable viscera it is of a granular texture. Age also has much influence upon the grease, for in the fœtus it is a kind of granulated gelatin, becoming gradually more solid till the springtime of life; and at the age of forty the fat attains its maximum. Then it is a species of greasy cachexy, and after that age gradually diminishes, as shown by the wrinkling of the skin, which it serves to distend. It is much softer in males than in females, and though in old age the fat lessens in quantity, it is hard and of a deep yellow color.

Its composition is, oleic and margaric acids in union with glycerin, and perhaps, also, a volatile acid, to which its smell, and the odor given off by individuals during perspiration, are attributed. The offensive odor of negroes, especially in warm climates, is attributed to the evaporation of this acid by the heat of the body; but it is rather owing to an acid peculiar to their fat. As this is a question, the settlement of which may lead to interesting results, I intend that it shall be the subject of future investigation.

Adipocire, formerly considered a distinct constituent of human fat, is in fact that body itself partially saponified by ammonia, a greater portion of its acids being free and uncombined. Traces of lime, phosphate of lime and other matters have been detected in adipocire.

It is generally found in graveyards, and those situations where ammonia has been generated by the decomposition of animal matter. It is a peculiar fatty matter, resembling in some of its properties a mixture of fat and wax, and hence its name from two Latin words, *adeps*, fat, and *cera*, wax. It is produced in moist situations, and can be prepared artificially. The fusing point is 125° F., and from it, without purification, candles can be made as hard, and void of smell as those from wax; but the conversion of dead bodies to this purpose has been an unprofitable speculation.

Of butter.—This fat exists in the milk of mammiferous animals, from which it separates, spontaneously, after a sufficient repose. The cream rising to the surface and holding the butter being skimmed off and churned at 55° F., yields about 4.5 of butter in every 100 parts. It is white or yellow, of consistency more or less firm, and of agreeable smell and taste. Milk is composed of butter, casein, sugar of milk, several salts and water, but in variable proportions. The butter exists in the form of very small transparent globules of nearly uniform size. The contrast between human milk and cows' milk, is shown by the following analyses, by Henry and Chevalier.

	Human milk.	Cows' milk	Asses' milk.
Butter - - - - - - -	3 55	3 13	0 11
Salts or mucous matter - - -	0.45	0 60	0 34
Sugar of milk - - - -	6.50	4 77	6 08
Cheese - - - - -	1.52	4 48	1 82
Water - - - - -	87 98	87 02	91 65
	100.00	100.00	100.00

Braconnot has announced a difference between butter prepared in summer and that made in winter, as regards consistency. The following analyses exhibit its composition in each season.

	Butter of summer.	Butter of winter.
Oily or fluid constituent	60	35
Firm or solid "	40	65

Butter is insoluble in water, and becomes rancid by exposure. Its composition is margarin, olein, butyrin, caprone, and caprine, and, according to some, stearin. By steaming butter for five or six hours, or melting it by fire, and keeping it at a temperature of 70° F. for one or two days, the stearin and margarin crystalize or granulate out, and can be separated by filtration. The liquid portion is then acted upon by alcohol, in which the butyrin, caprone and caprine, being more easily soluble, are taken up, and the olein left behind. The above three are in small quantity.

When butter is saponified with an alkali, and the soap thus formed is decomposed by tartaric acid, glycerin is eliminated, the stearic, margaric, and oleic acids separate in an oily stratum, and the butyric, caproic and capric acids remain dissolved in the water. By distillation, they pass over into the recipient, and can be obtained by neutralizing with barytes and then separating the salts by repeated crystalizations. These three latter acids are volatile, and to them is due the pleasant and peculiar odor of fresh butter. Chemically speaking, then, we must call butter the stearo-margaro-oleo-butyro, capro-capriate of the oxide of glyceryl (glycerin), or an admixture of stearate, margarate, butyrate, caproate, and caprate of glycerin. Occasionally there is also a minute portion of *vaccinic* and *caprillic* acids present.

With respect to making butter, the scientific explanation of the theory of churning is as follows: By agitation the globules of butter are broken, and made to unite together in a mass. The introduction of air during churning, aided by the heat of the cream or milk occasions the formation of lactic or acetic acid, and this coagulates the casein, and thus assists the separation of the butter. In summer, when the heat prevents the ready coherence of the butter, a quantity of cold spring water thrown in, after the buttermilk has formed, often effects the desired end. The temperature is thus depressed, and the butter rendered solid, while the air, contained in the water, aids in the formation of acid, and in coagulating the casein. The only thing, in a scientific point of view, necessary to be attended to after the separation of the butter, is to free it from buttermilk, or casein. If the casein be suffered to remain, putrefaction ensues, and the butter acquires a rank putrid taste. Its separation, therefore, is of the first moment.

The superiority of certain brands of butter, noted for the excellent flavor and taste which they retain for a considerable time, is more due to their freedom from casein than to any mystery in the mode of preparing them.

Fish oil.—Certain fish, particularly the *cetaceii*, yield a greasy fluid or oil, called or named according to that of the fish whence it is extracted. The greater part of the oils under this head are liquid at ordinary temperatures, and consequently are excessively rich in olein constituency. Spermaceti oil is obtained by melting and pressing the adipose pannicle of the *physeter macrocephalus*, and of all that species of fish. It is found in commerce, bleached and unbleached, the latter having a brownish appearance and disagreeable odor. Cooled to zero, it drops a solid constituent (spermaceti), somewhat varying in the different oils, and bearing the same relation to fish oil that stearin and margarin bear to lard, and the so called fixed oils. Sperm oil, for instance, yields spermaceti, and, when separated from it, is a clear limpid olein admixed with a little margarin. It saponifies with facility, requiring but 0.6 of its weight of potassa and five parts of water. The soap obtained is brown, rather soft, and completely soluble in water, and, when decomposed by an acid, yields chiefly the oleic and margaric acids, and perhaps, some little stearic, glycerin being eliminated. Chemically speaking, therefore, sperm oil is an oleo-margarate of glycerin, with perhaps a little stearate of glycerin, and a volatile acid, to which its smell is to be attributed. Sperm oil is constituted, according to Bérard, as follows:

Carbon	79.65
Hydrogen	14.35
Oxygen	6.00
	100.00

and hence is more hydrogenated and less oxygenated than the fixed vegetable oils.

Common whale oil, *dolphin* (*delphinus globiceps*) oil, and porpoise (*delphinus phocœna*) oil, each extracted by a process similar to that for procuring sperm oil, when exposed to a temperature of three degrees below zero, let drop a solid crystaline deposit, somewhat analogous to spermaceti, and called *delphinine*. The liquid oily portion separated by filtration from this solid, is inferior, for purposes of illumination, to sperm oil, but is similar to it in chemical properties and in composition.

Spermaceti.—This fatty body is found in the cavities of the bones of the cranium of the different species of cacholots, particularly the *physeter macrocephalus*, mixed with an oily fluid, from which it is separated by being pressed in bags. It is a brilliant pearly white solid, brittle, with a lamellated structure. At 60° F., its specific gravity is 0.943. It melts at 112° F. One hundred parts of alcohol 0.821, solve 3½, of which 0.9 are deposited on cooling in crystaline plates. Warm ether takes it up in large quantities. The spermaceti of commerce always is contaminated with more or less of a yellowish oil, which it retains with great tenacity, and by which its melting point is greatly lowered. The best means of purifying spermaceti is to treat it two or three times with boiling alcohol, in which, however, it is very slightly soluble, and then to subject it to one or more solutions and crystalizations in ether, until the temperature at which the spermaceti solidifies, reaches 120° or 121° F., when it may be regarded as perfectly pure. In this state, the name of *cetine* has been assigned to it, and Playfair says it is precisely the same in composition as crude spermaceti, so that the small quantity of oil which accompanies the latter is probably isomeric with the more solid fat.

Spermaceti is saponifiable, but with less facility than the other fats, and the soap thus formed by its union with an alkali, when decomposed by an acid, is resolved into oleic (?) and margaric acids, and a peculiar substance named by Dumas and Peligot *ethal*, taking the place of the glycerin of other fats. Spermaceti, chemically speaking, is, therefore, an oleo-margarate of the oxide of cetyl (ethal). Playfair, however, who made an extended examination of the matter, concludes that it is a margarate of ethal (oxide of cetyl), but doubts the presence of oleic acid, because, when spermaceti is distilled, it yields no trace of sebacic acid; consequently there is every reason to conclude that the other acid of which the quantity is extremely small, is certainly not the oleic acid.

Mr. J. L. Smith, who, by his examination of the subject, verified the assertions of Dumas and Peligot, that there is neither margaric nor oleic acid in pure cetine, avers, that its real composition is ethal and cetylic acid; both set free by saponifying the cetine and decomposing the resulting soap by a strong acid. According to this view, spermaceti then is a cetylate of ethal.

Playfair's analyses of pure spermaceti or cetine, though, as he says, made with every attention to accuracy, and often repeated, differ considerably from those of Chevreul, Saussure, and Bérard.

The following are the results:

	Chevreul	Playfair	Saussure.	Bérard
Carbon	81.560	79.72	75.474	81.000
Hydrogen	12.862	13.32	12.795	13.000
Oxygen	5.578	6.96	11.377	6.000
Nitrogen			.354	
	100.000	100.00	100.000	100.000

Delphinine, or *Phocerine.*—This body is found in the oil of the porpoise, which contains, besides, a small portion of cetine, and is dropped from it at low temperatures. It is, in all probability, a phocenate of glycerin, for, when saponified with an alkali, and when the resultant soap is decomposed by an excess of tartaric acid, in which it is soluble, it separates from the fat insoluble acids free from any impurity of oil with which it may have been mixed. By conversion into a salt of barytes, it is precipitated, and can be eliminated from its combination with it, by the addition of sulphuric acid. The oil of the *delphinus phocæna* contains, besides, some parts of olein and margarin.

Neats' feet oil.—This oil is the product of a continued boiling with water of the feet of beeves, deprived of their hoofs. The oil swims to the surface, and is skimmed off as it rises to the top. For the benefit of those who live where the raw material is plenty and cheap, I insert the details of the process, for when the fatty ingredient can be obtained from this source, it greatly favors the economy of the manufacture, and consequently enhances the profits of the savonnier.

After having filled a large cast iron kettle with beef feet, cover them over with water and heat the contents up to 140° Fahrenheit; then with a pitchfork withdraw the feet from the kettle, and let each one be dextrously struck with a hammer so as to separate the loosened horny hoof; this being done, they are replaced in the same kettle and water, and the contents are boiled for about eight or ten hours, taking care constantly to skim off the oil with a ladle as fast as it appears on the surface of the water, and to transfer it to casks. By sufficient repose, the water that may have been dipped out with it, settles and forms a lower stratum, which can be drawn off at the bottom through a small gimlet hole. The remaining oil is now filtered through woolen bags to separate impurities and a portion of stearin which deposits during cooling. The oil is then sent to market, where, as found, it is a yellowish liquid, and, because of its fluidity at a very low temperature, is useful for lubricating machinery. Like tallow, it is saponifiable, and consists of olein, margarin, and stearin, the first named being the predominant constituent.

A great deal of the neats' foot oil of commerce is an admixture of the grease produced by boiling together, indiscriminately, the bones, tendons and feet of the ox and sheep. A product thus obtained, makes an inferior article of oil.

CHAPTER XVII.

THE CONSTITUENTS OF FATS, THEIR COMPOSITION AND PROPERTIES.

No work upon the manufacture of soaps and candles is complete, without it gives a thorough knowledge of the constituents of the materials employed in the business. Since the works of Chevreul, Braconnot, and other later chemists, have appeared, we know that the oils and fats differ but little among themselves, except in the immediate principles which constitute them and in the proportional contents of each. Having already, in describing these oils and fats, individually, named their components, we now proceed to give a thorough idea of the composition and properties of each, so that the intelligent savonnier may always have at hand the necessary information to explain the rationale of his process.

Stearin.—This is the most solid constituent of fat, and takes its name from στεαρ, the Greek word for tallow. It may be best prepared from tallow or mutton suet, by melting them at a moderate heat, washing them with ether as long as anything is dissolved, or by mixing up melted suet with six times its volume of ether, and subjecting the mass when cold to a strong pressure. In both cases, the stearin remains behind; it is generally in pearly white scales, not at all greasy between the fingers, is insipid and inodorous, fusible at 143° F., easily powdered, and, on cooling, forms into a solid non-crystaline compact mass. It is insoluble in water and cold ether, dissolves in boiling alcohol and ether, and crystalizes out on cooling.

Stearin is the acid stearate of the oxide of glyceryl (glycerin), and by the action of strong bases or acids, it is decomposed and separated into its constituents; a similar decomposition is effected by heat.

1 atom of glycerin	$= C_6H_7O_5$	
2 atoms of stearic acid	$= C_{136}H_{132}O_{10}$	$= C_{142}H_{141}O_{17}$ = Stearin.
2 atoms of water	$= H_2O_2$	

Stearic acid.—This is freed from its combination with the glycerin, and obtained in a state of purity by saponifying pure stearin with potassa, decomposing the resulting soap by an acid, and running off carefully the upper oleaginous layer into a clean capsule, where it cools into a crystaline mass. It is white, inodorous, tasteless, and soluble in alcohol and ether, melts at 167° F., and gives, by cooling, very white interlaced crystaline needles. At a temperature of 212° F., anhydrous alcohol dissolves it in all proportions, but on cooling, drops it in large brilliant scales. This acid burns like wax, and is the material of which adamantine candles are made, and upon which we treat in full at Chapter 47. Its density when solid is 1.01, in the liquid state 0.854. Dissolved in alcohol, it reddens litmus paper, and when distilled with access of air, is converted into margaric acid, and another substance called *stearone.* Its composition, when crystalized, is $C_{68}H_{66}O_5+2Aq$, or $\overline{St}$ 2aq, which, by the action of nitric acid, or other oxidizing agents, receives another equivalent of oxygen, and is converted into $C_{68}H_{66}O_6+2HO$, which is the formula for margaric acid.

Stearic acid is bibasic, that is, it forms two series of salts. In the cold, stearic acid decomposes the alkaline carbonates only partially, so as to produce a bistearate and bicarbonate; but with heat it decomposes the same carbonates, completely. The neutral stearates of the alkalies are perfect soaps, soluble in 10 to 20 parts hot water; but by the addition of a large quantity of water, a part of the potassa becoming free, they are decomposed into acid stearates, which are deposited, and basic stearates, which remain dissolved. For the same reason a hot solution of a neutral stearate becomes gelatinous on cooling, from the separation of the acid salt. The bistearates contain one atom of water, and one atom of fixed base, and the neutral stearates, two atoms of fixed base. The alkaline stearates, only, are those soluble in water. Neutral stearate of potassa ($2KO+C_{68}H_{66}O_5$) is a firm soap when cold, soluble in 25 parts of hot water, but more soluble in an alkaline solution. When common salt is added to a solution of this soap, an exchange of bases takes place, and stearate of soda is formed—a harder and less soluble soap than its corresponding potash compound.

Stearate of ammonia.—This salt is, with difficulty, obtained by exposing stearic acid to an atmosphere surcharged with ammoniacal gas. The salt thus produced by absorption of ammonia is white, almost inodorous, and of an alkaline taste. It is soluble in warm ammoniacal water, from which it falls, on cooling, as a bistearate, under the form of small brilliant scales. Its composition is $NH_3, \overline{St}+2HO$.

Stearate of lime.—This salt is obtained by adding to a solution of stearate of potassa a solution of hydrochlorate of lime. Stearate of lime is precipitated, and can be collected on a filter and dried. Its composition is $2CaO,+\overline{St}$.

Margarin.—This solid fatty matter, mostly of vegetable oils and of human and goose fat, was so termed by Lecanu, from μαργαριτη, the Greek for *pearl*, on account of the nacreous lustre of margaric acid. It is prepared by evaporating the ethereal liquor which remained from the undissolved stearin in procuring that body from mutton suet and tallow, and dissolving the residuum in boiling alcohol.

On cooling it crystalizes out. But in these it exists only in minute proportions. Its most prolific source is olive oil, from which it is deposited on exposure to cold. It can be separated by pressure, and by solution in alcohol, and crystalization from it, it is purified. It melts at 118° F. Margarin, properly speaking, is the margarate of the oxide of glyceryl, and by the action of a strong base, is separated into glycerin and margaric acid. It consists of $C_{74}H_{74}O_{12}$.

1 atom of glycerin	=	$C_6H_7O_5$	$=C_{74}H_{74}O_{12}$ 1 atom of margarin.
2 atoms of margaric acid	=	$C_{68}H_{66}O_8$	
1 atom of water	=	HO	

Margaric acid.—This acid bears a great similarity to stearic acid, but it is different in the melting point, which is 140° F. On solidifying, it crystalizes, and when carefully heated, volatilizes without alteration. Its alcoholic solution has an acid reaction. It is obtained by macerating completely dry olive oil potash soap, in twice its weight of cold alcohol The oleate of potassa dissolves, whilst the margarate, which is insoluble, remains untouched. It is collected on a filter, washed with cold alcohol, the residue treated with 200 parts boiling alcohol, and then left to cool. The margarate crystalizes, and, as it always retains a little of oleate, it must be submitted to a second and third treatment with alcohol. The purified salt is then dissolved in warm water, and the solution decomposed by hydrochloric acid ; the acid then precipitates in white flocculæ, which are to be washed with warm water, and after having been dried, solved in alcohol, from which, by careful evaporation, it crystalizes in beautiful interlaced needles. It is also obtained pure by boiling for some minutes stearic acid with an equal weight of nitric acid (NO_5), of density 1.273 (32° Baumè). The mixture is then left to itself, and, after expressing the product which has become concrete on cooling, between folds of blotting paper, it is crystalized several times successively from alcohol, till its point of fusion remains constant at 140° F., (60° Centigrade.) The formula of margaric acid is $C_{34}H_{33}O_3+Aq$, or $\overline{Mr}$, Aq, and when mixed with lime, and distilled, carbonic acid is generated, and combines with the lime, whilst a volatile substance called *margarone* ($C_{33}H_{33}O$) is obtained.

Redtenbacher, in his examinations, discovered a remarkable source of margaric acid in the distillation of stearic acid. "The distilled product, though in appearance unchanged stearic acid, does not, in reality, contain any trace of it, being a mixture of margaric acid, of margarone, and of a volatile oily carbohydrogen." The reaction is as follows:—

4 atoms of hydrated stearic acid $C_{272}H_{272}O_{28}$	produce	6 atoms of margaric acid	$= C_{204}H_{204}O_{24}$
		1 atom of water	$= HO$
		1 atom of margarone	$= C_{33}H_{33}O$
		1 atom of carbonic acid	$= C\ O_2$
		the oily carbohydrogen	$= C_{34}H_{34}$

Margaric acid, like stearic, forms salts, with the bases resembling the stearates in their properties, but the acid being monobasic (?), there is but one class of margarates. This point is mooted by some, and it is averred, that in the formation of salts of margaric acid, sometimes both its atoms of water, and sometimes only one is replaced by another base. If it is bibasic, then its formula is $C_{68}H_{66}O_6+2HO$. Assuming the correctness of this position, the neutral alkaline salts, say, for instance, the neutral margarate of soda ($2N_aO+C_{68}H_{66}O_6$) when in solution, and largely diluted, precipitates acid salts or bi-margarates. The neutral margarate of potassa is rather soft, but when solved and then decomposed by a solution of hydrochlorate of soda (common salt), becomes, by an exchange of bases, margarate of soda, which is a somewhat harder soap.

Margarate of potassa.—The process for preparing this is the same as that for procuring the stearate of potassa, viz., by boiling together equal parts of margaric acid and potassa solved in ten times its weight of water. On cooling, the salt appears in granules, and after being separated from its mother water by filtration, is dissolved in boiling alcohol, from which it crystalizes on cooling, in slightly pearly spangles. In contact with water, it swells and forms a transparent jelly. If the water is augmented, the salt is decomposed and resolved into an acid or bi-margarate of potassa. The composition of the bi-margarate is $KO,HO+\overline{Mr}$; of the neutral margarate, $2KO+\overline{Mr}$.

Margarate of soda.—Substitute soda for potassa, and prepare this salt in the same way as the margarate of potassa. Its taste is slightly alkaline after some minutes. Heated, it enters into fusion. Is but little soluble in cold water, but dissolves in ten times its weight of water at 176° F. It is very slightly soluble in alcohol and ether. The action of water upon this salt is the same as upon margarate of potassa. The composition of bi-margarate is, $NaO,HO+\overline{Mr}$; of the margarate, $2NaO+\overline{Mr}$.

Margarate of ammonia.—Margaric acid unites either with liquid or gaseous ammonia. In the former case, dilute aqua ammonia should be used, and its action assisted by heat. On cooling, the salt deposits in small pearly scales, which always contain an excess of acid. It dissolves in slightly ammoniacal water. Exposed to the air, it loses a portion of its base, and becomes an acid salt, that is, a bi-margarate of ammonia.

Margarate of lime.—This is obtained by double decomposition of margarate of potassa and hydrochlorate of lime, both in aqueous solution. The resulting precipitate, when separated by filtration and dried, is white, insoluble in water, and composed of $2CaO+\overline{Mr}$.

Olein.—This body, constituting the most fusible portion of fats and oils, exists in the solid fats, however, in but comparatively small quantity. From its oily liquidity, it is also called *elain*, from ἔλαιν, the Greek for oil. Olein predominates in olive oil, and particularly in almond oil. It is prepared pure, when almond oil is dissolved in hot ether and the solution exposed to great cold; the traces of margarin crystalize out completely, and by evaporation the ether is separated from the olein. Pelouze and Boudet, as well as Redtenbacher, declare that there are two oleins, because the liquid portion of those fats which are not drying, but disposed to become rancid, differs from the liquid portion of the drying oils in solubility, and particularly in the transformation which it undergoes, under the influence of hyponitric acid (NO_4), into elaïdine and elaidic acids, while, under the same circumstances, the olein of drying oils experiences no sensible alteration. Olein remains liquid at 6° F., and at six or seven degrees below it crystalizes in needles. It is this article, perfectly pure, which watchmakers use for greasing their work. In constitution it approximates to the solid fats, and chemically speaking, is an oleate of the oxide of glyceryl, and when decomposed by a strong base, shows its composition as follows:

1 atom of glycerin (oxide of glyceryl)	$= C_6H_7O_5$	1 atom of olein or $C_{94}H_{87}O_{13}$
2 atoms of oleic acid	$= C_{88}H_{78}O_8$	
2 atoms of water	$= H_2O_2$	

Oleic acid.—This acid is obtained by saponifying olein with a strong solution of potassa, decomposing the oleate of potassa, thus formed, by hydrochloric acid, washing the oily layer which separates with warm water, and drawing it off by a syphon into a clean vessel. Oleic acid is an oily liquid of a faint yellowish tint, an acrid taste, and slight odor; it reddens litmus paper, and below 20° F., congeals into a mass of needle-like crystals. Its specific gravity at 60° is 0.898. It is insoluble in water, soluble in all proportions of alcohol, of 0.822 from which solutions it is separated by water. Ether also solves it, and it is decomposed more readily than any other of the oily acids. It combines with the bases to form salts, both insoluble and soluble. Those of potash, soda, and ammonia, are in the latter category, the two former of which are without tendency to crystalization, and are both soft, unless they contain an excess of acid. According to Varentrapp, its composition is $C_{44}H_{39}O_4+Aq.$*

When oleic acid is distilled, a portion of it passes over unaltered, but the greater part is decomposed, and some charcoal remains in the retort. The distilled products, are, sebacic acid, (which concretes on cooling,) and a liquid carbohydrogen, isomeric with olefiant gas. Sebacic acid is not produced by the distillation of any other fatty substance than oleic acid, and hence may be considered as characteristic of it. The decomposition consists in, that

2 atoms of hydrated oleic acid $C_{88}H_{80}O_{10}$	produce	1 atom of sebacic acid	$= C_{10}H_9O_4$
		3 atoms of carbonic acid	$= C_3O_6$
		carbohydrogen	$= C_{71}H_{71}$
		residual charcoal	$= C_4$

Liebig says, that, in the Giessen Laboratory, it has been observed, that the oleic acid, as obtained from stearic acid factories, produces upon rapid distillation, more than half of a fluid product, which, on cooling, becomes as hard as tallow, and upon expression produces 35 per cent. of margaric acid.

The action of nitric acid upon oleic acid results, according to Redtenbacher, in the production of the whole series

* Gotlieb's later analyses give $C_{36}H^{33}O_8+HO$ as the formula for pure hydrated oleic acid.

of volatile fat acids boiling below 300°. In this way he obtained the acetic, metacetonic, butyric, valerianic, capronic, œnanthylic, caprylic, perlagonic and caprinic acids. The general production of these volatile fatty acids from oleic acid, explains their occurrence in almost all fatty bodies. In the fat of hens, geese, snakes, and particularly human fat, Dr. Redtenbacher found, without exception, small quantities of volatile fatty acids. The most diffused appear to be the caprylic, capronic and valerianic acids. In a diluted state the caprylic acid produces on the organs of smell the same effect as the sweat of a healthy man, and it is probable that this acid is carried off with the perspiration during violent exercise. *Oleic* acid is a valuable material for soft soap, and, when mixed with two or three per cent. of palm oil, is applicable to the manufacture of hard soaps.

Oleate of potassa.—This is obtained by the saturation of one part of oleic acid with one part of potassa solved in five parts of water, aided by a moderate heat. Be careful to stir the mixture constantly during the reaction. The union being completed, after allowing to cool, separate the excess of potassa, by pressing between two leaves of blotting paper. It is then dissolved in ten or fifteen times its weight of boiling alcohol. Filter the warm solution; by this operation, the subcarbonate is separated, and the oleate, in a state of purity, can then be obtained by evaporation. I will merely mention, that it is necessary, before treating with alcohol, to leave the matter exposed to the action of the air for some days, in order that the excess of alkali may become converted into subcarbonate, a salt upon which the alcohol has no solvent action. The oleate of potassa is colorless, of a slightly bitter alkaline taste, and is deliquescent, and hence, soluble in water. One part of this salt, and two parts of cold water give a transparent jelly; with four parts water a syrupy solution, and, finally, when largely diluted, it is decomposed into a very soluble binoleate of potassa, and into potassa, which remains in the solution. With the exception of the salts of potassa, soda, and ammonia, all the soluble salts decompose it. Its formula has not yet been ascertained. This salt enters into the composition of soft soap.

Oleate of soda.—Substitute soda for potassa, and prepare this salt in the same manner as the preceding. In properties, in color and taste, it is similar to the oleate of potassa, though much less deliquescent, a characteristic which, in a measure, determines the difference between the two. Ten parts of water at 53° Fahrenheit, dissolve one part of this salt. It is less soluble in alcohol than the corresponding salt of potassa, 100 parts alcohol at 55° Fahrenheit dissolving but 4.84 of oleate. The action of acids and salts is similar to that upon oleate of potassa. Its composition has not yet been determined.

Oleate of ammonia.—Oleic acid readily unites with ammonia, and forms a salt soluble in water, which, when boiled, becomes thick or muddy, owing to the escape of ammonia; the salt thus passes into the state of binoleate, the solubility of which is much less.

Oleate of lime.—You procure this salt by mixing together the aqueous solutions of hydrochlorate of lime, and oleate of soda. Double decomposition takes place, and oleate of lime is precipitated as a white powder, which, when separated by filtration, must be dried. It is pulverulent and insoluble in water. Its formula, as known at present, is inaccurate.

Cetine.—This body is nothing more than pure spermaceti, from the crude article of which it is obtained, as directed at p. 141. It is soft to the touch, insipid, almost inodorous, brittle, fusible at 120° F., and does not redden litmus paper. It is insoluble in water, but soluble in

40 parts boiling alcohol, of 0.821, from which, on cooling, it crystalizes in white brilliant plates, or pearly scales. It fuses at 120° F., volatilizes without decomposition at 250°, but, subjected to dry distillation, is altered. Cetine is, chemically speaking, according to one authority, an oleo-margarate of the oxide of cetyl (ethal), and, by saponification with potassa, and decomposition of the resulting soap, its component acids are developed—a new base (ethal), taking the place of glycerin in other fats. Dr. Stenhouse thinks, however, after examination of the subject, that the ingredient which is considered as oleic acid, and which exists in but a very small quantity, is not that body; for spermaceti, when distilled, yields no trace of sebacic acid, and in this he is supported by J. L. Smith, who further opposes the assertion of the presence of margaric acid, as averred by Dumas and Peligot. Its formula is $C_{32}H_{32}O_{2}$. Pure spermaceti, with which it is isomeric, and, according to Stenhouse, identical, consists of

2 atoms of margaric acid	$= C_{66}H_{66}O_{8}$	=1 equivalent of spermaceti or $C_{206}H_{207}O_{18}$
1 atom of oleic acid	$= C_{44}H_{39}O_{4}$	
3 atoms of ethal	$= C_{96}H_{102}O_{6}$	

Mr. Smith affirms the composition of cetine to be ethal and cetylic acid, united as cetylate of ethal, and the *cetylic acid* is described as a colorless, inodorous solid, crystalizing on cooling, from fusion at 131°; distilling unchanged, insoluble in water, but readily soluble in alcohol and ether. Its formula is $C_{32}H_{31}O_{3}+HO$. It is formed from sperm with ethal, by the action of potassa. Therefore, if this composition is correct, what has heretofore, in the analysis of cetine, been considered as a compound of margaric and oleic acids, is an entirely different and definite body, viz., cetylic acid.

Phocenine.—This fatty body is prepared as directed at p. 140. It is an oleaginous fluid, of an odor resembling ether. It has a specific gravity of 0.954, has no action on litmus paper, is insoluble in water, very soluble even in cold alcohol; it is remarkable, however, that alcohol always renders it acid. It is doubtless an oleo-phocenate of glycerin, for by saponification with potassa, and decomposition of the soap thus obtained, it is resolved into

Acid phocenic, dry - - - -	32.82
Acid oleic, hydrated - - - -	59.00
Oxide of glyceryl - - - -	15.00
	106.82

If, as is stated by some authorities, there is also a proportion of cetine constituent, then the composition of the phocenine varies, and must be written accordingly.

Phocenic acid.—By saponification with potassa, the phocenine is transformed into glycerin, and a mixture of the oleate and phocenate of potassa. To procure the phocenic acid, decompose the mixture by tartaric acid. The fatty acids are then separated by filtration, the mixture is washed with water, and this water is united to the first. These waters retain in solution the phocenic acid, which is separated by distillation, the acid passing into the recipient with a certain portion of water Saturate the product with solution of barytes, and evaporate the salt to dryness. You now have a phocenate of baryta, of which mix 100 parts, dried and pulverized, with 33.4 parts sulphuric acid diluted with water. The sulphuric acid seizes on the baryta and sets free the phocenic acid which collects at the surface, and is to be decanted. Add to the residue about thirty-three parts of water; then another portion of acid is separated, and when syphoned off, can be mixed with that first obtained. This acid is then introduced into a retort, and distilled over a water bath. It passes over into the receiver with a small portion of water, on the surface of which it swims. The brown matter remaining in the retort is due to the decomposition of a portion of the acid by the air confined in the apparatus.

Phocenic, also called delphinic acid, is a volatile oleaginous fluid which remains liquid some ten or twelve degrees below zero. Its specific gravity is 0.932, and its boiling point a little above 212° F. It dissolves in eighteen parts of water at 86° F. Anhydrous alcohol takes it up in all proportions. Its composition is represented by the following formula, $C_{10}H_7O_3+HO$. It combines with the salifiable bases and forms salts, among which those of interest to the savonnier are two, viz., of potassa and soda, forming a part of the fish oil soaps.

The *phocenates of potassa* and *of soda* are obtained by direct combination of the phocenic acid with these two bases respectively. Both are of similar properties.

The *phocenate of lime* is soluble in water, and slightly deliquescent.

Butyrin.—It is to Chevreul that we are indebted for the discovery of this body, which derives its name from a Greek word βουτυρον, compounded of βους, *a cow*, and τυρος, *cheese*, or *coagulum*. As its name indicates, it is found in butter united with olein, stearin (?) margarin, caproin, caprin, and perhaps a little butyric acid. To isolate it, you melt pure butter and keep it for some days at a temperature of 68° F.; the stearin and margarin crystalize out while the others remain liquid. By means of alcohol the olein is then separated from the other fats, which are more easily soluble in that menstruum. Their further purification depends on successive solutions in alcohol, but none of them can be considered as having been obtained completely pure. Its point of congelation is 32° F It is more or less yellow colored, totally insoluble in water, but soluble in all proportions in alcohol. It is said that it can be formed artificially, by warming together a mixture of butyric acid, glycerin, and sulphuric acid, in proper proportions. If so, it is the first compound of glycerin reproduced artificially.

Butyrin, chemically speaking, is a butyrate of glycerin, with slight admixture of the caproate and caprate of glycerin. *Caproine* and *caprine* have not been obtained sufficiently free from butyrin, to be described.

When butter is saponified, and the soap decomposed by tartaric acid, stearic (?), margaric, and oleic acids separate, while the other acids remain dissolved. On distilling this liquor, the butyric, capric, and caproic acids pass over along with the water, and being neutralized by barytes, the three barytic salts are separated by repeated crystalizations. The elementary composition of butyrin is not known.

Butyric acid.—This oily limpid acid has the odor of rancid butter and a nauseous ethereal taste. Its specific gravity is 0.9675, and when put in contact with the tongue, produces, like the phocenic acid, a white spot. It is very soluble in water, alcohol, and ether. It is volatile, its boiling point being above 212° F. Its formula is $C_8H_8O_4$.

Butyric acid unites with bases and forms salts called butyrates.

Butyrate of potassa.—This is formed by direct combination of the acid with potassa. It has a sweetish taste and crystalizes in a confused mass. It is very deliquescent at 59° F.; 100 parts of water dissolve 125 parts without becoming saturated. Its composition is $KO+C_8H_7O_3+?Aq$.

Butyrate of soda.—This salt is less deliquescent than the above, but otherwise has the same properties. Its composition is $NaO+C_8H_7O_3+?Aq$.

Butyrate of lime.—This salt is soluble in water, and what is remarkable, a solution of 17.58 parts in 100 of water, becomes thick when heated, and resumes its liquidity on cooling. Formula: $CaO + C_8H_7O_3 + 10\ HO$.

Caproic acid.—This acid agrees in property closely with the butyric acid. It is a limpid oily liquid, having the odor of sweat. Its specific gravity at 71° F. is 0.922. It evaporates in the open air, boils above 212°, and is soluble in 96 parts of water at 44° F. Its formula is $C_{12}H_9O_3 + Aq$.

Capric acid.—This acid ($C_{18}H_{14}O_3 + HO$) crystalizes in fine needles, which melt at 60° F., and when liquid greatly resembles caproic acid in physical properties. It is dissolved by alcohol in all proportions, and is soluble in six parts of water at 68° F. Its odor is like that of the goat.

Hircin.—This derives its name from *hircus*, the Latin word for goat, because it is obtained from the fat of the goat, in which it is found along with olein forming the liquid portion. It is insoluble in water, and more soluble in alcohol than olein, and by saponification with potassa, yields *hircic acid*, of which but little is as yet known.

Cholesterin.—This is a fatty substance extracted from human bile and the brain, by ether. It is crystalizable, and very rich in carbon, as its formula shows = $C_{37}H_{32}O$ (*Couerbe*); and fuses at 279° F. It is generally classed by chemists among the fats, although it only resembles them in outward appearance, and is not saponifiable.

Palmitin.—*Cococinic* acid and *myristicin* are considered at pages 98 and 106. The formula of the hydrated palmitic acid is $C_{32}H_{31}O_3 + Aq.$; of the cocoic or cococinic acid, $C_{27}H_{26}O_3 + Aq.$; and of the crystalized sericic or myristic acid, $C_{28}H_{27}O_3 + HO$.

CHAPTER XVIII.

BASIC CONSTITUENTS OF FATS.

THE term base is applicable to any substance capable of combining with acids to form salts. Every salt, therefore, consists of an acid and a base, or, in chemical language, of an electro-negative, and an electro-positive ingredient. The term base is also extended to the electro-positive element of the binary compounds: thus sodium is called the metallic base of chloride of sodium. An organic base is a substance possessing the properties of the alkaline or metallic bases, as regards affinity for, and union with acids; but it differs from them in originating in the animal or vegetable kingdom. Those from the latter source are also styled alkaloids. The oxides of glyceryl and ethyl come under the head of organic bases.

1st. *Glycerin.*—This body was discovered by Scheele, and received the name of "sugar or sweet principle of fat;" hence, its present title from the Greek word γλυκυς, sweet. It is a product of the saponification of most of the fats and oils, and is obtained isolated as follows: Mix together equal parts of olive oil and powdered litharge diffused in water; stir well, and keep it over fire until all the water has evaporated. When it has acquired the consistence of plaster, add a new quantity of water, and remove it from the fire. Filter the supernatant liquor, and pass through it a current of sulphuretted hydrogen gas, in order to separate the lead. Heat slightly, to expel the excess of gas, then filter, and concentrate the solution over a water bath, and finish the evaporation in vacuo.

Thus obtained, the glycerin is a transparent, colorless, inodorous liquid. Its taste is very sweet. Its specific gravity is 1.127, at its greatest degree of consistence; but it cannot be obtained solid. It dissolves in water, and in alcohol, but is insoluble in ether. Thrown on burning coals, it takes fire and burns like an oil. It is decomposed by heat, yielding a peculiar volatile compound, called *acroleine.* It is not fermentable, nor subject to spontaneous decomposition. In its power of solving alkaline salts and other bodies, glycerin is nearly equal to water. Glycerin, chemically speaking, is the hydrate of the oxide of glyceryl (an hypothetic radical), and is expressed as follows: $C_6H_7O_5+HO$. Liebig has remarked, that glycerin probably exists in nature, combined with other acids besides those of fats; an assertion somewhat strengthened by the fact, that Gobley has found it, in the yolk of eggs, in the state of phospho-glyceric acid, combined with soda and ammonia.

Acted on by nitric acid, it is slowly converted into oxalic and formic acids. When it is mixed with twice its weight of sulphuric acid, they unite without being charred, but with the evolution of much heat; and form an acid compound, *sulpho-glyceric acid,* the formula of which is $C_6H_7O_5$, $2SO_3$, HO. With bases, it forms soluble salts.

2d. *Ethal.*—This body derives its name from the first syllables of ether and alcohol, on account of its analogy in composition to these liquids. It is obtained by saponifying spermaceti with powdered potassa, twice repeated, so that none of the spermaceti may escape decomposition. A lime soap is then formed by precipitation with solution of chloride of calcium. It is dried by a gentle heat, and the ethal extracted by ether. Alcohol must not be used, as it dissolves also a large quantity of the lime soap. It is advantageous to mix the lime soap with pounded glass, as it is thus prevented from adhering to the sides and bottom of the vessel when heated; and the ether also is enabled to act more equally on every part of the mass. The ethal first obtained, is boiled again with milk of lime, again extracted with ether, and repeatedly crystalized. It is a solid, colorless fat, without savor, crystalizable in thin semi-transparent spangles. It is fusible at 119° F., insoluble in water, soluble in alcohol, uninfluenced by the alkalies, and consequently not saponifiable. Ethal burns like wax, and volatilizes rapidly at 250° F. It is neutral to test paper. In spermaceti or cetin, ethal takes the basic position that glycerin holds in other fats. Ethal is, chemically speaking, the hydrated oxide of cetyl (a hypothetic radical), and is expressed by the following formula: $C_{32}H_{33}O+HO$. It is decomposed by nitric acid, and combines with sulphuric acid, and, when neutralized with bases, forms double salts.

CHAPTER XIX.

THEORY OF SAPONIFICATION.

THE word saponification is derived from two Latin words, *sapo*, soap, and *facio*, I make. Pliny ascribes the invention of soap to the Gauls, who are supposed to have made it from tallow and ashes. The German soap, however, was considered best, and hence, by a transposition of letters, the Latin term *sapo*, probably from the original German *sepe*, now written *seife*, has become *soap*. The title of soap is generally extended to all those products which result from a union of fatty and oily matter with the alkalies, alkaline earths, and oxides of metals, though it has now become, by common consent, only applicable to those combinations of grease or oil, with potash or soda, which, by their detergent properties and solubility in water, are appropriate assistants to the laundress in the washing of clothes, and other scouring operations. Oils or fats themselves, are neither soluble in water, nor miscible with it; but nearly all are made so by union with alkali. The process of effecting this connection is called *saponification*, and the compound produced, as above said, is *soap*—a solid when its base is soda, but of less consistence, and more soft, when potash is the alkali employed. We have already considered the composition of fats, and designated the properties of which each is constituted, distinctly pointing out those naturally in union with glycerin, and also that of which ethal is the base. We have mentioned, also, that some of the fats afford, by the action of alkali and distillation with water, a volatile acid, independent of their fixed component principles.

The suets and fixed oils are mostly composed of certain proximate principles. Those most commonly found are stearin, margarin, and olein. There are some oils which have constituents peculiar to themselves, but as they are similar in their relation to alkalies, we cite the above three as examples. Stearin, for the most part, characterizes animal fats; margarin, the vegetable and human fats; olein is present in both, though the olein of the drying oils is different from that of the non-drying oils, as may be seen on exposing each to nitric acid. Besides, this dissimilarity may be easily recognized, because one is more soluble in certain menstrua than the other. Chevreul, to whose untiring industry and consummate skill we are indebted for that store of knowledge in regard to fats, which has served as a basis for the improvements and additions of our modern chemists, has opened the way to a new era in discoveries relating to organic chemistry. It was he who taught us the composition of fats, and showed us that the stearin, margarin, and olein they contain, are not free and uncombined bodies, as was supposed, but positive compounds, each of its respective acid with a peculiar base, glycerin, which the discoverer, called the "sweet principle of fats." Hogs' lard, which according to this chemist, embraces each of these components, is incapable of *direct* combination with the alkalies, or, to speak technically, of saponification, but acquires the facility by the presence of water, which, it was supposed, occasioned a change proportional to its elements; converting the stearin, margarin, and olein into new bodies, which, from their acquired acid properties, were called adipose or saponic acids; the peculiar unsaponifiable body, glycerin, above named, being left behind as an incidental product. Hence, what are now known to be positive compounds of adipose acids with their natural base, glycerin, were considered as altered in ultimate composition by the action of water and alkali. Further researches have shown the fallacy of this belief, and it has been conclusively demonstrated, that the acids exist already formed in the fats, and are separated from glycerin, according to that law which regards the power of stronger bases over weaker ones. So, consequently, when potash or soda solution is boiled with oils, the saponic acids are abstracted, and the eliminated glycerin, as a dernier resort, unites with the water, and forms a hydrate. This accounts for the excess of weight which the fat acids, estimated together with the glycerin, are found to have acquired. This excess is due to the water entering into combination with the glycerin and also the fatty acids. To be more clear and explicit, let us suppose we have a portion of melted lard. We will select this fat because its composition is known to be olein, stearin, and margarin. It is equally well known, that olein consists of oleic acid and oxide of glyceryl (glycerin); stearin of the same base, and stearic acid; margarin also of that base, and margaric acid. Heat this melted lard, and add a *solution* of caustic soda. A commotion and milkiness immediately take place; the soda, by the law of might, has driven off the oxide of glyceryl, and appropriated the adipose acids to itself, leaving the glycerin to take up with the water as a hydrate. This is the exact principle upon which soap is made, for the union of these acids with the base is, in fact, a hard soap, whilst the hydrated oxide of glycerin remains in the water used for solving the soda. The soap, therefore, does not result from a direct union of the fatty matter with the alkali. This statement is confirmed by the decomposition of the soap above obtained, with a strong acid. Here, by the power of might above mentioned, the acid decomposes the soap by robbing it of its base, and leaves the fatty matter in a different state from what it was when first added, or, in other words, as the saponic acids, before spoken of, and called respectively the oleic, margaric, and stearic acids. These fatty acids possess properties different from those when originally combined with glycerin in the state of olein, margarin, and stearin. They are more ready to unite with alkalies; and this facility arises from the fact that they are uncombined, and have no base to overcome, as was the case in the first instance.

In all other saponifiable fats, the same reaction takes place, though with the production of different acids where different greases are employed; for instance, palm oil, &c. As an instance of the elimination of *ethal*, a natural organic base of some few fats, reference may be made to the saponification of spermaceti, the resulting soap of which, upon decomposition, yields its characteristic acid, the cetylic.

To speak chemically, therefore, a soap of lard or tallow is an oleo-stearo-margarate of potassa, if a soft soap and of soda, when hard. The titles of soaps from palm oil, spermaceti, &c., must be modified to correspond with the titles of their respective acids; for instance, the soap of palm oil is the oleo-palmitate of the alkali employed. It is unnecessary to multiply words upon this subject, as we consider it clearly explained by the preceding remarks. For a clearer exposition of the action of each constituent named, and of its characteristic properties, the reader is directed to the respective paragraphs in which they are considered.

CHAPTER XX.

UTENSILS REQUISITE FOR A SOAP FACTORY.

Steam series.—One of the most requisite fixtures for a soap laboratory, in these times of brisk competition, when every effort must be made to improve the economy of a manufacture, is a "steam series," an arrangement by which the soap is made with steam heat, instead of by the old method of heating the kettle with coal, or wood. The advantages of the steam process are manifold and evident, for, to say nothing of the saving in fuel, and the facility of readily arresting ebullition at the desired moment, its conveniences in other respects, requiring less attention, and, consequently, fewer hands, are important items which recommend it. An experience of some years has attested its superior merits, so that now-a-days, in all well-regulated factories, steam only is used for boiling the soap. On the adjoining page is a representation of the whole arrangement, consisting of three caldrons, one for white, another for yellow, and a third for palm, and the finer soaps. G designates the main pipe or feeder, which is attached to the steam boiler W, of the establishment. It is stationary, and generally fitted against the wall, immediately above the kettles. The boiling caldrons are partly of iron, and partly of wood,—the upper portion or curb A being of wood, well hooped round by iron rings, and the lower portion D of cast iron, and so shaped, that the worm can hug closely to the sides without loss of room, and that the "blow-pipe" can fit snugly to the bottom. For the convenience of drawing off the spent leys,

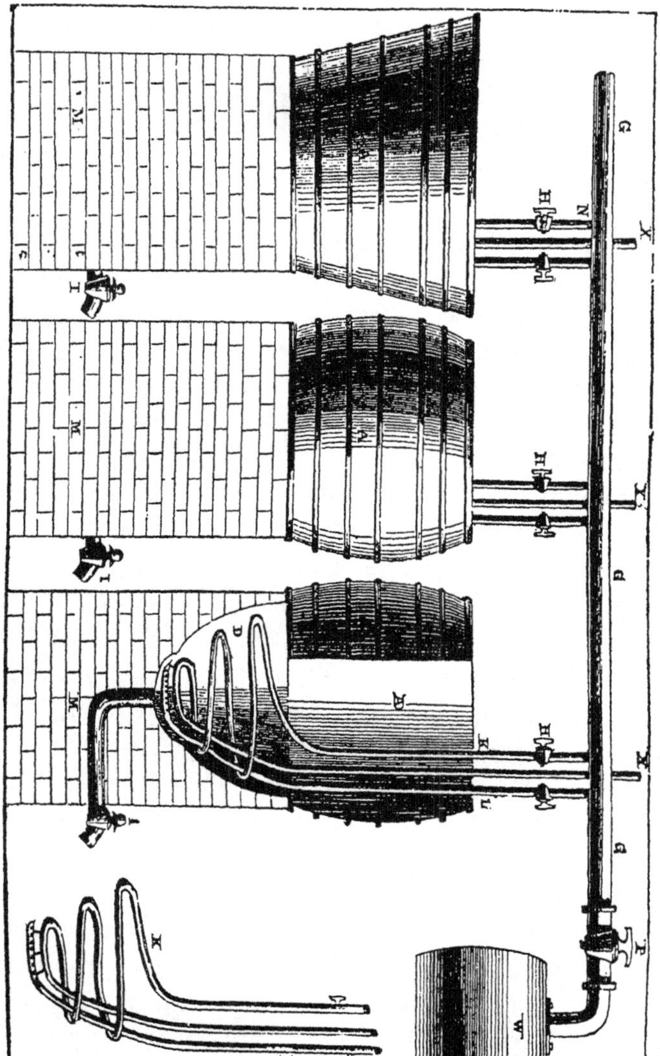

STEAM SERIES.

there is attached an *epine* pipe and cock I. Each of these kettles, setting upon a hollow pile of circular mason work M, is furnished with a welded wrought iron worm, which connects with the main feeder at N, and serves as the boiling medium of the soap paste. The steam is let on or off, by opening or shutting the cock H, and the waste steam is conducted through the other end of the worm X, which passes upward by the side of its inlet. and thence out in any convenient way through the wall of the laboratory. Also affixed to the main feeder, is another pipe with a stopcock attached, and leading immediately downwards to the bottom of the kettle, where it is affixed to a circular iron tube, pierced around its circumference with holes. It sets immediately below the worm, and is called the "blow-pipe," serving to give additional heat occasionally to the contents of the kettle, as well as to stir it up when necessary—an operation more effectually executed in this way, than by a crutch in the hands of a workman. The whole interior arrangement of the boiling pan is seen at the figure AD, the worm detached at K, and the "blow-pipe" at L. These kettles are worked much in the same manner as the ordinary fire caldrons, except that they require less attention. The charge of material is put in and melted by a rush of steam through both the blow-pipe and worm, the cock of the latter being shut off when it is necessary. We have inserted three caldrons in our figure. In large factories it is convenient to have this number; one, however, will answer in a small laboratory, though there will be necessarily a loss of time in cleansing it always, when the charge is to be changed from yellow to white soap. Messrs. Morris, Tasker & Morris, of Philadelphia, are particularly intelligent and expert in the construction of this and similar arrangements, and to their machine shops we recommend all manufacturers who intend ordering one of the above described convenient and economical apparatus. The curbs are made sometimes inclining inwardly, but the most economical shape is a slightly conical one, as shown by the figure in the extreme left. The cock P serves to shut off or let on the steam in the main feeder. Some manufacturers dispense with the iron bottoms entirely, and have their boiling cisterns or vats made wholly of wood well fastened by strong iron clamps.

Bugadiers or ley vats.—The bugadiers, in very extensive factories, are made of brick work, smoothly cemented within; but much the better material would be lead; for then one set of vats would answer for all kinds of soaps, as the ley prepared in them, being not acted upon by the metal, will be perfectly clean. Large oil casks lined with sheet lead, and cullendered at the bottom, are perhaps the best and most durable fixture of this kind that could be put up. Under each, should also be a lead-lined cistern, or reservoir, for the reception of the clear ley as it trickles through from the bugadier above:—or the lead-lined casks can be fitted near the bottom with a strong wooden cullendered diaphragm, also covered with metal. In this case there is also a cock fitted near the bottom of each cask, and through it the clear ley collecting in the lower part of the hogshead, between the diaphragm and the bottom, can be drawn off into the reservoirs below or for use, as may be wanted. Close by these bugadiers there must be a pump or hydrant, with its outlet spout conveniently arranged for a supply of water, in quantity as required. In small establishments, the reservoirs are not requisite, because the ley can be drawn off, as needed, immediately from the bugadier. There are generally several bugadiers to each laboratory, but the number depends entirely upon the amount of soap manufactured, and consequently the proportion of ley necessary for the steady prosecution of the work.

In a Marseilles soap house, there are four ley vats in each set, the arrangement being somewhat similar to that mentioned in this work, for the soft soaps. We describe it after Ure. No. 1 is the *fresh vat* which receives the fresh mixture of alkali and lime; the next one, or No. 2, being the *avancaire*, or an advanced stage. No. 3 is the small *avancaire*, being two steps in advance, and therefore, containing *weaker* liquor; and No. 4 is the *water* vat, because into that the water is directly introduced. Into No. 3, the moderately exhausted or somewhat spent leys are thrown. From No. 3 the ley is pumped into No. 2, to be strengthened, and in like manner from No. 2 into No. 1. Upon the lime paste, in No. 4, which has been taken from No. 3, water is poured, and the ley thus obtained runs upon the paste of No. 3, which has been taken from No. 2. No. 3 is twice lixiviated, and No. 2, once. The receiver under No. 1 has four compartments, into No. 1 of which, the first and strongest ley is run; into No. 2, the second ley; into No. 3, the third ley; and into No. 4, the fourth ley, which is so weak, as to be used instead of water, for lixiviation. The lime of vat No. 4, when exhausted, is emptied out of the window near which it stands, in which case the water is poured upon the contents of No. 3; and upon No. 2, the somewhat spent leys. No. 1 is now the *avancaire* of No. 4, because this has become, in its turn, the fresh vat, into which the fresh soda and quicklime are put. The ley discharged from No. 3, comes then upon No. 2, and after having been run through it, is thrown upon No. 1.

When, however, the iron vats are adhered to, as is the case in some factories, there must be a separate set for the finer qualities of soap, as it is requisite, especially for toilet soaps, to have the ley perfectly clear and colorless. On this account, both the bugadiers and reservoirs should be of material not easily injured by the leys. Deal wood would answer tolerably well, but as it is rapidly destroyed, and imparts a slight yellowish tinge to the leys, (very prejudicial to the whiteness of soaps,) a better substitute should be preferred. Laugier, the distinguished Parisian soapmaker, uses empty iron-hooped brandy pipes lined with sheet lead.* The price of these pipes is but $1 50, and when lined with lead, they will not cost as much as a cedar vat of the same size; besides, they neither affect, nor are affected by the leys, though they may have remained in them a month or more. Besides, they preserve the causticity of the ley much better than the wooden cisterns, and even when worn out, still possess an intrinsic value, as old lead, whilst the others, when once laid aside, are valueless for any purpose.

Soap frames.—These are rectangular cisterns, into which the soap, after it is properly formed, and separated from the ley, is conveyed in buckets from the boiling caldron. The soap, as it goes in, is pasty, but becomes more firm and consistent after a few days. The room in which these frames are set should be airy, and expressly appropriated to the purpose, and so constructed as to admit of every facility in drying the soap. It is called the "frame room," or "drying loft." This room should in winter be of a moderate temperature, and cool during the summer months, for too rapid or too slow congealing of the paste would be equally improper. The frames, of which a portion are set apart exclusively for the white and toilet soaps, and another quantity for the yellow soaps, are most generally made of wood; they consist of a pile of uniform rectangular enclosures, measuring internally about forty-five inches by fifteen inches, the thickness

* The shape to be preferred is that of a cone, and exactly similar to the body of a funnel. This form should always be adopted in the construction of *new* leech vats.

or height of each being about six inches. These frames are very smooth, and so jointed as to fit closely together. When piled one upon the other to the height of five or more feet, they form a tight square well to receive the boiled soap paste. Those for the marbled or mottled soap should be each of about 10 feet length, and 5 breadth, and 13 or 14 inches height. They are tightly bound together, by clamps and screw rods. Cast iron frames, as a substitute for the above in the manufacture of yellow soap, are of late introduction, "such frames being composed of five rectangular plates, one for the bottom, two for the ends, and two for the sides, so arranged and fastened together, as to form a well or reservoir of the same length and breadth as the wooden frames, and about five feet in height. These frames are put together and taken apart with more facility than those of wood, and the good conducting power of the iron considerably accelerates the cooling and solidifying of the soap."

Caldrons or boiling kettles.—In smaller factories, where the limited means and capital of the proprietor will not justify an outlay for a steam series, the old mode of boiling the soap by the naked fire, is employed, and in order that nothing may be left unsaid, we proceed to give a description and drawing of the kind of kettles most advantageous for the purpose. The size of the caldrons should be proportional to the amount of soap intended to be made at each boiling. The bottom pan can be of cast

Fig. 2.

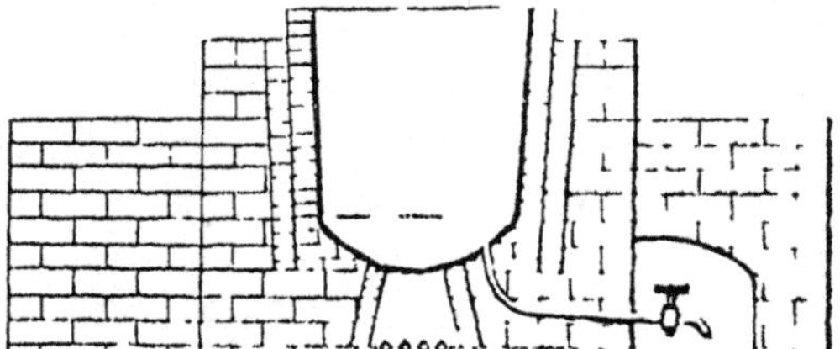

iron, but in England they prefer Swedish wrought iron plate. This bottom pan is consolidated in brick masonry, and is so built around, that the heat acts solely upon its bottom. Fig. 2 shows one of these caldrons. Should there be several, they are placed on a line with each other, and over a furnace beneath. To the caldron a tube of about two inches diameter is adapted, which serves as an outlet for the waste ley which remains under the boiled paste. At the mouth of the furnace is an arcade; at the bottom of this vault iron bars are placed to hold the fuel which is to heat the caldron. The arrangement of the mason work is generally, however, left to the skill and ingenuity of the bricklayer. These soap pans or kettles are cast with a flange at their top, so that, when necessary, an adjunct cylinder of wood, in the shape of a cone, may be fastened to them. This is called the *curb*, or upper part of the caldron. It is nothing more than a hollow cone of wood iron bound, made to fit the flange of the iron kettle. It can extend as high as desired, and is made of wood, so as to save the cost of metal, and the mason work necessary to enclose it. The cones stand erect, but they should be strongly and tightly fastened, and jointed to the lower pan. In this way you can enlarge your kettle extensively, at much less cost than if the whole caldron was of iron, and consequently entirely enclosed with mason work.

The minor utensils are a long-handled poker and a rake, for arranging the fire in the furnace, and withdrawing the coals. A wooden instrument, formed of a square

Figs. 3—4.

piece of board with a long handle, (Fig. 3,) and called a *mottler*, and used for stirring the paste in the kettle, when it is to be marbled.

A *round headed plug* (Fig. 4) of wrought iron; it should be a little crooked, and of about an inch diameter along its length. At one of its ends is a conical butt, which, when wrapped with tow, makes a sort of valve.

These instruments are fitted to the ley reservoirs, and answer more conveniently than cocks for drawing out the liquid. They are put from withinside; to open the channel in order to give vent to the liquid, give them a push inwardly, and draw them towards you again when the vent is to be closed.

Two *iron shovels*, with wooden handles, for general convenience.

One *heavy iron spade*, for shoveling and mashing the barilla and soda.

One *hollow shovel*, for throwing the mixture of lime and alkali into the bugadiers.

One *sieve*, for cleansing the lime of dirt and lumps.

A *plane*, to smooth the surfaces of the bars of white soap.

A sort of *iron-toothed scribe*, (Fig. 5,) to trace the sizes into which the soap in the frame is to be divided.

Fig. 5.

Copper pans, (Fig. 6,) for dipping up the ley and the oils from the reservoirs.

Fig. 6.

Small coppels, or basins, to draw the soap from the kettle.

Wooden buckets or *pails*, (Fig. 7,) of one foot diameter, and eight inches height, in which to carry ley, oil or water.

Fig. 7.

A *brass wire*, (Fig. 8,) with which to cut the bar soap into tablets.

Fig. 8.

Servidous or *copper buckets*, with handles, for conveying the boiled soap paste from the kettle to the cooling frames.

Iron ladles, pierced, for dipping up the paste from the kettle, and putting it into the *servidou*.

Fig. 9.

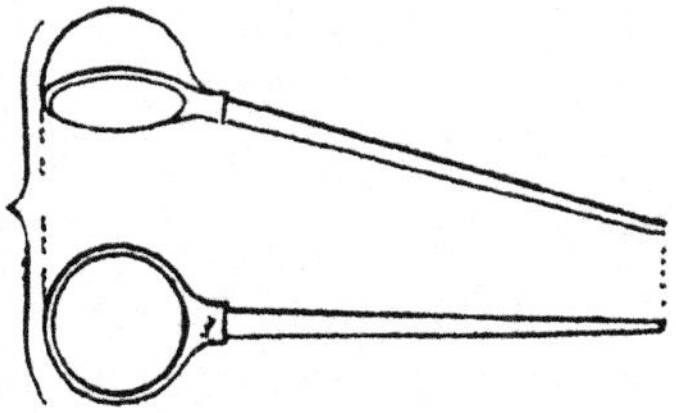

A *copper basin* with a handle (Fig. 9) for ladling the soap out of the caldrons, or conveying water to the lime when it is to be slaked.

A *press*. Fig. 10 very clearly exhibits the construction of this press. It has two spiral springs, A and B, by which the cake of soap is immediately expelled from the

Fig. 10.

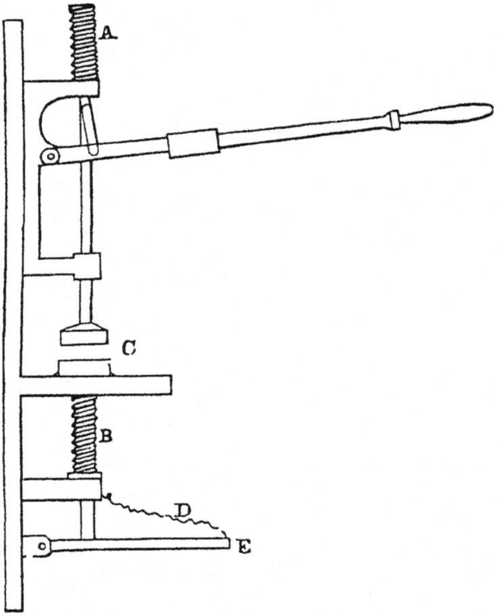

box C as soon as it is pressed. The workman knocks it off with the tablet that is to take its place, and so the pressing goes on successively, without that delay in the removal of the blocks of soap which would be necessary without some such arrangement. D is a rope suspending a wooden rod E, which serves as a support to the bottom of the die during the pressure. The box C can be suited to any size desired for the tablets. It is screwed on, and can be removed to make way for one of other proportions. This is a great convenience, because the fashion now is to press all kinds of soaps into blocks; the finer grades into fancy shapes, and the commoner sorts in pound lumps, bearing the impress of the maker's name given by the die which forms the top of the box C. This box also is movable, and its place may be supplied, at will, with a device suitable to the change of fashion or the taste of the savonnier. As they are very economical presses, it would be more convenient to have two, one for toilet soaps, and another of larger dimensions for the yellow soaps.

Culbertson's brick pressing machine, constructed upon a diminished scale, might also be applied to this purpose. The patentee resides in Cincinnati. I form my opinion from a model which was exhibited in operation at the Franklin Institute last winter. Its rapidity of movement and efficacy, it seems to me, would make a most advantageous substitute for the press, in forming tablets of hard toilet soaps. A small machine will scarcely be more expensive than a press, and as it throws out six blocks every rotation, the economy of time and labor is a valuable consideration in its favor. It is doubtful, however, whether it would answer as well for pressing soaps while in their *pasty* state; it is more applicable for the pressure of *dry* soap.

CHAPTER XXI.

THE SYSTEMIZED ARRANGEMENT OF A SOAP LABORATORY.

AFTER having mentioned the utensils requisite for the manufacture of soap, we now proceed to describe the different stages of the process. The manufacture should be systematic, and conducted in consecutive rooms, each properly arranged for its destined use; for instance, there should be separate store rooms for receiving the raw materials, and a perfectly dry and close room for storing the potash and soda. In this room should also be placed a large thick flat stone, upon which the workman reduces to powder the saline ingredients for the leys. He first crushes the larger lumps to smaller, with a heavy iron spade, and these, in their turn, are further powdered with a flatter and broader shovel. The injury to the health of the workman, owing to the alkaline dust which he draws into his lungs, has rendered necessary some arrangement which would obviate this disadvantage. This is effected by an apparatus called *bocard*, (Fig. 11,) which is propelled either by steam or hand.* This bocard consists of a certain number of pestles *b b b b*, which are placed above the mortars *a a a a*. An examination of the figure will obviate the necessity of a more extended description. The same power also gives motion to a drum-formed sieve (Fig. 12). A is a wire

* The bocard is more useful when the alkaline material is in lumps too large to be ground between stones. If it is in small clots, the better and more speedy way is to pass it through an ordinary mill. A suitable apparatus for this purpose can be had of the proper size at any millwright's.

cloth, the spaces in which are larger or smaller, according to the degree of fineness which it is desired that the substance shall have. B C is the axis of the drum, upon which is fixed a pulley; D D a wooden case, in which the drum partially enters; it is intended to prevent any loss of the material, in its passage from the sieve to the

Fig. 11.

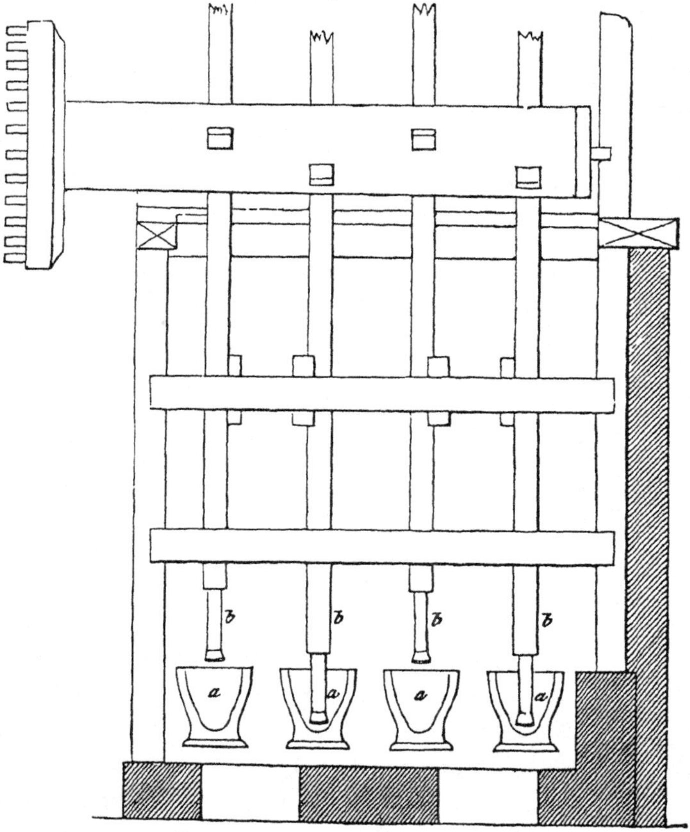

tub E. By this means the process is accelerated, and goes on regularly, and with the saving of one work`hand. By the side of this alkali room, called, by the French soapmakers, *picadou*, is another room, or what would be preferable, a separate building, in which the lime is kept. This building must be perfectly water tight, for the heat, generated by the accidental contact of water with the contents of a lime house, has occasionally ignited and destroyed the whole factory. Next to this lime building, or forming a part of it, is a room in which the mixture of alkali and lime is made, preparatory to its being thrown in the bugadiers or leech vats.

Around the boiling caldrons is a large area, into which the mouths of the furnaces, ranged along in a straight line, open. It should be bricked around and paved, with a staircase in the middle, by which to ascend and descend with facility. We have already said, that

Fig. 12.

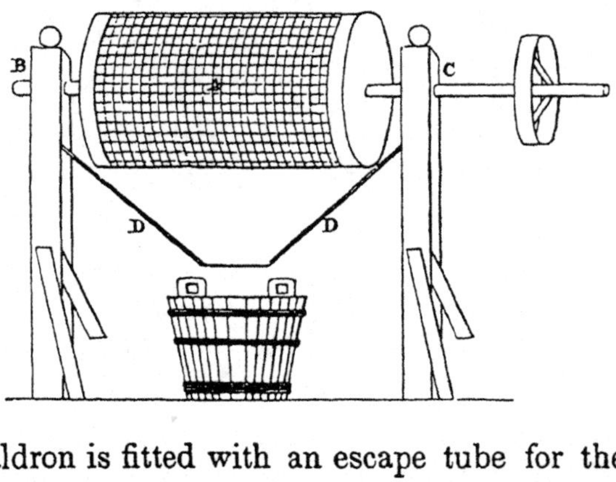

each caldron is fitted with an escape tube for the waste ley. This tube, instead of having a cock attached to it, can be more conveniently stopped and opened by one of the round-headed plugs, (Fig. 4,) with a sufficiently long handle; the cock is apt to become glutted, and is difficult to be worked. At the end of this waste pipe there should be a large bucket furnished with a very fine sieve; at its outlet, to retain what soap may run from the kettle with the ley, without interfering with the exit of the latter liquid. The second stories of the soap laboratory are generally partitioned off into four divisions; one for drying the soap, another for cutting it and pressing it in tablet forms, another for packing, and the remaining apartment as a warehouse for its sale.

In factories where the boiling of the soap is accomplished by steam, that power is at the same time made subservient to other purposes; in this way the economy of the establishment is much advanced. For instance, besides other advantages, there is this: the leys can be made with hot water, as it is very easy to fit up a cistern adjoining the bugadiers, which can be kept filled with water and constantly heated by the waste steam from your engine; by a little power it may be pumped up as wanted, into the bugadiers.

We have given the description of a steam series at p. 166. There the steam is introduced directly into the material. A too great temperature, however, is not desirable for some soaps; in that case, it is better to apply the steam upon the outer surface of the kettle. The arrangement then is different, as shown by Fig. 13. A

Fig. 13.

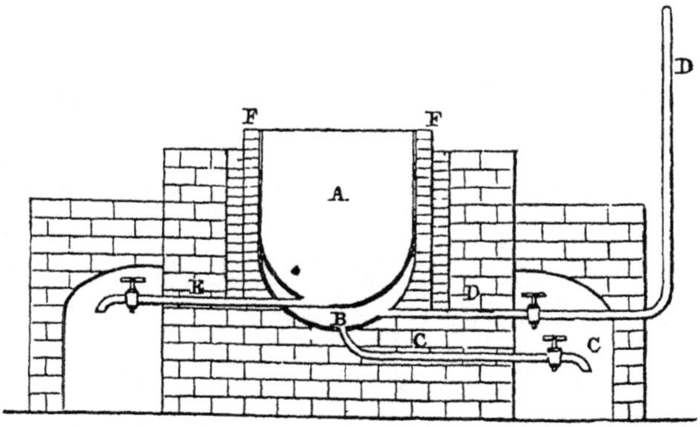

is the interior of the cast iron kettle, surrounded by brick work. B is the outer cast iron caldron, which should fit to the inner kettle so tightly as to prevent any escape of steam. D D is the tube leading from the steam boiler, and conveying the steam to the kettles. It is fitted with a cock, which is opened or shut, according as the steam is to be let on or off, to accelerate or retard the boiling of the soap. C C is the tube by which the condensed vapor is discharged; the cock in this tube can be left slightly open so as to operate as a safety-valve, when one of these necessary appendages is not affixed to the apparatus. The tube E is the discharge-pipe of the caldron.

The brick work F F is similar to that for other furnaces. The steam boiler, besides furnishing the power for this apparatus, can be made to work a small engine, by which the expense of several hands may be saved.

CHAPTER XXII.

PRELIMINARY REMARKS ON THE PROCESS OF SOAP MAKING.

It has already been asserted that olein, margarin, stearin, butyrin, cetin, hircin, phocenin, and the like, when treated with alkali, undergo saponification; or, in other words, are converted into soaps, composed of the respective acids of these fats and the alkali employed; that their natural base glycerin, or ethal, is eliminated in the state of a hydrate. As the proportion of these constituents varies in the different oils and fats, it is evident, that the latter are not equally applicable to saponification. Oils require a larger proportion of potash than of soda, one part of potash being capable of saponifying only from four and a half to five parts of oil. The consistence of all fats is proportional to their solid constituency, viz., their amount of stearin or margarin; and such as are rich in these two ingredients, are most fitting for the hard soaps, whilst those in which olein predominates are more proper for the soft soaps. According to Darcet, Lelièvre, and Pelletier, the following is the order in which the oils and fats are to be ranged, in regard to their facility of saponification:—Olive oil; almond oil; the animal fats, such as the suets, butter and the like; the oils of colza and navette; the oils of beech nut and of poppy, but mixed with olive oil or the greases; the various fish oils, also in mixture with the fats or olive oil; the oil of hempseed; nut, and linseed oil. In those localities where olive oil is indigenous, that is the ingredient most used for making soap; but in places where it is not manufactured, lard and suet are mostly substituted. The soaps are of two kinds: the *hard* soaps, made of hard fat and soda, and the *soft*, from the oils and potash. The first are called soda soaps, the latter potash soaps. They are veritable salts, the result of the union of the fatty acids with alkali. The term *savonule* is given to those formed with volatile oils; that of *ammoniacal* soap to the product from the action of ammonia upon the oils; and the designation of *plaster* to such compounds as proceed from the union of the metallic oxides with the oils.

The hard soaps of suets are either white or marbled; they are purer, and particularly applicable for washing the finer articles of dress. The soft soaps are more caustic than the solid soda soaps. They can be converted into hard soaps by the addition of salt, (chloride of sodium,) which, by an exchange of its base, renders them soda soaps. The process by which this change is produced, is called "cutting" the soap. We have already made known the constituency of fats and the rationale of the process of soap making, and have also in the proper place explained the parts acted in saponification, by the greasy body alkali, water and salt. Therefore, merely mentioning that those oils deprived of their mucilage yield soaps of mean consistency and impaired tenacity, we quote some appropriate remarks of Liebig, from the translation by *Booth.*

"Potassa and soda soaps are readily soluble in hot water and alcohol; the addition of a quantity of water to the aqueous solution produces a precipitation, the neutral salts of stearic and margaric acid decomposing into free alkali, which remains in solution and acid stearate and margarate of alkali, which precipitates in the form of pearly crystaline scales. Potassa soaps are more soluble in water than those containing soda. Stearate of soda may be regarded as the type of hard soaps, and when in contact with ten times as much water, it suffers no striking change. Stearate of potassa forms a thick paste with the same quantity of water. Oleate of soda is soluble in ten parts of water; the oleate of potassa dissolves in four parts, forms a jelly with two parts, and possesses such a strong affinity for water, that 100 parts absorb 162 in a moist atmosphere. Margaric acid acts similarly to stearic. It follows from this, that soaps are soft in proportion to the oleates, and hard in proportion to the stearates and margarates they contain. Soda soap exhibits a peculiar behavior to common salt; it loses the power of being penetrated by or dissolving in a solution of salt of a certain strength, and this remarkable action is an important condition in its manufacture, on which depend the separation of all free alkali and oxide of glyceryl, its content of water, and the state in which it is brought into market.

"If a piece of common hard soap, cut into pieces, be put into a saturated solution of salt, at ordinary temperatures, it swims upon the surface without being moistened, and if heated to boiling, it separates without foam into gelatinous flocculæ, which collect on the surface, and upon cooling unite into a solid mass, from which the solution flows off, like water from fat. If the flocculæ be taken out of the hot fluid, they congeal, on cooling, into an opaque mass, which may be pressed between the fingers into fine laminæ, without adhering to them. If the solution be not quite saturated, the soap then takes up a certain quantity of water, and the flocculæ separate through the fluid in boiling. But even when the water contains $\frac{1}{100}$ of common salt, boiling produces no solution.

"If the soap be boiled in a dilute and alkaline solution of salt, and suffered to cool, it again collects on the fluid in a more or less solid state, depending on the greater or less concentration of the solution, that is, on the quantity of water taken up by the soap. By boiling the dilute salt solution with soap for a considerable time, the watery flocculæ swell up, and the mixture assumes a foaming appearance; but still they are not dissolved, for the solution separates from them. The flocculæ, however, have become soft and pasty, even after cooling, and their clamminess depends more or less upon the quantity of water they have taken up. By still continued boiling, this character again changes, and in proportion as the water evaporating renders the solution more concentrated, the latter again extracts the water from the flocculæ; the liquid continues to foam, but the bubbles are larger. At length a point is attained, when the solution becomes saturated; before this, large, iridescent bubbles are observed to form, and in a short time all froth disappears; the liquid continues boiling without foam, all the soap collects in a translucent mass on the surface, and now the solution and soap cease to attract water from each other. If the plastic soap be now removed and cooled, while the solution is pressed out, it has become so solid as scarcely to receive an impression from the fingers. In this state it is called *grain-soap* (*Kernseife*).

"The addition of salt, or its solution to a concentrated alkaline solution of soap in water, precipitates the soap in gelatinous flocculæ, and the mixture behaves precisely like solid soap boiled with a dilute solution of salt. Carbonated and caustic potassa act exactly like salt by separating soap from the alkaline fluid in which it is absolutely insoluble. The application of the above to the manufacture of soap is evident. The fat is kept boiling in an alkaline ley until all pasty matter disappears, but the ley should have only a certain strength, so that the soap may be perfectly dissolved in it. Thus tallow may be boiled for days in a caustic potassa ley of specific gravity 1.25, without saponifying; if the ley be stronger, a partial saponification takes place, but being insoluble in the fluid, it floats upon the surface as a solid mass; by the gradual addition of water and continued boiling, at a certain point the mass becomes thick and clammy, and with more water a kind of emulsion (*seifenleim*), is formed, which continued heating renders perfectly clear and transparent, if a sufficient quantity of alkali be present. In this state it may be drawn into long threads, which, on cooling, either remain transparent, or are more milky and gelatinous. As long as the hot mass suffered to drop from a spatula exhibits a cloudiness or opalescence, the boiling is continued, or more alkali added. When excess of alkali is present, the cloudiness arises from imperfect saponification or want of water; the former is shown by dissolving a little in pure water, which becomes perfectly clear when the whole is saponified; if the ley contain lime, the mixture is also clouded, but the addition of carbonated alkali instantly clarifies it."

"In order to separate the soap from water, free alkali and oxide of glyceryl, a large quantity of salt is gradually added to the boiling mass, on each addition, waiting until it is dissolved; the first addition increases the consistency of the mass, while each successive portion renders it more fluid, till it loses its threading character and drops from the spatula in short thick lumps. As soon as the congelation is complete, that is, when gelatinous flocculæ separate from a clear watery liquid, the fire is extinguished, the soap suffered to collect on the surface, and cooled either on the liquid or ladled out and allowed to get solid. In the former case it is impure from water, free alkali and other impurities of the ley, and is therefore illy adapted for commerce, although sufficiently good for domestic purposes. As in other chemical operations, a precipitate is purified by boiling it in a fluid in which it is insoluble, so soap is purified by a solution of salt rendered alkaline.

"When the saponified fluid is made with potassa, the

salt (chloride of sodium) operates in a twofold manner; it dissolves in the pasty liquid and decomposes, forming on the one side, chloride of potassium, and on the other, soda soap. When potash ley is employed in soap-making, the first salting requires more than twice the quantity of salt. In the preparation of potash soaps, a concentrated potassa ley is employed for separating the soap. The saponification of fats is not completed by the first treatment with weak leys, and the subsequent repetition of fresh leys, besides purifying, also renders saponification more perfect."

CHAPTER XXIII.

THE PROCESS OF SOAP MAKING—EACH STAGE SEPARATELY DESCRIBED.

THE manufacture of soap is divided into nine principal operations, which are consecutively recorded as follows:

Preparation of the leys.—The deficiency of knowledge heretofore as to the mode of preparing the leys of suitable strength, has been a great hindrance to any improvement in the manufacture of soaps. It is positively requisite that the ley should be perfectly caustic when added to the oil, because, in the state of a salt that is in chemical combination with any other body, the saponification is not complete. We now know that 100 pounds of olive oil require about 54 parts of soda, of 36° alkaline strength, and that three parts of the soda are rendered caustic by one part of lime; for it must be recollected, that the soda exists as a subcarbonate in the commercial sodas. As it is, therefore, a salt or connection of carbonic acid and soda, the lime, when added, causes its decomposition by a superior affinity for the carbonic acid. Thus is formed an insoluble carbonate of lime, which precipitates and leaves the soda free, and deprived of its acid, in a state of caustic solution. It is, therefore, easy to ascertain what quantity of lime and oil the different species of soda will neutralize, since it is in direct proportion to the quantity of soda employed. For example, we give some calculations which can serve as a basis.

Each alkalimetrical degree indicates a hundredth of soda, or, if one soda marks 36 degrees by the alkalimeter, then it requires the third of its weight of slaked lime to be rendered caustic; if it marks 30 degrees, one-sixth less of lime; at 24° two-sixths less, and so on progressively.

After these examples, you can easily calculate not only the proportions of lime for the intermediate degrees, but also the quantity of oil that a certain amount of alkali can saponify; for it is evident that if 100 parts of olive oil neutralize 50 parts of soda at 36°, it will require 108 at 18°, or 216 at 9°, &c. The preliminary operations, in the fabrication of soap, consist, then, at first, in being assured of the degree of strength of the soda, so as to proportion the lime to it ratably. In preparing the mixture, the lime should be first slaked, and then thoroughly intermingled with the soda ash, and thrown into the bugadiers, and covered over with straw. The vat being nearly filled, water is let on its contents gradually, and allowed to percolate slowly through the mass. In about twelve hours, the outlet cock is opened, and the ley which has run through, is let off into the reservoirs. This ley marks 20 to 25°, and is called the first ley. Before the application of the alkalimeter, the strength of this first ley was determined empirically, by its ability to swim an egg upon its surface. Upon the same vat add a new quantity of water;—this, when it has passed through, is the second ley, and marks 10 to 15°. By adding again, you get a third ley of four to five degrees: but the residual mixture is still unexhausted, and will bear a new quantity of water, which, when drawn off, may serve either to dilute the first ley, or answer as a first water for fresh materials. The *marc*, or, as it is called, leeched ashes, makes an excellent manure for meadow, and for this purpose commands brisk sale at a fair price. Be careful to use only the prescribed quantity of lime, for when a great excess is employed, useless expense is incurred, without any increase of the causticity of the ley. It will be better, however, to use too much than too little, else a portion of the soda may be left undecomposed, and consequently useless. It is a well ascertained fact that the ley made from black ash contains more or less sulphur; this is a source of inconvenience in the process of soap making. To avoid it, always give the preference to the best quality of "soda ash" or "barilla;" this is rather free of it when properly calcined.

For common soaps, the ley is made from the ashes of hard woods, such as oak, hickory, ash, &c.; but as the soaps formed of this ley are soft, and of a potash base, it is necessary to convert them into hard soaps by the "cutting process," spoken of in another place.

The water used, should be as pure as possible, and in setting the vat let its bottom be covered well with straw so as to prevent the stoppage of the holes by the falling in of the larger particles of the mixture. Throw the first bucket of the mixture in lightly, and spread it well over the bottom, and scatter throughout the mass thin layers of straw so as to prevent a too great compactness, which is a hindrance to the percolation of the water. Let on the water in detached portions, always leaving a vacancy of a foot at the top of the bugadier, so as to allow room for the swelling of the materials. Keep the vats supplied with water, as they need it.

In England, according to Ure, they lixiviate the ground soda, of whatever kind used, by stratifying (instead of mixing) it with lime in cylindrical cast iron vats, from six to seven feet wide, and from four to five feet deep; the lowest layer consisting of unslaked or shell quicklime. The vats have a false bottom perforated with holes, and a lateral opening under it; this is closed commonly with a wooden plug, similar to the *épine* of the French soap pans, by which the leys trickle off clear and caustic, after infiltration through the beds of lime. The quantity of lime is proportional to the carbonic acid in the soda.*

* Tennant & Co., of Scotland, the extensive and celebrated manufacturers of soap and soda ash, have issued to their customers, printed directions as to their method of preparing caustic soda leys. I am indebted for a copy, to Messrs. Trimble & Johnson, their sole agents in this city. It is as follows:

"Proper cylindrical cast iron vats being provided, from six to seven feet in diameter and four feet deep, having a circular aperture of about one inch in diameter, at the bottom of each vat, stopped with a wooden plug—sometimes a false bottom perforated with small holes is used—but in general a good setting or filter is made by setting five square bricks inside of the plug hole, one on the edge on each side, two over, and one at the end, leaving sufficient space between the bricks for the ley to run, the box formed by the bricks to be covered with rye or wheat straw, over which is a layer of cinders.

"A layer of fresh burnt lime, say five measures of 112 pounds each, to be laid equally over the bottom of the vat and over the setting or filter, a few gallons of water to be thrown upon the lime, until it begins to slake or fall; this layer to be covered immediately with six hundred weight of soda ash, the next layer four measures of lime slaked as before, the fourth layer the same quantity of soda ash, the fifth layer, lime as before, and the last layer, the same quantity of alkali.

"After standing two hours, the vat is to be stanched by filling it with water or weak ley of a former vat; this to be done gradually. After standing about fifteen or sixteen hours, the plug is to be gently loosened so as to allow the ley to run off or trickle clear and caustic after infiltration through the beds of lime,—this is called the first runnings. As soon as the ley ceases to run, the plug is to be tightened, and the vat again covered with water, and after standing a sufficient time, to be run down as before; this is the second runnings, and worked together with the first runnings in the soap pan, is an excellent ley, and works freer and better than if used separately. After the vat is run dry, it is to be turned over into another empty vat. covered with water, and again run down,—this ley is very weak, and is seldom worked in the soap pan, being used instead of water, to stanch or fill up the strong or first set vats. As soda ash is not all equally soluble, it is sometimes necessary to turn the contents of the vat over a second time in order to obtain all the free alkali,—experience and care can only teach this. The receivers for the ley are generally much smaller than

2d. *Empatage.*—This operation, in which a magma is produced by the mixture of the oil and alkali, is considered as the most delicate and important of the whole series. Upon the success of this operation depends the profit of the savonnier; for if the empatage has been badly managed, the employment of a greater quantity of ley, without increasing the product of soap, is required. By way of illustrating the consecutive stages of the manufacture of soap, we will show the course by which "OLIVE OIL SODA SOAP" is prepared. Upon the manufacture of this soap, so generally used throughout France, and considered by Laugier as the best sort, much has been written; but the ignorance of savonniers, in matters directly appertaining to their art—for instance, as to the theory of saponification, and the requisite degree of strength of the leys—has made a trade, which is really scientific when well understood, a mere empirical occupation. We were ignorant for awhile concerning the mode of making this soap, as well as of other important matters, though, by a little attention and intelligence, they might have been explained long before. We now know that 100 pounds of olive oil require about 54 parts of soda at 36°, and that three parts of this alkali require one of lime to render it caustic.

In the manufacture of soap, the first thing is to be assured of the degree of strength of the soda. Well, then, supposing you have leys prepared as directed at page

the vats, but it is preferable to have them of the same size, it being at all times desirable to have a sufficient supply of strong caustic ley.

"Should the ley in the course of the process of boiling the soap, (what is termed) close with the materials, and will not separate, a small quantity of common salt, thrown with care into the boiling soap, will effect a separation; but this is always to be avoided if possible.

"The ley may be taken out of the vat with a pump or syphon; a third running may be taken from the first vat to staunch with."

190, you commence by pouring into the boiling caldron, to a third of its capacity, a mixture of three leys, the strength of which should be 8° to 9° if the oil to be used is rich in solid constituent, and of 10° to 11° if olein is a largely predominant component. This precaution is necessary, because olive oil is often mixed with lard oil; the latter contains much stearin, which it drops at cold temperatures. You then heat the contents of the boiler to ebullition, and at that point add about 720 kilogrammes of oil. The reaction is such, that a magma is immediately formed. Now decrease the heat by opening the doors of the furnace; as soon as the empatage has taken place, dose with weak leys, and have the precaution to stir the mass constantly during the addition, in order that it may have thorough contact with the whole. The *mottler* or *redable* is applicable for this purpose. The mass should remain homogeneous; the oil should neither rise to the surface, nor the ley descend to the bottom. If oil will present itself, it is then necessary to add strong ley to promote the empatage. The weak ley also can be used for this purpose, but the workman must judge which is more suitable to the capacity of the caldron. If the ley is in excess, you must add a new quantity of oil. The workman must stir the paste constantly, upon any addition of new material. The operation requires from 18 to 20 hours. Some savonniers, to facilitate the operation, throw in the scrapings or the waste of soap already made; by this they gain some four or five hours in the empatage. The signs by which an excess of soda is recognized, are, the liquidity and transparency of the paste. If the oil is in excess, it rises to the surface. This defect is often occasioned by the presence of a large quantity of marine salt in the sodas—a salt which has the property of retarding the empatage. If it is contained in considerable quantities, the empatage is entirely prevented from taking place. This accident renders the use of soap scraps indispensable. When native sodas were used by the soap-makers, this inconvenience frequently happened, but the employment of the purer manufactured soda, which is rarely contaminated with common salt, prevents its recurrence. When the empatage has been badly conducted, the consecutive operations will be of difficult management.

3d. *The relargage.*—Having in the preceding process used a large quantity of water, which was required for the ample saponification of the fat, the next step is to withdraw it when it is no longer needed. This operation is effected by adding those leys that contain marine salt; having poured these in, you then boil the mixture for about 15 or 20 hours. The fire being extinguished, the compound is allowed three or four hours' repose before the epinage, or the settled waste ley is drawn off. During the relargage, the workman, standing on the top of the brick work, should continually stir the contents of the caldron. When the mass opens in different places, the relargage is complete, and you then proceed to the next step.

4th. *The coction, or final boiling of the paste.*—Again add to the caldron, ley charged with marine salt; boil the mixture gently, and take care to scrape from the sides of the kettle any adhering soap, so that all portions may have the benefit of this process; by this the mass acquires more consistence.* After some hours' repose, the settled waste ley is again drawn off. Again, pour into the caldron, ley of 20°, and keep up the fire until the paste boils; by this it acquires a still greater consistence.

* In making the ley that is to be used for the coction, the materials should be mixed as follows: 64 measures of soda ash, 26 of slaked lime, and 4 to 6 of common salt; moreover, the ley itself must be strong for this process.

After continued boiling for about three hours, give it sufficient time to settle, and again draw off the waste ley. Repeat this same operation once more, but with strong leys, taking care to continually stir the paste while being heated, in order that the lower strata may be so mixed with the upper as to form a homogeneous mass. The fire should be so managed as only to produce a gentle ebullition. It is then that the soap commences to acquire firmness. That the coction may be perfect, it is requisite to repeat this operation four or five times. If, during these successive boilings, a crust should form at the surface, be careful to mix it well in with the paste. In these last operations, the vapor escaping upwards from the caldron is apt to spatter, or throw out portions of the paste, and as their temperature is very hot, the workman should so place himself as to avoid being hit by them.

When the odor of violets is recognized, and the soap has no longer the smell of oil, and when, by pressing a little of the paste between the fingers, it scales without adhering to them, the coction is complete. In winter it should be of eight or ten hours' duration; in summer ten or fifteen hours will be requisite; however, the length of time generally depends upon the quantity of paste to be operated upon. As soon as the coction is perfected, the fire should be withdrawn, the mass allowed one or more hours' repose, and the waste ley again drawn off. The soap thus obtained, is more or less bluish, and sometimes blackish, contains but 0.16 of water. Darcet has observed, that the color proceeds from a combination of a greasy substance with sulphoxide of iron and alumina, formed during the empatage, and dissolved by the soap.

5th. *The madrage.*—The madrage or mottling is the coloring of the soap, and Thenard's explanation of it has taught us the theory of the "marbled soaps." The alumina proceeds from the kilns in which the sodas are made, and becomes dissolved during the leeching of the alkali. The sulphuretted hydrogen comes from the sulphuretted sodas in the ley, and is set at liberty the moment the empatage is made. As to the oxide of iron, that proceeds from the impurity of the materials employed, and is held in solution by the hydro-sulphuret of soda. When the leys do not contain sufficient oxide of iron to give the aluminous soap a blue coloring, a sufficient quantity is added, in the form of a solution of copperas or sulphate of iron, after the empatage. In all these instances, the oil, it appears, combines almost instantly with the alumina and oxide of iron, thus producing a soap, becoming colored only at a boiling temperature.

6th. *The cooling and solidifying.*—When the soap paste has remained in the caldron long enough to have partially cooled, it is ladled out with the pierced iron ladles into buckets or pails, from which it is poured into the cooling frames, before described. These frames are set upon a bottom plate or plank so formed, that the ley which settles can run off into a reservoir arranged to receive it. After two or three days, when the ley has drained off, and the soap is hard, the iron screw fastenings which bind the frames together are unloosened, and each frame is lifted off the soap, which is left in one solid mass of the size of the capacity of the well. This mass is then cut with a brass wire, into blocks of three inches thickness, and then again subdivided into bars which are to be piled away upon slat work frames, intersecting each other so as to leave openings for the passage of the air. The soap then dries, and the saline solution with which it is still partially impregnated, exudes through the pores to the surface.

It is said that the soap acquires a finer grain, and becomes softer, by adding a small quantity of water as soon as it is put in the frames, and well agitating the paste with a long wooden paddle or crutch, until nearly cold.

7th. *The division of the blocks.*—The soap being dry, and in a state to be taken from the frames, a fact known by the gentle pressure of the finger making no impression, it is cut into equi-sized blocks, as marked out by the dentier. The teeth of this instrument are near or distant to each other, accordingly as the blocks are desired of small or large size. The thicker the bars the longer time required to dry them; however, the blocks should be cut up in bars of different sizes to suit the market for which they are intended.

8th. *The drying.*—The bars and tablets of soap being put into form, are then carried into the drying room and arranged upon the slat frame work, where they should remain until sufficiently dry.

9th. *The boxing and packing.*—The boxes in which the soap is packed are of common Jersey pine, and of sizes proportional to the length of the bar and the number which they are to receive.

The present fashion is to divide the bars of the yellow as well as the white soap into pound lumps, each being impressed with the name of tl e manufacturer. This is a neat way of sending the article to market, and therefore it is generally preferred by purchasers. The additional expense is very little, and is soon compensated for, by the increased sales which will be produced by this improvement upon the old mode of roughly packing the soap in irregular shaped bars of indefinite weight. The packing boxes for the white soap should be lined with white paper.

CHAPTER XXIV.

HARD SOAPS.

THIS term is applied to those soaps with a soda base, in contradistinction to the soft soaps which are made with potash. The hard soaps are firm, and are the kind used for the toilet, and domestic purposes; the extreme causticity of the soft soap renders it fit only for rough use. The hard soaps made from the oils, excepting palm and two or three other oils, are mostly a mixture of oleate and margarate of soda, whilst those made with animal fats, in addition to the foregoing, contain also a large portion of stearate of soda, which may be regarded as the type of hard soaps. Hence the hardness of soda soaps is proportional to the amount of stearate and margarate they contain, and the softness of the potash soap in a ratio corresponding with the amount of its oleic component.

Germany is one of the localities where soap was first manufactured. Nearly all of the soap product of that country is made from grease and potash, but with this difference, that the soap thus formed of potash is "cut," to speak technically, but, more properly, decomposed by the addition of common salt, (chloride of sodium,) the chlorine of which seizes on to the potassa of the soap and leaves its soda in compensation to the fatty acids, an actual soda soap being thus created by an interchange of bases. The method most usually followed in Germany, is the following.

Soap of potassa converted into soda or hard soap;—the "cutting" process.—Take one part of potash to two of grease. Some persons use ashes, of which they make leys according to the rules heretofore prescribed, by adding the fourth of their weight of lime and of salt, a handful to the pound of grease, or about 30 per cent.

When the solution is sufficiently saturated, it is poured into the caldron, and its combination with the grease effected. Agitate the mixture, whilst pushing it to boiling, and keep it at this temperature for six or eight hours, stirring during the interval until the liquid becomes clear; it is then run into moulds of desired form. The red color of common soap is given by bole Armenian, whilst peroxide of manganese is used for the brown tint. The ley which separates from the soap contains muriate and subcarbonate of potassa. By evaporating and calcining it, you obtain a product called "soapmaker's waste," much used in the European glass works.

In Bavaria they use indiscriminately the suets of mutton, beef, &c. At Vienna, principally tallow. The varieties of soap made in Germany are numerous, but the two principal sorts, as found in commerce, are *gemen seife*, common soap, and *mandel seife*, almond soap.

The first is prepared with ordinary materials; the latter is rendered very light and flotant by lively agitation. The first quality soap is made with the best grease, to which one-third of lard is added, an addition which gives beauty and brilliancy to the soap. When it is desired to make the soap still lighter, after the chloride of sodium has been added, and the ley well separated, it must be transvased into another vessel, that it may become frothy.

That the value of the oils and fats in the manufacture of soap may be correctly known, we give the result of some experiments made by Darcet, Lelièvre and Pelletier, relative to this point. The process detailed is the same that was followed for the essays.

1st. *Soap from olive oil.*—To proceed to the saponification of olive oil, we prepared leys with three leys of the soda of commerce and a ley of lime; the strongest or first ley; the second, of less force, and a third, or weak ley. We put into a small copper basin, three pounds of olive oil and a pint of weak ley of 5°, boiled the mixture, stirring carefully, with a wooden spatula, during the ebullition, and adding, from time to time, ley of the same degree. After four hours' boiling, we poured in of second ley, marking 6°, and continued to use it for about two hours; then we employed a small quantity of strong ley at 15°. After an hour's ebullition, the soapy matter became thick, and its incipient separation apparent; we then added two ounces common salt, to complete the separation, and afterwards withdrew the basin from the fire, and when the soapy paste seemed to be congealed, lifted it out, with a skimmer, into a tureen. Having emptied the basin, and well cleansed it, we replaced the soap in it with a small quantity of water, and when it reached the boiling point, added what remained of the ley of 15°. Having allowed it to boil an hour, we separated the soap in the same manner as before; threw away the small quantity of liquor which was at the bottom of the basin, and again replaced the soap in it, with a pound of water, and when it was perfectly liquified and united, run it into a small frame. Taken out the next day, and weighed, the block equaled six pounds ten ounces. It was sufficiently solid, but contained too much water to be saleable; three pounds of oil should produce but three pounds and some ounces of soap. The excess of water, however, was not slow in evaporating, for the same soap, upon exposure to the air, weighed, two months after, but four pounds fifteen ounces. It was then much drier, and perfectly solid, of excellent odor, resembling that of Marseilles soaps. By further exposure in a dry place, it underwent a still greater desiccation.

2d. The soap of *sweet almond oil* was, in consistency, next to the olive oil soap. The soap was very white, uniform, and of an agreeable odor. Its weight was five and a-half pounds, and after two months diminished to four and a-half pounds.

3d. The soap obtained from *colza* (*brassica campestris*) *oil*, with soda, is less solid, and yellowish gray. It retains the odor of the oil, and does not acquire the hardness of the preceding, and can also bear a greater quantity of water.

4th. The *oil of rapeseed* soap is of a grayish yellow, more consistent than that of colza oil, and of the peculiar odor of the oil from which it is made.

5th. *Oil of beech nut* soap is of a dirty gray, softer than the preceding, and of the odor of the oil. It is greasy, pasty, and clammy, and when exposed to the air, becomes yellow. By associating these oils with proportionate quantities of suet, mixtures will be formed of which soaps of proper quality and consistence can be made.

6th. The soaps made from *poppy oil* are of a dirty gray, without any disagreeable odor, of a medium consistence, and clammy. They become yellow in the air, and, exposed to cold, become soft at the surface. This oil, mixed with the greases, furnishes a soap which much resembles that of olive oil.

7th. *Hempseed oil* soap is green in color, very pasty, and so soft, that the least addition of water renders it liquid. Exposed to the action of air, it loses its green color, exteriorly, bleaches, and then takes a brown color.

8th. *Nut oil* soap is of a yellowish white, with but little consistence, greasy, and clammy; it takes a brown color by exposure.

9th. With *oil of linseed*, a whitish soap is obtained, which becomes yellow by contact with the air. It is greasy, pasty, and clammy, of a feeble consistence, strong odor, and, by a slight addition of water, is reduced to a thin paste.

10th. *Sperm oil* soap is of a dirty gray color. It retains the odor of fish oil, becomes reddish brown, and of some firmness, but much less than the olive oil soap.

11th. Soap prepared from *fish oil*, is but little different from the preceding.

12th. Soap of *cod liver* oil only differs from the above in having a more unpleasant smell and less consistency.

13th. *Suet* soap with soda base, is white, solid, and slightly smelling of suet. Exposed to the air, it dries, and becomes brittle.

14th. The soap prepared with lard is very white, solid, inodorous, and is specially used by the perfumer for making toilet soaps.

15th. Soap prepared from *rancid butter* is white, and approaches in properties, suet soap, but retains an odor of rancidity.

16th. With *horse fat* a soap is made resembling that from suet.

17th. Soaps made from castor oil and spermaceti are emollient, and particularly applicable to toilet purposes.

18th. Palm oil, as soap ingredient, holds a rank only inferior to castor oil and spermaceti.

The soaps from drying oils are most generally soft and flabby.

COMPARATIVE TABLE

Of the quantities of soap obtained from three pounds of oil, saponified with sub-carb. soda rendered caustic.

Names of the oils or fats.	Color of the soaps.	Quantity obtained as taken from the frame.	Loss of weight in	Time.
Of olive	white	7 lbs. 10 oz.	5 lbs.	2 months.
“ sweet almond	“	5 “ 11 “	4 “ 6 oz.	“
“ colza	lemon yellow	5 “ 14 “	5 “	15 days.
“ rapeseed	white	6 “ 8 “	5 “	20 days.
“ beech-nut	dirty gray	5 “ 4 “	4 “ 13 “	2 months.
“ poppy	gray	4 “ 8 “	4 “ 6 “	1¼ month.
“ hempseed	green	5 “	4 “ 14 “	15 days.
“ nuts	deep yellow	4 “ 7 “	4 “ 6 “	“
“ linseed	yellowish	5 “	4 “ 12 “	1 month.
“ sperm	dirty gray	4 “ 12 “	4 “ 10 “	15 days.
“ fish	reddish brown	4 “ 11 “	4 “ 8 “	1 month.
“ cod liver	dirty gray	4 “ 14 “	4 “ 12 “	15 days.
“ suet	white	8 “ 4 “	6 “	2 months.
“ lard	“	8 “ 3 “	5 “	“
“ rancid butter (unsalted)	“	11 “	7 “	“
“ horse fat	“	9 “ 8 “	6 “	“

White soap.—The olive oil soap, treated of at p. 194, having arrived at the point of coction, as described, can be converted either into white or marbled soap. When you wish to obtain it as white soap, thin out the paste by a mild heat, with a little ley, and covering the caldron carefully, leave it to settle. The result is, that the *alumino-ferruginous* soap, which is blackish, ceasing to be soluble at this temperature, precipitates to the bottom of the caldron. Then remove the upper stratum of white soap with the cullendered ladle, and run it into the cooling frames. White soap contains more or less water, according to its state of dryness. Generally, however, that made at Marseilles consists of

Soda - - - - - - -	4.6
Fatty acids - - - - - -	50.2
Water - - - - - - -	45.2
	100.0

It is easily seen from this, that white soap can be obtained whenever the soda leys which you use are very pure.

Marbled or *mottled soap.*—We have already said that the soap, having arrived at its proper degree of coction, is of a more or less deep blue color, and that it contains but 0.16 of water. Now when you wish to make marbled soap, the requisite water must be added so that the *alumino-ferruginous* soap may separate itself from the white paste and centre or assemble in large or small veins, in a manner to form a blue marbling upon a white ground. According to Thénard the separation of this body is comparable to a kind of crystallization, for that it shall be complete, it is necessary that the soap be properly thinned out with weak leys, and prevented from cooling either too rapidly or slowly. If it is too much diluted, and if it cools too slowly, all the marbling falls to the bottom and the soap is white; if too rapidly, it is in small granules. This process, therefore, is founded upon the lesser solubility of the *alumino-ferruginous* soap at a low temperature, and upon the inability of its solution to retain it, and its power of separating therefrom at a certain density.

If the soap, when it has attained to perfect coction, is not colored almost blue, then add in solution of copperas until the shade shall be sufficiently deep. The quantity of copperas requisite is about 18 ounces to 450 pounds of oil or fat. It is dissolved in the weak ley, and this solution poured in during the coction of the soap, and before it has acquired too much consistence. By varying the proportion of copperas (sulphate of iron), you can produce a blue of light or deep tint. This salt, by becoming peroxidized in the air, gives a rusty color to the exterior of the bars of soap, and the shade can be deepened, by the addition of colcothar, (brown peroxide

18

of iron,) into a tint, called, technically, *manteau rouge*, or *manteau Isabelle*, which upon the deep blue produces the appearance of marble. This manteau rouge can be given as well before, as after the marbling; it suffices to thin out the oxide of iron with water, and in adding it to the soap to incorporate it well. The soap is then put in the frames to cool, and cut into bars, as are other soaps. It is always harder, more uniform in its proportions, and contains less water than the white soap. This is easily explicable, since the coloring materials which constitute the soap are precipitated upon the addition of more than the requisite quantity of water; hence the proportion is limited.

The marbled soap is composed, according to Thénard, of

Soda - - - - - - -	6.0
Fatty acids - - - - -	64.0
Water - - - - - -	30.0
	100.0

Soda which contains sulphurets is preferable for making the mottled or marbled soaps, whilst that free of sulphur is most fitting for white soap.

The rationale of the process is, therefore, that the dense sulphuretted liquor, in passing through the pasty mass of soap, causes the marbled appearance, by the abstraction of alkali, by the sulphuric acid of the copperas and the elimination of the protoxide of iron, to become more or less brown colored by further oxidation, thereby producing a variety of tints; whilst a portion of the sulphur, falling in with hydrogen on its way, escapes in the form of sulphuretted hydrogen.

Marseilles soap.—The process for making soap at Marseilles, is as follows: For a boiling of fifty quintals of white soap, it requires in summer 40½ barrels of oil, whilst in winter 40 barrels suffice. The difference is owing to the larger quantity of ley, necessary to produce a coction of soap in winter, and also to the greater thickness of oils when cold made. While the ley is running through the bugadiers, the savonnier puts into the caldron 40 barrels of oil, which, if even clear and limpid, must be yet purified by a gentle fire under the kettle, so as to produce a slight ebullition of the oil; if, on the contrary, it is turbid, then whilst in ebullition you must add thereto two barrels of ley, and boil the mixture more actively. If the oil is of that kind which is so thick that it runs from the barrels with difficulty, it is necessary to increase the fire and prolong the ebullition with the ley. When a precipitate has formed at the bottom of the caldron, then the oil has become clear and limpid, and swims upon the ley, and is to be drawn off, by means of a syphon, into suitable reservoirs. The caldron, being afterwards well cleansed, one half of the oil is again poured in, and the fire rekindled thereunder. Now pour in eight buckets of strong ley, or, if preferable, half and half—that, is four kettles of the first strong ley, and four kettles of the second. The oil bubbles and boils with the ley which has been added, and the savonnier must be attentive, during the ebullition, to observe exactly and minutely the movements, for it is upon the remarks made at the commencement that he decides what is necessary to follow; nevertheless, he adds the remaining half of oil. As soon as the charge has advanced to a paste, it spatters, and throws out jets of pap in such a manner as to exhibit throughout the mass oily spots; this is a sign that the paste is thirsty, and that the ley heretofore added has been consumed.

It is then an evidence that ley is needed, when thick vapors ascend from the boiling matter, or when the paste, which is upon the oil, remains sunk and motionless; and four buckets, of the same as first added, are now poured in. But in adding it, be careful to distribute well throughout the mass, because if it all goes in one place, the cold ley precipitates to the bottom of the caldron, rarefies and spatters, and throws the paste against the sides of the kettle. By diffusing it over the surface, when pouring it in, it heats and rarefies gradually before reaching the bottom, and without bubbling or throwing out the mass. The four kettles of ley having been added, the savonnier is again attentive to the movements of his operation. When the same indications as before present themselves, another four buckets of strong ley must be thrown in. Continue supplying this ley, gradually, as wanted, and little by little, until all the oil is converted into paste.

When no more uncombined oil is apparent, then the empatage is complete. If the materials of the ley used are good, this process generally takes about thirty hours; but when the leys are weak and defective, nearly two days elapse before the paste is completed, and even then, only after much fuel and ley are consumed.

To know if the paste is good, homogeneous and perfect, the savonnier takes a kind of wooden spatula, three or more feet long, and one and a-half inches wide, and dips it in, and then draws it out and cools it. If, upon examination, the paste is well united, white, and without defect, and shows no particles of uncombined oil, he then pushes up the fire, and maintains an ebullition during the whole day; for, though the paste is complete, the soap is not yet perfected.

When the savonnier observes, by the escape of thick fumes, that all the ley is consumed, he adds ten more buckets, and always of the strongest ley. The paste which was thick, then becomes soft, and in this state is called *varre*. Whilst one workman revives the fire in the furnace, another adds the ley to the caldron, from time to time, until the whole ten buckets have been poured in. Some pretend that the soap is made handsomer and more profitable by dosing the oil in the commencement entirely with strong ley; this, however, is a mooted point among soap boilers, each adopting his own views in practice; and each avowing, that though a good soap can be made by such and such a method, yet his own is much the preferable plan.

When the above quantity of strong ley has been consumed, that which requires 36 to 48 hours, according to the quality of the material of which the ley is formed, then the paste *flaques,* to speak technically, or sinks and remains motionless in the caldron, an indication that its ley is exhausted. When a paste flaques but feebly, it ejects its mushy particles against the side of the caldron, and then the fire must be moderated. Whenever it is void of weakness, it appears perfectly free; and if it does not boil readily, must be dosed with five or six buckets of that ley which is left in the caldron when the soap is transferred to the cooling frames, and which is generally drawn off and preserved for this purpose, though, as is seen, its use is not always necessary. If the charge, after being boiled over a moderate fire, for two or three hours, is observed to have contracted, it is then to be dosed with fifteen buckets of the second ley. It then melts, and is converted into red paste; but this redness changes half an hour afterwards, and becomes white, an occurrence by which it is known that the soap is not perfectly *leyed.* Increase the fire, in that case, in order that the combination of the ley may be effected and the paste become of more body, and until its humidity has been dissipated, a fact easily determined by rubbing a little of the paste with the finger in the palm of the hand, and examining it. If this test does not prove the paste of the requisite qualifications as to beauty, freedom of water, and perfection of coction, then pour in three buckets more of strong ley, which have been and should always be reserved for such a contingency as this. After three or four hours' boiling with a reasonably hot fire, again serve it with fifteen buckets of the *second* ley. The paste again commences to become red, and as the coction is better promoted by this *abreuvage*, redouble the fire and boil briskly until the mass mounts nearly to the edge of the caldron, taking care now to stir it well with a long rod, so as to admit a portion of air to the mass; otherwise it will boil over. After the ebullition has ceased, the paste is generally in a state to be liquefied, but before proceeding to the granulation and separation, lend an ear to the following remarks:

When the paste is liquid, that is, has been boiled for half an hour, it becomes white and free; by continuing the fire, it dries up and becomes granular; now when the workman observes, that by reason of the weakness of the materials, the continuation of the fire fails to grain it, he then doses with three buckets of strong ley; an addition which seldom fails to put it in that state. If, by continuing the fire, it is seen that the paste splits all over, or parts asunder, the workman takes in his hand a sample to examine its state of coction, and whether it is perfect or not; he also tastes the ley in the kettle with the paste, and if this has but little taste, the paste is then drenched with strong ley, in order to its liquidation: if on the contrary, the ley is of a strong and piquant taste, then instead of drenching with strong ley, substitute pure water. It is at the termination that the workman must be most attentive, for it is in accordance with his judgment upon a sample of paste kneaded between the fingers, whether the ley to be added should be strong or weak, and when the fire is to be augmented or lessened. These operations are repeated four or five times, until it is seen, that all of the oil has been "tied" by the salt, and that the water thereof has sufficiently evaporated.

When thick vapors are emitted, it is a sign that but little water remains under the paste, and it is dosed with ley to prevent its falling to the bottom. If the intention is only to melt the paste in order to continue its coction, then weak ley must be used, for the strong ley will granulate it anew.

When, by employing weak ley, the paste becomes too soft, the fire must be increased. Such are the different treatments given to the paste, the whole process of which is technically styled *the liquidation*. It is conducted in accordance with the savonnier's personal observations of the paste itself. It is impossible to give a minute description; a general idea is all that can be expected. Finally, when the paste perfectly separates from the ley, and it is well perfected, it is allowed to repose for a day or two in the caldron, and after having sufficiently cooled, transvased into the frames. I will only remark, that upon the manner of managing the liquidation depends the quantity of soap obtained, and, consequently, the profits of the savonnier. In a well-conducted factory, with two sets of hands, one for day and the other for night duty, in two caldrons, three boilings of soap can be made per week. Some manufacturers use a certain per centage of poppy oil with the olive oil. It has been supposed that 40 barrels of oil have been used at one boiling, but the quantity can be increased or decreased according to the amount of soap required to be made. Forty barrels should furnish 50 quintals of soap, at an expense of 18 to 19 quintals of materials to form the ley and about 70 quintals of wood.

The preceding process fully verifies our charge as to the empiricism of the art previous to the explanation of the chemical principles applicable thereto, and the presentation of a method by which to determine the strength of an alkaline solution. What were the means used for this purpose heretofore, and even still, by the more ignorant of the craft? Why, the density of leys was distinguished in proportion to the ability of an egg to swim upon or in them. The ley was weak, when the egg sunk to the bottom; of medium force, or second ley, when the egg only partially descended; and of full power, or first ley, when the ley was strong enough to support it on its surface. Our previous remarks upon alkalies and leys show the fallacy of such a mode of operating.

The process of saponification is of duration, longer or shorter, according to the quality of the raw materials. It generally takes some hours of boiling, for the material to *ley*, become white, and thicken. This ebullition is kept up for about eight hours, adding from time to time of weak ley, and four or five hours after, the second ley, and from that passing on to the strong ley. It is important to keep a constant watch over the operations, for the least negligence or inexactness can vary the results.

One of the latest improvements in the manufacture of soap is the application of streams of air to the materials, whilst in the course of saponification. The air is forced below their surface by means of a pump, and this operation is said to facilitate and improve the process. Below is the method.

Yellow soap, Dunn's process—the arrangement being applicable for other soaps.—Into each of the ordinary boiling caldrons, a circular ring of pipe (of about one and a-half inch bore, perforated with small holes) is fixed in the well of the kettle just below the flange or joint, keeping it sufficiently off the bottom to allow a stirrer to be used to scrape the bottom when necessary. The circular ring of pipe is supplied with atmospheric air from a cylinder blast or other suitable forcing apparatus; this circular ring being connected with such forcing apparatus by means of a pipe attached thereto, and rising up to the top of the kettle, where it is furnished with a stop-cock and union joint for the purpose of connecting or disconnecting the parts of the pipe within and without the copper. For a clean yellow soap, put into the caldron 90 gallons of leys of specific gravity 1.14, made from strong soda ash. The fire being kindled, the caldron is charged in the usual way with, say 2050 pounds of grease, and as soon as the ley is hot and on the boil, or nearly so, the blast is set in action, keeping up a good brisk fire, so as to continue the materials in the kettle as near ebullition as possible. When the leys are exhausted, as is easily known, more ley is gradually added until the grease, oil, or fatty matter is killed. Then add 550 pounds of resin, a bucketfull at a time, with more ley occasionally, until 300 gallons of the strength above mentioned have been used, keeping the blast in action the whole time if the fires draw well; but if not, it is advisable to stop the blast for a short time, before adding the resin, to allow the contents of the kettle to approach to ebullition. When the whole of the resin is melted, and completely mixed with the soapy mass, and the strength of the leys taken up, stop the blast and give a brisk boiling to the contents of the caldron, and let it rest that the spent leys may separate and settle, which being now drawn off, the soap is then brought to strength on fresh leys, as in the ordinary process of soap boiling.

During the operation of the blast, the soap must be kept in what is technically called "an open or grained state," and for this purpose, salt or brine is to be added when necessary. Experience proves that it is better not to make a change of the ley during the operation of the blast, where ley of the strength before mentioned is used, but if weaker ley is employed, one or more changes may be made, as is well understood. It is found desirable that the soap should be kept in what is called a weak state, during the action of the streams of air through the materials; otherwise, the soap is apt to swell up, from the air hanging in the grain; and this is found troublesome to get rid of, requiring long boiling. If dark-colored materials are used, it is well to keep the blast in operation three or four hours after the resin is melted, provided the soapy mass is kept weak and open or grained. When a charge is to be worked upon a nigre, such nigre should be grained, and the spent ley pumped, or drawn off as usual, and the fresh charge added in the manner beforementioned, using less ley in proportion to the quantity and strength of the nigre, taking care not to turn on the blast until there is sufficient grease present to make the nigre weak.

"*Yankee*" *soap.*—This is the ordinary yellow soap, with an increased proportion of resin; the nigre or nigger of each boiling of this, as well as of all the species of yellow soap, being separated and sold as a second quality to the better portion or upper stratum of soap. The different qualities of both Yankee and yellow soap are made by varying the proportions of resin ingredient.

English soap.—We have already stated, that in England they make the soaps mainly of grease. As their mode of procedure is a little different from those of other countries, and as their soaps are much esteemed, we describe it in detail, commencing with

Yellow soap.—"The essential difference between yellow soap and all other kinds, is due to the presence in the former of a considerable quantity of resin, which, as before observed, is capable of uniting with alkalies to form a detergent compound, soluble in water. Besides the resin and common fat or inferior tallow, a portion of palm oil is now used. The usual proportions of palm oil and resin are three and a-half parts of the former to one of the latter: if made with a large proportion of resin the soap is soft and dark colored. The extreme proportions are, three parts oil to one of resin, and nine parts of oil to two of resin. Since the compound of resin and soda does not separate, like soap, from an alkaline solution or concentration, it would be improper to have the resin present in the copper, while the palm oil is being brought through all the stages of the saponifying process, as in that case, a portion of the resin would be removed with the spent leys. When the saponification of the palm oil is nearly complete, viz., at the last charge of ley, the resin should be introduced in a state of coarse powder and well mixed with the soap by agitation. The mixture is then boiled for some hours (a small quantity of ley being occasionally added, if necessary, to preserve an excess of alkali), until the soap is fully formed, when the ley may be withdrawn, and one or two waters, or else weak fresh leys, applied successively with agitation for the purpose of washing and purifying the soap. The resinous scum being removed for another operation, the soap may be conveyed into the iron frames to solidify. Yellow soap, which contains a large proportion of water, has just the same appearance as that which contains only a small proportion. When anything like 50 per cent. is present, it has been added while the soap was in the frames, for the purpose of fraud." But to the English plan, without palm oil:

Charge the caldron with 2000 pounds of tallow or soap grease, about 600 pounds of resin, and 150 to 175 gallons of ley, (of soda, and marking 10 to 20,) and when the whole is melted, heat up the mixture to ebullition, being careful to stir all the while to prevent the adherence of the resin to the bottom and sides of the caldron. If the mass seems disposed to become bloated or swell out, lessen the fire. This boiling should continue but two or three hours, because of the facility with which the union of the fat and alkali is effected. After six hours' repose, the exhausted ley is drawn off, and fresh substituted, and the whole again boiled for three hours more. Allow another repose of six hours, and again draw off the spent ley and renew it with fresh. The boilings are thus continued from day to day until the soap shall have acquired consistency, a fact determined by taking a sample and when cool, squeezing it between the thumb and finger; if hard thin scales are formed, it is finished, or nearly so; if greasy, clammy, and soft, it is on the contrary, not perfect, and you must then add ley, and give it another boiling. In the first case, you give a brisk boiling to the paste and then put out the fire. Cool the soap by adding three buckets of ley, and two hours after, draw off the liquor. Now throw in six or eight buckets of water and boil briskly, stirring the mixture until the soap is melted; then, with a wooden spatula, taking a little of the boiling paste, hold it up and observe whether it runs from the ley, clear; if it does, add water to the caldron and continue the boiling. If it does not run from the ley, too much water has been added already, and there must then be poured in half a bucketfull of strong solution of common salt.

The most delicate part of the operation is that of boiling, and should, therefore, command the particular attention of the savonnier. That the boiling or cooking may be complete, the soap should, when the spatula is held obliquely, not run off, but shake and disperse tremulously in a manner as would a body of gelatinous consistence. It is then that the fire may be withdrawn, and the soap regarded as finished.

If you wish to give a pretty color to this soap, you can add about 20 pounds of palm oil, and after two days run it into frames, whence, after a week or less, it can be taken and cut into bars. In France, they make yellow soap by the same process as for the other kinds of common soap, adding at the commencement of the operation, fifteen to twenty per cent. of resin. Some factors make a soap with but five to ten per cent. of resin, which being at a lower price than the pure grease soaps, has a more extensive sale. Hence its preference for all domestic purposes, and by its property of solving better in sea water than other soaps, is well adapted for use on shipboard. A common brown soap is made, as are the preceding, but by increasing the quantity of resin, and using the commoner kinds of grease, such as soap fat, bone fat, &c. It is called "Yankee soap."

Pure white soap.—English process.—Charge the caldron with 190 gallons of ley of specific gravity 1.090, and 2000 pounds of best melted suet. When the mixture becomes milky, increase the fire and keep it at gentle ebullition for two hours, then put out the fire and after two hours' repose, draw off the ley. Continue the operations as directed for the preceding, using the same care in examining the point of coction. If it forms, by pressure between the fingers, clean solid scales, cool the magma by throwing in some buckets of ley, but soon after draw it off clear. After the fire has been withdrawn, turn into the caldron eight or ten buckets (an English bucket is nine or ten gallons) of ley, and after fusion make a trial of the paste with the spatula. If it runs from the ley, add water;—if it does not run or seem disposed to, boil it a little more, adding a bucket of water, containing a third of its weight of common salt in solution, in order to effect the separation of the soap from the water. When this separation appears to be perfect, after half an hour or more repose, the water which contains the greater part of the alkaline ley remaining from the first boiling, and generally of a very deep bottle green color, is then drawn off. Replace the fire under the caldron, and turn in about eight buckets of water, and when, after a continued boiling, the incorporation is complete, examine with the spatula if the soap runs from the water. If it does, add water in small portions until it ceases, and the pasty liquid moves tremulously like melted gelatin. To end the operation give a good boiling to the contents of the caldron, and unless the soap has a bluish tint (in which case it should have another washing) extinguish the fire entirely, carefully cover the kettle and leave the whole rest for a day or more, and then ladle it off into the frames.

Soap from human fat.—This fat gives hard soaps by a process similar to that used for saponifying the ordinary fats. These soaps dry quickly, and become yellow.

Beef marrow soap.—This more properly belongs to the class of extempore soaps, as the marrow is generally worked into soap by the "cold" plan. To 500 parts of beef marrow add 250 parts by weight of solution of caustic potassa, 36°. Stir the mixture with a wooden spatula, and keep it at a continued heat until it forms a mass soluble in water. In this state thin out with 2000 parts of boiling water and add 1000 other parts holding solution 180 parts of common salt, stirring well together the whole, and then allowing it to repose. After some time, transvase the soap into frames that it may drain, and then mould it in porcelain, or iron, or tin forms.

Soap from rancid butter.—If the butter is salted, boil it in water to abstract the saline matter, and then put it over the fire to evaporate all moisture. Take three pounds thereof, and treat it with leys in manner as before indicated. The butter saponifies very well, and when it has been converted to soap, can absorb a sufficient amount of water, without being impaired in solidity by cooling. It is thus that Darcet, Pelletier, and Lelievier obtained a soap, very white and slightly odorate, of rancid butter. As taken from the frames, the soap thus obtained weighed eleven pounds, but in two months it had decreased to seven pounds.

Soap from horse fat.—This soap is made by the usual process. It has no disagreeable odor, and is of a white color and very consistent.

Cocoa nut oil (butter of cocoa) soap.—There is very little of this kind of soap made in the United States; a small lot occasionally finds its way here from England, but its disagreeable smell limits its sale, and that too at a low price; besides, the oil even there has been used but to a limited extent in the manufacture of white soap, because it is rather difficult of saponification, and yields a soap which requires for its separation from alkaline leys, a much larger quantity of saline matter than the suet and palm oil soaps. Below is a modification of Sturtevant's process.

The amount of ingredients for a batch of cocoa soap consists of 2072 pounds cocoa nut oil, 168 pounds olive oil or tallow, 84 pounds common salt, 375 gallons of soda ley, 24° Baumé, and 60 gallons potash ley of 20°. The cocoa nut oil is first deprived of its unpleasant smell by boiling it in a wooden vessel, by the aid of steam, with six pounds sulphuric, and twelve pounds muriatic acid to every ton of oil. The purified cocoa nut oil and the olive oil are first heated in the soap pan, and ten gallons of soda ley being added, the mixture is caused to boil, after which another ten gallons of soda leys are added, and this addition is repeated at intervals, until the whole of the soda ley has been introduced in the pan, the contents of which are kept at the boiling point. When this mixture has boiled for half an hour, the potash ley is added to it in the same manner as the soda ley, and the boiling is continued for about ten minutes. The salt is now slowly sprinkled over its surface, and after it has boiled for half an hour longer, the fire or steam by which the soap pan is heated, is withdrawn, and the soap allowed to cool; it is afterwards finished in the usual way.

Palm oil soap.—Of all soaps for toilet use or for washing the skin, there are few better than the palm soap, its emollient properties, agreeable violet odor and consistency rendering it even superior to the Windsor or olive oil soap. But to possess these properties, it must be correctly made, and of bleached oil, not, however, whitened by nitric acid, as its goodness is thus impaired both as to quality and odor. Mr. Graves of Philadelphia, one of the most intelligent savonniers in this country, has, by skill and perseverance, brought this soap to an advanced state of perfection, and corresponding with that which entitled all of his products to the compliment of two silver medals at the late exhibitions of the Franklin Institute.

Palm soap consists of palm oil (bleached by any of the methods before given) and suet saponified by soda ley over fire, and in the usual manner of making suet soap. The proportions are one pound of palm oil to four pounds of suet for the *demi palme*, and about three pounds of oil to one of suet for the *palm* soap.

When the materials are boiled to the consistence of paste, the separation or "cutting" is effected by additions of strong caustic ley, being careful, however, to manage properly, lest the odor of the palm be too much altered. The paste having been separated, it is transvased into the cooling frames.

Mr. Walterson of Manchester, in 1844, obtained a patent for a process for making soap in general, and particularly adapted to palm soap. His method consists in a mode of mixing or combining the oils or fatty matters employed, with caustic soda and in subsequently adding water for the purpose of converting such mixture or produce into soap. The inventor says, that the soap is thus rendered purer and more efficient, and that a considerable economy of time is effected in the operation, the soap being sufficiently hard for sale or use in the course of a few hours, instead of several days, as is the case under the ordinary process of manufacturing. The ingredients used are animal or vegetable oils, either separately or mixed, caustic soda ley of 22 per cent. strength, and water as free from earthy and metallic impurity as can be. The manufacture is as follows:—for PALM SOAP.

Take 784 pounds purified or bleached palm oil, and as soon as it has become fluid in the caldron, add thereto, gradually, 407 pounds more or less of liquid caustic soda, of strength above named, taking care to mix the ingredients thoroughly by stirring. The heat must then be increased, and the mixture constantly moved about to prevent it from caking on the bottom of the pan. After continuing the operation, and at the end of three or four hours, the mixture assumes a whitish appearance, and by continuing the heat, the whole of the aqueous part is entirely evaporated, and the mass reduced to a perfectly dry state. The heat is now increased, and in a short time the mass again becomes of liquid form and changes to a brownish color, which indicates that the combination of the oil with the alkali is effected. The fire is then speedily withdrawn, and the stirring is continued so long as any fear of scorching is apprehended. When this is over, the first part or day's operation is concluded, and the pan may be left to cool. The second part of the operation consists in breaking up the product, now in a solid form, into powder. To this add about 45 gallons pure water, and thoroughly agitate the mixture for about half an hour. The heat is then applied, and the contents of the pan raised to ebullition, and kept so for about three hours, during which time the stirring and agitation must be continued. So soon as the evaporation has been carried to the desired extent and the soap appears of the proper consistency, it is allowed to cool gradually. The whole contents of the pan, (no leys or waste accruing from the operation,) while yet in a liquid state, are now put into the ordinary frames and left to cool. The day following, the contents of the frame will be found hard enough for cutting into bars, &c., in which state it is ready for sale or use.

Windsor soaps, (English.)—This celebrated soap was formerly made exclusively with mutton suet, but at the present day the savonniers use a mixture of it and olive oil or lard in the proportion of 20 to 35 per cent. By this addition the soap gains in quality, and the saponification is more perfect; though, it is true, the olive oil slightly detracts from its whiteness. The Windsor soap, originating in England, is now made in France, of excellence equaling at least the best of imported manufacture. The English classify Windsor soap into several species: viz., the plain Windsor, violette, benzoin, palm, and rose.

The *plain Windsor* is made with lard, and perfumed with caraway oil. The *violette* is composed of one-half lard, a third, palm oil, and the remainder of spermaceti; —the essence of Portugal, and oil of cloves forming the perfume.

The *benzoin* is similar to the Windsor, except that an hour after it is taken from the fire, 10 pounds flowers benzoin are added to every 200 pounds soap. If the soap is to be made into powder, then the benzoin flowers must be increased to 15 per cent.

The *palm* Windsor is made as are the above, with palm oil, and a small proportion of lard. To increase the odor, add a little essence Portugal, and oil of cloves. If this soap is intended to be powdered, add 5 per cent. of benzoin. It thus becomes a most valuable soap.

The *rose* soap is the plain Windsor, colored with vermilion or oxide of iron, and aromatized with essence of roses; but the perfume must not be added until the soap has been poured into the frames, else loss will accrue from evaporation.

Windsor soaps (*French*).—So long as the French adhered to the old English plan of making Windsor soap with only mutton suet, its quality was inferior; it became rancid and yellow by exposure. By their present method of mixing with the suet a certain quantity of olive oil or lard, the product is superior.

The addition of olive oil is preferable, but lard answers every purpose and the proportion should be 25 or 30 per cent. The advantages gained by this admixture are these:—the soap remains unaltered for a long time, does not give the disagreeable odor of suet, and becomes more perfectly saponified by reason of the excess of olein in the oil forming a compensation for the large amount of stearin in the suet, and thus inducing a more ready and perfect union of the alkali and fats.

When the soap quits its waters, and the paste in separating from the leys, becomes clotted, that is, forms into granules, that instant you must cease the fire, in order to facilitate the complete deposition of the leys. This operation requires at least 12 hours, at the end of which time, and while the soap is still hot, entirely liquid, and perfectly neutral, pour into every 2000 pounds of paste, 18 pounds of the following mixture.

12 pounds oil of caraway		Laugier's proportions.
3 " "	lavender (fine)	
3 " "	rosemary	

Then stir the mass well, and incorporate the essences thoroughly throughout every part of it, taking care not to reach to the ley that has settled at the bottom. After two hours' delay, ladle it off and put it in the cooling frames. Twenty-four hours generally suffice for the solidification of the whole mass, which is then to be cut and divided into bars and cakes. This mode of perfuming is pursued for large quantities of soap. When the amount does not exceed several hundred pounds, it is better to dip out the paste and put it into a water bath caldron, (Fig. 14,) and when heated sufficiently, to stir in the essences. In this way the incorporation is more complete, and the danger is avoided of mixing in a certain quantity of ley, by accidentally stirring too deeply into the caldron containing the paste, after its completion.

Fig. 14.

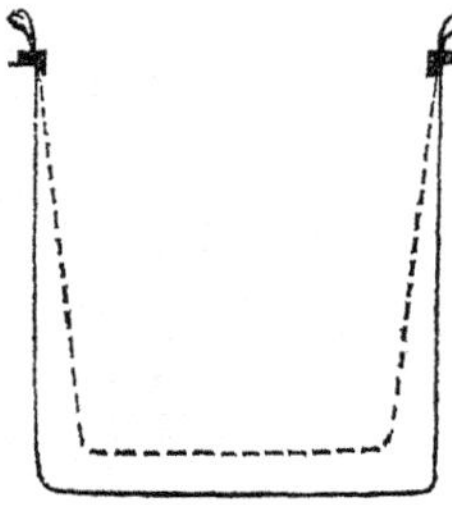

Annexed are the analyses of several soaps, as made by Ure and others.

Castile soap.—Specific gravity 1.0705.

Soda - - - - - - - - -	9
Oily fat - - - - - - - - -	76.5
Water and coloring matter - - -	14.5
	100.0

English imitation of Castile soap.—Specific gravity 0.9669.

Soda	10.5
Fat of pasty consistency	75.2
Water, with a little coloring matter	14.3
	100.0

Perfumer's white soap.

Soda	9
Fatty matter	75
Water	16
	100

Glasgow white soap.

Soda	6.4
Tallow	60.0
Water	33.6
	100.0

Glasgow brown resin soap.

Soda	6.5
Fat and resin	70.0
Water	23.5
	100.0

London cocoa nut oil soap.—This soap is called "marine soap," because it can be washed with in sea water. It is sufficiently solid, but dissolves in hot water with great facility. It consists of

Soda	4.5
Cocoa nut butter	22.0
Water	73.5
	100.0

Poppy nut oil hard soap.

Soda	7
Oil	76
Water	17
	100

Marseilles soap, according to D'Arcet's analysis, consists of

Soda	6
Oil	60
Water	34
	100

English curd soap.

Fat (tallow)	52
Soda	6
Water	42
	100

Ordinary "*Savon vert*," by Thénard's analysis.

Potassa	9.5
Greasy matter	44.
Water	46.5
	100.0

Semi-hard soap, made in Scotland for washing woolens, and called "*economical soap*."

Potassa	11.5
Solid fat	62.0
Water	26.5
	100.0

CHAPTER XXV.

EXTEMPORANEOUS SOAPS.

THESE soaps form a class sometimes designated by the name of *little pan* soaps, because they are not generally prepared in very large quantities. They are made without boiling, and the leys used must be in strength of 36° Baumè, (specific gravity, 1.324.) Experience has proved the disadvantage of soaps factured in this way, for, independent of their excessive consumption of ley, they are extremely hard, and so difficultly soluble in water, as to give but a weak lather. The extempore soaps* are, when made of pure materials, white, and it is true, that containing only so much water as is necessary to their constitution, they are ready for market soon after having been finished; but still there are other drawbacks, which should discard their use where quality is preferable to beauty. Moreover, after some months they become so changed, as to have partially lost the properties of soap. The process of manufacturing them is founded upon the ready affinity of oily matters for ley of the strength above named, for after several hours contact of these two materials, there results a solid of some consistence, assisted in

* Nearly all the *toilet* soap made in this country is factured by the "cold process." Experience and competition have enabled a proficiency on the part of our manufacturers which more than partially relieves their soaps from the disadvantages above mentioned. This article, as now made either by Hauel or Roussel, has all the efficiency of true soap, and a specimen I have lately examined from the laboratory of Mr. Rogers of this city, is entirely unobjectionable.

its formation by the elevated temperature attendant upon the chemical union of the oil and the alkali. The finest specimens of this class of soaps are made of lard, the management of which for the purpose is as described in the following recipe.

Soap, entirely of lard.—To prepare this soap, weigh out 112 pounds of lard, and 56 pounds of caustic soda ley at 36° After having melted the lard by a gentle heat, add half of the ley, stirring it well in, and continuing the agitation, without allowing the mixture to boil. This first operation should be considered as the empatage. When all appears to be well incorporated, then pour in, little by little, and during constant stirring, the other portion of the ley, and be careful not to augment the temperature above 149° Fahrenheit. The paste should be well united, homogeneous and of some consistence. Pressed between the fingers, it should be mild and unctuous, without being greasy. Attained to this state, it is run into the frames, and in two days ought to have acquired sufficient solidity to be cut up into tablets and pressed. The perfume must be added whilst the soap is yet in paste. This soap is of a very brilliant whiteness and flinty consistence.

Sweet almond oil soap, (*medicinal soap.*)—This soap, by reason of the high price of the oil, is only manufactured for the toilet or medicinal purposes, and consequently the choice of the best materials must be insisted on. Select oil of almonds, free of rancidity, and pure carbonate of soda. Dissolve the latter in water, and add the third of its weight of hydrate of lime, frequently stirring the mixture during an interval of some hours; after which filter the clear ley, running through and concentrate it by evaporation until it marks 36°; then take 12 parts thereof to 25 of oil, placing the ley first in the vessel and gradually incorporating therewith the oil, added in small quantities, at a time, and well and continually stirred in until the mixture has the appearance of a soft grease. In two or three days its consistence becomes such, that it can be turned into porcelain moulds. When placed in a temperature ranging between 68° and 72° Fahrenheit, it acquir e in a month sufficient solidity to be taken from the moulds. The temperature of the ley should be 10° to 15° Cent., but if you wish to prepare the soap quickly, it suffices to place the mixture over hot coals, being careful to prevent the concentration of the ley by additions of water proportional to the quantities evaporated.

This *amygdaline* soap, well prepared, is beautifully white and of a very mild odor. It becomes so hard, that when dry it is easily pulverizable.

The other oils are but seldom used for soda soaps, because they saponify less easily and give a product less solid than olive and almond oils.

Hawes' soap, (without boiling.)—The invention of Mr. Hawes has for its object, the intimate combination and admixture of the soap ingredients by mechanical means, and without boiling, and the inducement of perfect saponification of the fat with the alkali without recourse to that high degree of heat requisite in making soap by the usual methods, and at a great cost of fuel. To what success he attains may be judged from the following description of the process as furnished by himself.

"I take any given quantity of tallow, say two and a-half tons, and having melted it, keeping the temperature as low as possible, I mix it with the quantity of alkaline ley which is required to completely saturate the tallow and convert it into soap, and such mixing I perform by mechanical means, and the apparatus or machinery I employ is hereafter described. I use the ordinary ley of soap boilers, preferring that made from the strongest and purest alkali.

"The saponification of the tallow or other fatty matter, may be ascertained by the absorption or combination of the tallow or fatty matter with the ley, care having been taken in the first instance to use sufficient quantity thereof; or about 20 gallons of ley made from strong alkali, and of the specific gravity of 1.125, to every 100 pounds of tallow. It is necessary to state that the proportion of alkali varies with the different fats and oils. The combination of the fatty matter and ley may be effected in an ordinary boiling caldron, with the addition of a machine to produce an intimate admixture and the minute division of the tallow. The whole apparatus is represented by Figs. 15, 16, 17. It consists of an upright shaft A (Fig. 15) from which arms, *a a a a a*, radiate to the sides of the caldron B. This shaft, either permanently or temporarily

Fig. 15.

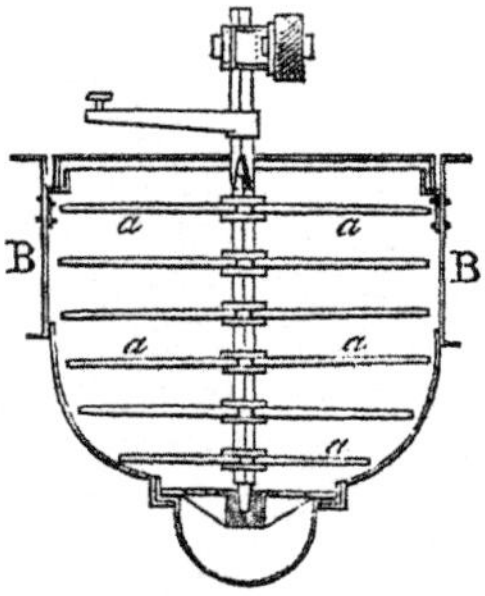

fixed in the copper, may be of wood or iron. The mode of fixing the apparatus and the materials used first, will depend on the nature of the caldron, and the convenience of the manufacturer. An oscillating motion, or rotatory motion may be given to the shaft and connected arms by any of the ordinary methods of communicating mechanical power, or a cylinder may be employed with a shaft C (Fig. 16) passing through it horizontally, and from which arms, *c c c c c c*, may radiate, when a rotatory motion will thoroughly incorporate the fatty matter and the ley.

Fig. 16.

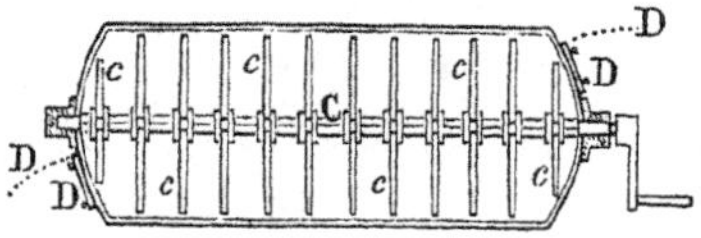

"The size of the cylinder, for two and a-half tons of tallow, will be about six feet in diameter, and twelve feet in length. It must be provided with convenient doors D for charging and emptying. Motion being communicated to the machine, and the caldron having been previously charged with the tallow, the ley is to be gradually added thereto, and in a short time every particle of the fatty matter will be brought into intimate contact with alkaline ley, and by such means saponification will take place. The stirring is continued for about three hours, or until the tallow appears completely saponified, as is indicated by the mass thickening, after which it is allowed to stand from one to four days, according to the quantity of paste.

Fig. 17.

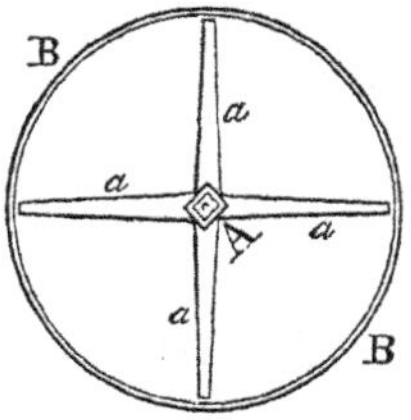

"Should a cylinder be used, then immediately upon its being charged with tallow, at a temperature just high enough to keep it fluid, the ley is run in and motion communicated to the shaft, and continued from three to four hours, or less time, if the mass becomes thick sooner. As the benefit of this process arises mainly from the saponification of the ordinary materials in a comparatively cold state, it is desirable, as soon as the mass thickens, and the ley is absorbed, that the cylinder should be emptied, and the contents turned into an ordinary caldron, preparatory to being finished and converted into yellow soap, by the addition of resin; or into mottled soap or white soap by the operation of finishing leys, as at present practised by soap boilers generally. By this transfer from the cylinder to the ordinary caldron, time is allowed for the combination of the tallow and alkali to become perfect."

The above process is described as for tallow, but the inventor declares it as well applicable to the oils or other fats usually employed as ingredients of soaps.

Macquer's soap.—Take two parts of good soda, and one of lime, and boil them together in an iron kettle with twelve times their weight of water. Filter the liquor, and concentrate it by evaporation until it weighs one ounce three drachms. Dilute the ley with just one ounce of water, and mix one part of it with two of olive oil, in a glass vessel, and stir the mixture constantly with a wooden spatula. It soon thickens, takes a white color, and in seven or eight days you obtain a very firm white soap, differing nothing from that described by Baumè as "*medicinal soap.*"

With the oils of linseed, hempseed, &c., soaps are obtained in the same way, but of inferior quality.

Extempore soap, by Darcet, Pelletier and Lelièvre.—Place in a shallow wooden or stone vessel, six pounds of olive oil, and one and a half pints of caustic ley 8°, and agitate the mixture lively with a paddle for 15 minutes at least. Then add one and a half pints of ley at 18°, and agitate again for an hour, after which pour in three other pounds of this last ley and renew and continue the agitation until the material has acquired a good consistence. After two or three days' repose, it must be malaxated in an open vessel with a wooden pestle, and then put into the frames. In a few days it is fit to be taken therefrom, and in a month or more afterwards, is sufficiently firm and solid for use.

If oil of rapeseed, or colza is substituted for olive oil, then the leys must mark 20° instead of 18°.

We have inserted these processes, because it is the author's desire that there shall be in this work a representation of every class of soaps. Perhaps, too, these improvements, imperfect and objectionable in themselves, may lead the thinking savonnier to the means of supplying their deficiencies, and thus establishing a new and economical mode of manufacturing soaps; if such an end is attained to, then will the results accomplished, be a recompense for the money and time spent in the experiments and essays which may be induced by what has been written.

CHAPTER XXVI.

SILICATED SOAPS.

These soaps are admixtures of silex with the paste, either mechanically, in the form of white granules of sand, or, as silicate of soda, (liquor of flint,) well stirred in with the soap whilst still hot and pasty. The peculiar characteristic of these soaps is their great detergency and causticity, but it is a question, which experience is about determining adversely, whether, the disadvantages in its use, owing to its gritty hardness, are compensated by any other desirable qualities. Under this head are also comprised those soaps, admixed with pipe clay, and such like materials, and though not strictly silica soap, still having that material as a part of their composition, properly belong to the class of which we speak. It is doubtful whether such additions serve any beneficial purpose, except to the manufacturer, who increases his profits by what cannot be called else than a cheap adulteration. We commence first in order with an article now in prevalent use styled "sand soap."

Sand soap.—This soap has a grayish color. It is heavy, flinty, rough to the touch, feels like sand stone when the hand is passed over it, and when rubbed between the fingers, it abandons its sandy constituent, the granules becoming apparent to the sight. It is made by melting any of the white soaps, and while in paste, thoroughly incorporating therewith seven or eight per cent. of fine sifted white sand. As soon as it has entirely cooled, it is taken from the frame and cut into tablets or moulded into balls.

Silica soap, (*from liquor of flints*.)—This soap in appearance differs entirely from the preceding, being mild to the touch, smooth like the ordinary soap, and does not at first sight betray its adulteration, but when washed with, it has the action of fine powder, is somewhat caustic and leaves upon the skin a fine deposit. Otherwise, it is very detergent, cleanses perfectly, and imparts a fineness to the skin, which, however, becomes, after a short time, dry and rough. We give the method followed for making this soap, by Mr. Sheridan, the inventor.

A quantity of calcined flint or quartz is reduced by wet grinding with horizontal stones to a nearly impalpable state. To one part of such ground flint or quartz, thinned out with about 20 per cent. of water, are added two parts of caustic soda leys, of strength 20° according to Baume's soap-boiler's hydrometer. The mixture being thoroughly incorporated, is to be boiled about eight hours, during constant stirring, until it becomes a homogeneous mass, having the appearance of saponified matter, an experience soon accustoming the eye to recognize this state. The mixture is then ready for application in the process of soap making, and is technically termed "detergent mixture."

When the ingredients for making soap have undergone the usual process, and being perfectly saponified, are in the proper condition to be cleansed, (a term familiar to soap makers,) they should be placed in a pan or vessel, and dosed with the detergent mixture; progressively added, a pailfull at a time, care being observed after each addition to stir or "crutch" the contents, so as to intimately blend the whole together. The detergent mixture when added, should be in temperature as nearly equal as possible to that of the paste. For soft soap the detergent mixture should be prepared with potash ley.

The quantity of this mixture to be added to a given

amount of saponaceous matter, depends on the strength required in the soap being manufactured.

In curd soap, equal quantities by weight of each, will answer best; in yellow soap, about one-tenth more of the detergent mixture may be used; and in soft soap, two-tenths less; but the savonnier can better inform himself how much a soap will receive by taking a number of uniform sized small frames, say of half-pound capacity, and putting in each, a portion of the ordinary materials of soap perfectly saponified and ready to be "cleansed," and intimately blending with each separate measure of soap, different quantities of this detergent mixture, and allowing the same to cool. In this way are obtained several samples of soap, each having different quantities of the detergent mixture, and from them the savonnier can judge as to the quantity that should be used, to obtain the soap of quality desired. The proportions, however, given in the preceding method, are about right, and supersede the necessity of any test trials. Well, then, to proceed, the contents of the pan having been thoroughly crutched,—if hard soap, the paste is poured in the ordinary frames, and if soft soap, into tubs.

If the detergent mixture is to be made of caustic soda leys and *sand*, which is nothing more than an impure silex, the sand should be previously washed, and then mixed with carbonate or subcarbonate of soda, in powder, in the proportion of one part by weight of sand to three parts by weight of the soda and calcined to fusion in a reverberatory furnace. The product of this calcination is drawn out into water and dissolved, it being similar to that body known as Dobereiner's soluble glass. By injecting therein a current of carbonic acid gas, the silicic acid (silex) is precipitated, and the liquid in a state of carbonate is drawn off and evaporated to dryness for use again, and the precipitate collected together, mixed and boiled with the proportion of caustic soda or potassa leys, as before mentioned, until the matter attains to the appearance of saponified matter. It is then sufficiently boiled. The detergent mixture thus prepared is added to the soap in paste, as before directed.

Dunn's silicic soap.—The difference between Mr. Dunn's silicic soap and that by the preceding process, is only in the mode of its preparation, the silica in the former instance, either alone or in combination with caustic alkaline leys, being made to unite with the soap under steam pressure. Mr. Dunn, the author of this process, also declares it as applicable to all other kinds of soap, even where silica is not an ingredient, and with the advantage over the usual mode of boiling soap materials, of effecting a more perfect union of the ingredients, in a shorter time, with less waste, and at a diminution of expense. We insert below, the method in detail, which, with an accompanying drawing, we doubt not, will convey a full idea of its operation and efficacy.

You take the materials for soap in the usual proportions, say for *yellow* soap, 7 cwt. of tallow, 3 cwt. palm oil, 3 cwt. of resin, and 140 to 150 gallons caustic soda leys, specific gravity 1.16, or containing 11.5 per cent. of real soda, and place the whole in a steam boiler such as is represented by Fig. 18. The boiler should be furnished with a man hole, safety valve, and all the ordinary appendages of such an apparatus, with a thermometer plunged into a mercury chamber. There should be a feed pipe as at A, and a discharge pipe as at C, through which the soap may be discharged into a pan or frame as at D. The fire being kindled, the pressure on the valve should be such as to allow the temperature in the boiler to rise gradually to about 310° Fahrenheit. When it has remained at this height for about an hour,

Fig. 18.

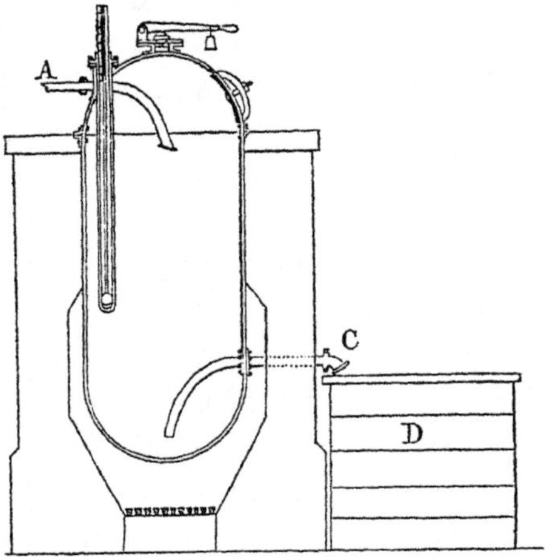

the ingredients may be discharged from the boiler into the pan or frame, and allowed to cool down, when the process of saponification will be found to have taken place.

When silica is to be added, it must be put through a preparatory process, which is as follows: Crushed flint or quartz mixed with caustic soda or potassa ley, in the proportion of one cwt. of silica to 100 gallons of ley, of specific gravity 1.16 is placed in a steam tight boiler, with apparatus, such as above described, and the whole is heated to a temperature of about 310° Fahrenheit, and kept under steam pressure of about 50 to 70 pounds to the square inch, for about three or four hours, when it is discharged and cooled down, and a silicate is thus obtained, of potassa or soda, according to which alkali has been used, in solution, and this solution is added in the proper per centage quantity to the soap paste in the caldron or frame, after the saponification is complete, and before it has cooled down.

To the above invention in its application to ordinary or silicic soaps, a gentleman by the name of Guppy has proposed certain improvements, such as the introduction of stronger leys, and in separate portions into the boiler or steam-tight vessel, to be injected from a reservoir by a force-pump, properly appropriated and arranged, and in connection with both the boiler and reservoir.

For every 24 pounds tallow, 10 pints caustic soda leys of specific gravity 1.12, are added to the boiler, and the same heated up to 300° Fahrenheit, and by means of a force-pump about 30 pints of soda ley, marking specific gravity 1.20 to every 24 pounds tallow, are then injected or thrown in, and the mixture maintained for two hours at 300° to 310° Fahrenheit. At the end of that time the saponification will be complete, a fact determinable by drawing out samples through a try-cock fitted in for the purpose. The stronger leys are kept at hand in a special reservoir, and from thence drawn by the pump, through pipes suitably connected, and forced in through other tubes.

The advantages gained by this mode of operating seem to be a saving of time and fuel, but whether these expectations are realized in practical experience must be determined by the essays of the intelligent and enterprising savonnier.

Davis's alkalumino-silicic Soap.—This soap is a patent invention, by which, as the patentee says, the cost of the soap is diminished, whilst its detergent and required properties, instead of being impaired, are much improved. The plan consists of a combination of fuller's earth, pipe clay and pearlash, with the soap as soon as it is poured into the cooling frames. When pearlash or soda is employed, it is necessary that they should be calcined and then ground together with the clay and earth so as to form as intimate a mixture as possible. In this mixed state they are incorporated with the soap. To every 126 pounds of soap already made and in paste, take

56 pounds of fuller's earth, slaked or dried, 56 pounds of dried pipe clay, and 112 pounds of calcined soda or pearlash, all reduced to powder and sieved as finely as possible, and thoroughly incorporate the whole by stirring or crutching. The mixing must be very perfect, and done as quickly as possible before the pasty soap cools. To obviate any objection against the use of this soap for washing white linens, a modification of the above process is proposed, by which the use of fuller's earth is entirely omitted, leaving the proportions then for every 120 pounds of soap, 112 pounds of dried pipe clay, and 96 pounds of calcined alkali. A soap procured by these quantities, the patentee says, is useful for general purposes at sea, and for washing white linens in salt water.

For washing white linens in fresh water, the process is still further modified by using to 112 pounds of soap, 28 pounds of dried pipe clay and 36 pounds of calcined soda; and as a toilet soap, either for fresh or salt water, by employing 28 pounds of fuller's earth, slaked or dried, and 20 pounds of calcined soda to 112 pounds of perfumed curd soap.

It is a matter of doubt whether the addition of silica or silicated materials to soaps really enhances their detersive properties;—if it does, the advantage must be due to the mechanical action of the finely divided silica, and not to any detersiveness that it may possess. In such case, then, the abrasive power of the grit renders it unfit for washing cloths, and consequently is a serious counterbalance to any desirable properties it may otherwise possess. It would be false economy, therefore, to buy a soap of cheap price, when the saving in cost between it and soap of good quality is to be nullified by the damage it occasions to the clothes which are washed with it.

CHAPTER XXVII.

PATENT SOAPS.

UNDER this head are included all of that species called "*economic soaps.*" We do not make any comment upon either the merits or demerits of these soaps, because the process of preparing them and the nature of their constituents will enable the savonnier to judge of the qualities of each soap respectively. They are mostly the subject of foreign patents, and are presented here with a view of showing at least the attempt at improvement;—if defective themselves they may furnish a hint by which to be perfected. They are all of recent announcement.

Snell's (*dextrine*) *Soap.*—This soap is nothing more than ordinary soap admixed with a greater or less amount of vegetable matter, obtained either from potatoes or other sources. This plan, it is said, is preferable to the old method of combining farina with alum and soda, in the manufacture of soap, because of its insuring the perfect incorporation of the admixture with the saponified fats and oils, whereas when they are previously acted on by alum and soda, their tendency to union is partially impaired. The vegetable matter or dextrine and fibre, as Mr. Snell calls it, is obtained from potatoes as follows: The washed potatoes are reduced to as fine a pulp as possible by a revolving cylinder furnished with parallel plates of steel or iron. These plates have fine teeth like saws, the teeth rising just above the surface of the cylinder, that revolves rapidly under a hopper, which receives the potatoes. The pulp falls on sieves or screens below, and passes over the upper sieve which is a plane of five feet breadth and eight feet length, slightly inclined, with four bars the whole length and one or two cross-bars, covered with a wire cloth of thirty holes to the inch. At the top and about midway down the sieve are pipes containing water, to be placed so that the water shall run from the pipes through small holes, the whole breadth of the sieve, the finer parts of the potato flowing through the meshes of the sieve, with the water and the coarser parts running over the sieve into a vat. The finer parts (dextrine) which fall on a wood plane placed under the second sieve, (but inclined in a contrary direction thereto,) are received in vats, where they are repeatedly washed over sieves of finer wire cloth until cleansed from all its impurities. The grosser parts (fibre) are washed over coarse sieves. The finer parts having precipitated, are taken up in buckets and are fit for use;—the coarser parts remain in a vat covered with water until wanted, or the dextrine may be dried by any convenient means, and in the dry state kept till required for use. Three to 350 pounds mixed with 150 pounds of water to every ton of saponified tallow or oil are the usual proportions. This admixture of water and dextrine, having stood an hour, should be further treated with 600 pounds of boiling water, as hereafter described. The white soap should be made of tallow, and by steam heat, and when as much alkali has been added as can in the usual way be taken up, and when the soap is in a fit state to cleanse in a frame, as is well understood by soap-makers, to each ton of saponified matter in the caldron, take 450 to 500 pounds of dextrine in its wet state, added to which are 600 pounds of boiling water, or enough to make it into a thick paste, and pour it by pailfuls into the soap, stirring well the mixture all the time. When the mixture is completed, and has been heated, it is fit to be cleansed into a frame for cooling and cutting up in the usual way. Before the dextrine is added, the spent ley must first be pumped or drawn off from the caldron. When yellow soap is made with resin, in the usual way, and is what is called "fitted," it should be heated with steam, and dosed with dextrine in manner and proportion as above directed. If no resin is put in, a larger quantity of dextrine may be used, and with or without fitting the soap.

In making common soap, either yellow or brown, by using the fibrin mixed with as much water as will make it a very thick mush, the product will, especially when containing an excess of alkali, be improved in quality and color, and waste and consume much slower in hot water. The preference given to steam for heating soap to be thus mixed, is owing to the liability of the dextrine to become carbonized over a naked fire. The patentee has named the ingredient which he uses, dextrine; doubtless he means starch, for the former term is very frequently but inappropriately applied to the latter article. Starch, before it can become dextrine, must be peculiarly modified by the action of acids or heat, and converted thereby into mucilaginous starch. As regards the fibre, this probably acts upon the skin and in washing, as bran does, by reason of certain detergent properties which it possesses.

Normandy's (salinated) soap.—This is another of the admixed soaps, the sophistication being sulphates and carbonates of soda and potassa, added whilst the soap is yet in paste. The advantages gained are said to be hardness and economy. For eighty pounds of soap, the dose is twenty-eight pounds of sulphate of soda, and four pounds carbonate of potassa, or two pounds of each of the last two materials. Or, if the substances are used singly, then only thirty-two pounds of sulphate of soda for eighty pounds soap; fifteen pounds only of sulphate or carbonate of potassa for eighty pounds of soap, and ten pounds only of carbonate of soda for eighty pounds soap. The application of these well pulverized materials, alone, or in admixture, is made, to the soap, while still in a liquid or hot state, and just as it is finished and ready to be taken from the boiler into the frames. The whole is well crutched and incorporated until the mass is perfectly homogeneous.

It appears from the composition of this soap that it might be a valuable one for marine uses, as the excess of soda which it contains would neutralize the *hardness* of sea water.

Gwynn and Wilson's soap.—The fat used in the manufacture of this soap, is that which has previously been hardened by sulphuric acid. The influence of oil of vitriol is in proportion to the quantity added. Where but a very small ratio is used, as in the case of bleaching the oils by this acid, then its power is diverted either by a superior affinity for the coloring matter, or for the watery particles of the oil, holding in solution the mucilaginous constituent which keeps in suspension the coloring and foreign matters. In the first case the oil is clarified by the separation of the coloring matter in an altered state: in the latter instance the coloring and other matters cease to be soluble as soon as their water is abstracted by the acid and consequently precipitate. In larger quantities, however, the sulphuric acid is a decomponent of the fats, and upon the changes thus induced is founded this process, the rationale of which will be explained in some remarks concluding the explanation of the mode of procedure, which is as follows:

"Ten tons of palm oil, whale oil, or other oily matter,

are put into a wrought iron vessel, provided with a perforated steam worm, through which steam is admitted until the matter is heated to 350° Fahrenheit. The fatty matter is then run into a tank, formed of brick, lined with lead and sunk in the ground; the tank is furnished with a pipe for the introduction of steam, but which does not terminate in a worm, as the sediment from the fat would be likely to choke a worm. The cover of the tank is made of wood, lined with sheet lead, and has two man holes closed by an oil joint about eight inches deep; through the cover a pipe passes, connected with a high shaft to allow the escape of offensive vapors; 2000 pounds concentrated oil of vitriol, of 1·8 specific gravity, are poured into the tank, the temperature of the mass being at the same time carefully watched by means of a thermometer, immersed in the mass, and not allowed to exceed 350° Fahrenheit. The heat is regulated partly by slackening the speed at which the acid is poured in, and partly by lessening the quantity of steam admitted into the tank. The admission of steam is continued throughout the operation, but so soon as the whole of the acid has been added, the fire employed for heating the steam is extinguished, and after that, the steam is continued for about four hours. The steam is heated after it leaves the boiler, by passing through pipes placed over a fire. At the expiration of four hours the steam is shut off, and a large pump introduced through a pipe in the cover of the tank, then, as soon as the vapors given off will allow the pump to work, the product is pumped into a wooden vessel, lined with lead and provided with a steam worm. In this vessel the fatty matter is washed by means of free steam, with about half its bulk of water, for two hours, and then allowed to rest for twelve hours."

The product thus obtained can be made into soap by

the usual method, either alone or combined with softer and cheaper materials.

The proportion of acid may be reduced one-half and still yield a good product for soap making.

We think that the change in this process may be attributed to the abstraction of the glycerin of the fats by the sulphuric acid, and the consequent elimination of the fatty acids, which in a free state, as is well known, are much more solid and less fusible than when in the form of palmitine or palmitate of glycerin, and olein or oleate of glycerin. Of course, as oleic acid is naturally fluid at ordinary temperatures, its presence, in large proportion, would be a hindrance to the perfect solidification of the palmitic acid; so, if it is not changed by the acid into a firmer body, a part of it must have united with a part of the sulphuric acid, to form sulpholeic acid, which is soluble in and taken up by the water used. This, however, would decrease the amount of product and render the process unprofitable. We shall allude to this subject again in another part of the book, mentioning, however, here, that the excess of alkali required, renders this plan of making soap, from fat hardened in this way, rather expensive.

CHAPTER XXVIII.

ANDERSON'S IMPROVEMENT UPON THE MODES OF MAKING SOAP.

In making yellow or resin soap, the process is that of boiling the materials upon successive portions of alkaline ley until they become, technically speaking, "strength," after which the curd so produced is subjected to a purifying operation, or "fitting," which is thus performed. The ley used in boiling to strength is carefully pumped off from the soapy paste, and the latter is diluted with weak ley or water and boiled to that requisite consistence well known to soap boilers; after two days' repose, it settles out of the perfect soap, and there is found below it a certain substance (inferior or second quality of soap) called "niger" or "nigre." The perfect soap is cleansed or lifted off carefully from this niger into the cooling frames;—the niger itself being added to the next operation, or concocted into an inferior yellow soap. The process followed in making curd soap is to boil the materials upon successive leys until they become "strength," after which, and when the curd is of proper consistence, it is cleansed off carefully into the cooling frames without further treatment.

In making curd soap, Mr. Anderson says, no fitting process has hitherto been practised; therefore these details are entered into in order to facilitate the description of the invention, which is as follows. When curd soap is boiled to strength and subjected to the "fitting" process used for yellow soap, there separates from it a peculiar substance analogous to the niger of the yellow soap, and that, by removing this niger, as it shall be called, and boiling the remainder of the materials into curd soap, a product is obtained, of better quality than could be without this operation. The niger removed is particularly fitted for making a mottled soap, and to this purpose it is applied. The invention starts from that point of the usual mode of making soap where the materials are boiled to "strength," and "ribbon out" well on the finger; but at this stage, instead of boiling out the head and finishing as heretofore practised, commence the improved process by pumping out or drawing off the strong ley on which the materials have been boiled, and treat them with successive portions of weak ley or water, and boil them together, until they assume the appearance of a fitting yellow color. This condition being arrived at, stop the operation, and allow time for the niger to be deposited, say twenty-four to forty-eight hours. To separate this niger, either pump it off from under the purified compound, into an adjacent kettle, or carefully remove the upper stratum of "clean soap" with basins or buckets. Either way, the material thus freed must now receive the proper finishing ley for curd soap, which is added, and the whole boiled until the soap becomes of sufficient consistence to be cleansed off into frames.

When the ingredients operated with or upon, are very impure, the above fitting or purifying process is repeated one or more times, in which case or cases, after separating the niger from the materials, as described before, add to the residual partially purified goods, a ley of moderate strength only (instead of the finishing ley for curd soap), and boil, taking care that no "head" is formed; then pump or draw off this ley, and again treat the materials with weak ley or water, until sufficiently diluted, so as to perform the fitting process; after which, allow time for subsidence, then separate the niger, add the finishing ley, and boil to a curd as before described.

The treatment of the niger and its conversion into mottled soap are effected by adding to it in an ordinary caldron, the usual finishing ley for a mottled soap, and boiling until it is fit to be transvased into frames. The quantity of niger obtained from one fitting or purifying operation is not enough to be conveniently boiled by itself; therefore it should be saved, and six or more batches operated upon at a time. Occasionally, a portion of bone fat may be added, and the soap finished as an ordinary charge of mottled soap.

CHAPTER XXIX.

SOFT SOAPS.

The causes which create the difference between soft and hard soaps, have been made apparent by the valuable researches of Chevreul. In speaking of the oily or fatty acids, it was mentioned that their combination with potassa resulted in the formation of salts of a mucilaginous or gelatinous aspect when in contact with water, whilst the salts or compounds which they formed with soda, unlike those of potassa, are not softened by water; and hence it is that soaps with potash base are mostly soft. It having been proved before that the stearate of soda offers more resistance to water than margarate of soda, and that this latter salt resists much more than the oleate, the result is that the hardness of soaps is in direct proportion to their stearic and margaric constitution and their softness, in accordance with their amount of oleic acid. It is then evident that if the greasy body employed in the fabrication of soaps of soda, is rich in stearin, the soap will be hard, but if olein predominates, the soap will be of softer consistence. It is easily conceivable then, that you can imitate a given soap by adding to the oils rich in olein, a certain quantity of fatty matter, which is rich in stearin. For instance, you can obtain a soap presenting the qualities of that made from olive oil, and such mixtures are frequently put into practical application by the savonniers.

Soaps with potash base, made either with the fats or

oils, are soft, and more or less pasty. There are two kinds of this soap, met with in commerce; 1st, those made with oils and fats, and called *savons verts*, or green soaps, because, as made on the continent, they are usually of a natural or artificial green tinge; 2dly, those prepared with lard and potash, of a finer quality, and styled toilet soaps. The theory of the reaction of potassa upon the oils and fats is the same as that of soda, but the mode of fabrication is entirely dissimilar. The oils used are those of hempseed and poppy, colza and rapeseed. Fish oil is also used, but its disagreeable odor renders it unfit for any but the commonest quality of soap. For the soft soap, all kinds of soda must be discarded, and potash exclusively substituted.

In summer, to every 1500 pounds of potash, 1200 to 1300 pounds fresh caustic lime are requisite; but in winter, the quantity is slightly increased. To set the vat, the potash must be well bruised and mashed, in manner before directed for the soda. The lime is prepared separately, by throwing water upon the proportional quantity, in a heap or pile, and then leaving it remain for about a half day, more or less. (Upon the proper preparation of the lime and its quantity depends the goodness of the leys.) The lime being well slaked, it is mixed as thoroughly as can be with the potash. In order to prevent the adherence of the mixture to the tools, it is best to dust them over with charcoal powder. At the same time, it will be as well to add a little also to the mixture, that it may not become too massy and compact, and thus prevent the infiltration of the water. This mixture thus well made, is filled into the last vat. These vats, 1, 2, 3, 4, 5, are of lead, or else of mason work, sealed and plastered with hydraulic cement, and interiorly, forming a capacity proportional to the amount of ley required for the operations of the establishment. To each of these vats, and placed beneath, there is also a cistern.

These cisterns, 1, 2, 3, 4, 5, are of the same breadth as the vats, but in order that there may be greater facility in drawing off the ley which collects in them, they are ranged under a shed immediately beneath the vats. Their depth is of no importance, though the larger they are, of course, the more ley they will contain. They should, however, have six feet of extent, below the bottom of the brick or stone vats, that they may hold all the ley which percolates through.

The capacity of the fifth vat should be much larger than that of the others, for it serves as a reservoir to the strong leys,—such as are to be used for the soap. For the convenience of the workman, it ought to be placed very near to the caldron. The cisterns are generally of the same materials as the vats, and carefully made, else the caustic ley liquor will eat its way through, and cause a leak. To avoid the danger of these inconveniences, perhaps it will be better to use the lead cisterns spoken of, as before. They are worked as previously directed, under the chapter on leys. By the mode already given, you also ascertain the strength of the ley running through.

Although the water which is poured in the vat No. 5, dissolves the greater part of the salt contained in the material, yet there still remains a small quantity. To extract this residual portion, when all the water has run through into the cistern, throw the contents of vat No. 5 into vat No. 4, and treat it with the same quantity of solvent as at first, but of that liquor which has been drawn from cistern No. 3.

Recommence the same operation, until having reached the earths in vat No. 1, then, as that has no preceding cistern, its contents must be *displaced* with ordinary water. The choice of water is a matter of importance; that called *hard* water, and in which soap is difficultly soluble, is not suitable;—the purer it is, the better; the distilled, or rain water being preferable to every other.

When the new water which has been poured upon the vat No. 1, has run through into the cistern No. 1, the mixed material has been washed five different times, and consequently is pretty well exhausted of saline ingredient. It is then thrown out through a window, placed for convenience in the building, facing vat No. 1. For cold and sandy soils, these leeched materials make an excellent manure, and are coming into use for that purpose, thus affording a small revenue to the savonnier for what was not long since considered as useless, and in the way.

The progress of the water is contrary to that of the alkaline material, that is to say, the fresh compost is always thrown in vat No. 5, whilst the new water is always poured into vat No. 1. Thus it is seen that the material is washed and leeched five times consecutively, before it is exhausted of saline constituents, and reciprocally, that the water, before having reached cistern No. 5, where it is a ley sufficiently strong for making soap, has passed five times upon the material; hence the force or strength of the ley increases as it progresses from cistern No. 1 to No. 5. That the order of operation may be systematically kept up, as soon as vat No. 5 is emptied, it is refilled with fresh compost, as directed heretofore. Thus are prepared the leys used in the manufacture of soft soap, and in manner, similar to that described before, for the hard soaps. As regards the oils, they require no previous preparation, and the soap is boiled in caldrons. The larger their capacity the better, as it is always more economical to use roomy pans; and to be properly proportional, their diameter should always be greater than their depth. As the soap, when boiling, swells and mounts up, the kettle or caldron should be only half filled with material. The quantity of ley requisite, is entirely regulated by its strength. As a general rule we can rely on four to three, that is to thirty tons of oil, say forty tons of ley.

The ley should always be added gradually, and in small quantities at a time by spreading it superficially over the whole mass. As soon as the two clear and fluid liquids, separately put in, unite together, they thicken and sometimes boil peaceably, but at others, swell up into froth and foam. In such latter case, it must be thrashed with a stirring pole and dosed with some measures of ley to allay the commotion and prevent loss by the running over of the materials. The caldron requires constant care and watching, whilst over the fire, and it is the art of the savonnier to know how to conduct the operation to success, without any baulking through deficiency of knowledge as to the mode of rectifying the mischances that will occasionally happen. If too much ley has at the commencement been put in, no union is effected; if the leys are too strong, they seize on the oil too rapidly, and in place of growing thick, the mass becomes clotted. This is remedied by adding some ley of the first cisterns, which is weaker. On the contrary, if the leys are too feeble, the combination ensues so tardily, that a part of the superabundant water of the ley must first be evaporated before a perfect union of the oil and alkali is effected. In this case, the waste of time and material is much more considerable.

The vivacity of the ebullitions depends mainly upon the manner in which the fire is regulated. No precise rules can be given for the management of the coction; the intelligence and experience of the savonnier must supply any deficiency. When the combination is perfect, and the great foamings have passed, and the froth has subsided, the pasty mass should become clear and transparent, and free from clots or granules. This is known when a small sample taken out and cooled, appears of the proper consistence, and free from opacity. This point arrived at, there remains only to be given to the material its proper finishing. The savonnier ascertains the completion of the finishing by an examination, from time to time, of cooled samples. The instrument used for this purpose is a ladle, with which a small portion is dipped out, and run upon a glazed tile, and exposed to the air. At each time the ladle is dipped into the caldron, the contents should be stirred up so as to disperse the foam and froth. If the test sample thickens on cooling, has a brown color, is fine grained, and is not ropy when worked between the fingers, then the finishing is at its maturity, and the fire must be withdrawn from the furnace.

To allay the foaming, and render the cooked soap fit to be barreled up without endangering its causticity or detergent quality, empty into the caldron about a ton of ready made soap;—this, in becoming melted, by that law of caloric which asserts a general tendency to equilibrium of temperature, abstracts a portion of the heat of the recently made soap, and thus makes the degree of that and its own, uniform. The foaming having ceased, the soap is then ready to be racked off into barrels.

In making this soap, it is better that the boiling should be too much than too little protracted. Soap which is not sufficiently cooked, changes and spoils, whilst the only disadvantage an excessive boiling produces, is a diminution of the quantity of product. The usual time requisite for a charge, is six or seven hours, but varies with the strength of the leys, the temperature of the atmosphere, and incidental circumstances.

The above is the mode of facture pursued at some

places, but it is very empirical; yet there is no reason why the manufacture of soft soaps should not be based upon rational principles, as well as that of hard soap. Instead of being at the expense of working a series of vats, the material can be as well leeched in the same manner as has been directed for the sodas; the strength of the ley estimated by alkalimetrical computation, and the operation of boiling and making, conducted in the usual manner, until all the oil has been added. This mode of operating, however, produces a soap not unlike an unguent with excess of oil, scarcely transparent, and of a dirty white color. The fire should be carefully managed, and the contents of the caldron, stirred constantly to the bottom with long wooden paddles, and new leys added, successively stronger. When the saturation of the oil takes place, the soap becomes transparent, and the operation is continued until the coction is perfect.

The difference between the mode of facturing soft and hard soap is easily perceptible

In the potash or soft soaps, the whole aim is to produce a reaction of the potash upon the oil, by constantly stirring and holding the soap solved in the ley.

In the soda or hard soaps, on the contrary, the soap is separated from the ley, even before all the oil has been neutralized.

According to Thénard, soft soap is a perfect soap, dissolved in an alkaline ley, and is composed as follows:

Potassa - - - - - - - -	9.5
Greasy matter - - - - - - -	44.
Water - - - - - - - -	46.5
	100.0

Soap made with the oil of colza is a little bluish. The green color which is sometimes imparted to the soaps, comes from a solution of indigo in alkali, applied in suitable doses, the yellow of the soap forming a green with the blue of the indigo. Fish oil soap has a dark brown color, and that made from refuse fat and oil, a dirty white shade. When the process has been properly conducted, no sediment remains at the bottom of the caldron.

The soap is filled into barrels, through one of their heads, which must not be fitted and coopered in again, until the soap has entirely cooled, and for this purpose, the barrels are as soon as filled ranged in a row. Sometimes it is twenty-four hours before they are ready to be headed up.

Soft soap is always of a pasty consistence, and, if even excessively boiled, does not harden, but becomes scorched and dry. The greater solubility of these soaps, economy, more powerful detergent action, render them particularly applicable for scouring floors and wooden utensils, and for washing woolen goods.

Crown soap, first quality.—For eighteen barrels, prepare 400 gallons of ley, with good hickory or oak ashes and lime, and put a third of this ley in the caldron, and then add fifty-two pounds of suet and as much of lard. When the whole is melted, pour in seventy gallons of olive oil, and leave the liquor in repose for two hours;—kindle the fire anew and turn nineteen gallons of ley into the caldron. As soon as ebullition commences, add from time to time a little ley in order to allay the frothing. Continue this addition until the liquor in the caldron has been reduced one-half. At this time examine whether the soap has been dosed too little or too much with ley. This test, or proof, should be made frequently during the saponification. It is merely to withdraw a sample from the kettle, upon a spatula and to examine it. If it becomes whitish, and falls in short pieces, it is too alkaline, and requires oil; if, on the contrary, ley is needed, it drops in long ropy strings. If it is proper, that is, deficient neither in ley nor oil, the sample should not be viscous, nor too white or transparent. Then the fire must be extinguished, and the soap run off into barrels. It may be as well to say that after the second time the fire is kindled, the soap should be kept in lively ebullition until its preparation is well advanced; arrived at such point, it must be carefully managed until the soap has acquired its degree of coction.

Crown soap, (second quality.)—For this soap take 286 pounds of suet or tallow, ley 135 gallons, sperm oil 80 gallons. Place in the caldron, first, ninety-four gallons of ley and the tallow, and when the latter is melted, add the oil and put out the fire. Two hours after kindle anew, add in nineteen gallons of ley, and carry the whole to boiling, and keep it so until the soap becomes half made. Then dose with nine gallons of ley, and finally resume and continue the ebullition, having care to add successively the other nine gallons of ley, to complete the facture of the soap.

Ure, in his Dictionary of Arts and Manufacture, gives the process followed in Scotland for manufacturing soft soaps. He says that these soaps are usually made with the oils, but that tallow is frequently added to produce white and somewhat solid grains of stearic soap in the transparent mass, called "figging," because the soap then resembles the granular texture of a fig. This appearance, however, is sometimes not evident until two or more weeks after the soap is made, and in warm weather is not at all visible.

The potash leys, as used for soft soaps in Great Britain, are made perfectly caustic, and of at least two different strengths; the weakest being of specific gravity 1.05, and the strongest 1.20 or even 1.25. Here is the process alluded to above.

Savon vert.—Two hundred and seventy-three gallons of whale or cod oil, and 400 pounds of tallow are put into the soap pan with 250 gallons of potash ley, in strength such that one gallon contains 6,600 grains of real potassa. Heat being applied to the pan, the mixture froths up very much as it approaches the boiling temperature, but is prevented boiling over by being beat down on the surface. Should it soon subside into a doughy-looking paste, it is to be inferred that the ley has been too strong. Its proper appearance is that of a thin glue. There should be now introduced about 42 gallons of a stronger ley, equivalent to 8,700 grains of potassa per gallon, and after a short interval an additional 42 gallons; and thus successively till nearly 600 such gallons have been added in the whole. After suitable boiling to saponify the fats, the proper quality of soap will be obtained, amounting in quantity to 6,400 pounds, from the above proportions of materials.

Soft soap, as made in this country, differs very little from the preceding in its mode of manufacture;—460 pounds good soap result from the saponification of 200 pounds oil by 72 pounds potash, in leys, at 15° Baumè. It is said, that "a fair criterion of the quality of the soap is furnished by a peculiar phenomenon observable during its cooling. A little opaque zone forming around the test portion taken from the caldron for an essay, indicates perfect saponification and is called *strength*, the absence of which denotes a deficiency in strength, and when the zone vanishes soon after having been seen, the soap is said to have *false strength*. When it occurs in the best form, the soap is perfect and may be secured in that state by removing the fire, and then as before directed, adding some good soap of a previous round, to cool it down and prevent further change by evaporation. No waste leys are pumped or drawn off as is the case in the manufacture of hard soap.

Oils require for their complete saponification at least one part of potash to every four parts of oil, and the resultant soaps contain about fifty per cent. of water mechanically admixed. The general composition of soft soap may, however, be considered as of alkali one part, fatty acids, four to five parts, water and other impurities, five to six parts.

When hempseed oil is used, the soap has a green color, and hence the name *Savon vert*, the French expression for green soap.

CHAPTER XXX.

OF THE CONVERSION OF SOFT SOAPS INTO HARD SOAPS.

It has already been stated that the consistence of soaps is more especially dependent on the alkaline base used in their preparation, and that those formed with soda are hard, whilst the potash soaps are soft. Soft soap is a definitely formed compound of the fatty acids with potassa. If, in its preparation, too little ley has been used, then acid salts are formed, but by the addition of more ley they become neutral. The (greasy) acid salts of potassa are apt to precipitate and adhere to the bottom of the soap pan in thick masses. So also the neutral, soft soap, (greasy salts of potassa,) may be precipitated from its solution in water or weak ley, by the addition of a strong ley; hence the necessity, in making soft soap, of using ley in a certain state of dilution. A knowledge of the composition of soft soap, therefore, shows, that their decomposition should ensue, whenever they come in contact with bases which have a stronger affinity for their acids. An every day exemplification of this is seen when attempting to wash in hard waters, or those containing muriate and sulphate of lime, &c. A decomposition always occurs, the potassa taking up with the sulphuric and muriatic acids of the lime, whilst that body goes to the liberated fatty acids, and forms insoluble salts or soap of lime, which swims in flocculæ throughout and upon the water. So also a potassa soap is decomposed when common salt, which is, properly speaking, a chloride of sodium, is added to it. Pelletier and others recommend the proportion of six pounds of salt to every three pounds of oil, and its solution in water, to be added little by little and gradually to the soap, and during continued ebullition. The new compound, or soda soap, which is produced, soon separates from the ley, and the operation is terminated in the usual way.

There is a triple reaction in this process; thus, the soap of potash consists of the oleate and margarate of that alkali;—on the other hand again, salt solution is composed of soda and muriatic or hydrochloric acid. Now when these two come in contact, an interchange of bases takes place, the hydrochloric acid seizing on the potassa of the soap and forming a muriate of potassa, which remains in solution, whilst the oleic and margaric acids, set free, take to the soda displaced by the potassa and form an oleate and margarate of soda, the union of which two produces the hard or soda soap.

A knowledge of this process and its rationale, enables the use of potash and ashes in those distant localities where soda is either not to be had or else is too costly for employment in the manufacture of soaps.

Ure gives the following analyses of different soft soaps. In one hundred parts

London manufactured soft soaps yielded

Potassa	8.5
Oil and tallow	45.
Water	46.5
	100.0

Belgian soft or green soap,

Potassa	7.
Oil	36.
Water	57.
	100.

Scotch soft soap,

Potassa	8.
Oil and tallow	47.
Water	45.
	100.

A rapeseed soft soap,

Potassa	10.
Oil	51.66
Water	38.33

An olive oil (gallipoli) soft soap—(slightly carbonated,)

Potassa	10.
Oil	48.
Water	42.
	100.

A semi-hard soap from Verrier's, for fulling woolen cloth,

Potassa	11.50
Fat	62.
Water	26.5
	100.00

CHAPTER XXXI.

OF THE FRAUDS PRACTISED IN SOAP MAKING, AND THE MEANS OF DETECTING THEM.

WHEN the grasping cupidity of a savonnier prompts him to practice a fraud upon his customers by means of improper admixture with the soap, he merits the mortification and punishment that should be meted out to all evil doers. It is an indirect robbery of a man's pocket, for when he buys soap for pure, which in fact contains one-third adulteration, he is actually cheated out of fifty per cent. of its real value. Consequently, as a protection and in justice to the honorable of the craft, whilst mentioning the more prevalent sophistications, we also give the modes of detecting them. In cases, however, where the combination of certain substances with the soap, have a tendency to improve its qualities, there is no fraud, properly speaking, and we shall so consider it. The most usually practised deceptions are the excessive absorption of water to which the paste is subjected; the mixture of foreign bodies, such as chalk, starch, fecula, leeched sodas, pipe clay, salt, &c., and sometimes, in case of imperfect saponification, a certain quantity of uncombined fatty matter. We will speak of each of these adulterations and imperfections separately. If the soap contains an excess of water, the proportion can be determined by two methods. The first consists in taking a certain quantity of soap reduced to small thin shavings, weighing out four or five grammes, and exposing it to a current of air heated to 212° Fahrenheit, or in an oil bath, until it ceases to lose weight. The dry substance is then weighed;—the difference between its present and former weight indicates the quantity of water evaporated. If it be a soft soap, it is weighed in a counterpoised shallow capsule. Marbled soap well prepared should contain for every 100 parts but thirty of water, and the white soap but forty-five to forty-six parts of water. The savonnier who sells soap, surcharged with water, keeps it in a solution of salt. The absorption is such that a white soap, as taken from the frame, when left to soak for a month, so augments in weight that 100 parts of dry soap, which should retain but thirty-eight parts of water, absorbs 154 parts. It is impossible to introduce an excess of water into the paste of marbled soap, for if it has not a certain tenacity retentive of its ferruginous material, this precipitates, and you get only white soap. The fraud can, nevertheless, be practised in another way, by plunging the soap, as soon as taken from the frame, in the solution of salt, and is detected in the same way as above directed for white soap. The mottled soap should contain but twenty parts of water per 100 of dry soap. That met with in commerce retains 42.85 parts per 100, and it can be made to take up as high as eighty-three parts and even more. In this state it is so soft that it can be easily kneaded between the fingers;—a quality which condemns it. The other method of determining the proportion of water is to decompose the soaps with hydrochloric acid; that is, provided they are entirely soluble in water. Weigh out 100 grammes of the soap to be examined, dissolve it in 1000 grammes of water, and add to the solution some hydrochloric acid. It is decomposed, the greasy matters rise to the surface, and must be collected upon a filter, and when well washed and dried, are carefully weighed, deducting that of the filter. On the other hand, unite the wash waters, evaporate them to dryness, and take the weight of the residue, which indicates, in the form of a chloride salt, the quantity of alkali contained by the soap. The difference between the two weights, subtracted from that of the soap employed, expresses the quantity of water. The marbled soap, well prepared, should give by analysis, per 100 parts, six of soda, (expressed by ten parts of chloride salt—as obtained in that state by evaporation,) sixty-four parts fatty acid, and thirty parts water. The white soap must yield 4.6 of soda, or in the form of chloride 7.5 parts; fatty acids 50.12 parts; water 45.2 parts. By comparing the results obtained, with the above, it will be easy to determine the fraud. The first method, however, being the most simple, practicable and expeditious, is much the most preferable for general use in the arts and commerce.

To estimate the amount of foreign matter, after having weighed ten grammes of the soap, previously reduced to thin shavings, treat them with boiling alcohol. If the soap be white and without admixture, the portion remaining undissolved is very minute, and a mottled soap of good quality does not leave, when operating on ten grammes, more than ten centigrammes, or about one per cent. If, however, there should be a sensible amount of residue from white soap, or more than one per cent. from mottled soap, some accidental or fraudulent admixture may be suspected. The exact amount of this undissolved matter must be ascertained by weighing it. To determine the nature of the materials composing this residue, it must be then treated with cold water, which dissolves out the salts, soluble therein, each of which is recognizable by certain chemical reagents, and can be quantitatively determined by evaporating the liquor to dryness. This duty properly belongs to the chemist, so it is unnecessary to give the minute methods of detecting and estimating each salt separately, for it is only necessary for the purchaser to know that the soap is adulterated without regard to the nature of the fraud. If the residuum is insoluble in hot water, then add boiling water. The presence of starch or fecula occasions a more or less thick gelatinous mush. The filtered solution, upon the application of tincture of iodine, in such case becomes violet blue. The insolubility of the residue in either cold or hot water, indicates that it is lime or pipe clay. If lime, it will be entirely soluble in hydrochloric acid, from which a white precipitate falls upon the addition of oxalate of ammonia. What remains untouched by water, acid and alcohol, is pipe clay, that being the kind with which it is usually adulterated.

If the soap contains any uncombined or non-saponified greasy matter, decompose its solution with hydrochloric acid;—the fatty acids as well as the unaltered fatty matter, assemble together at the surface of the liquor. Collect them together upon a filter, wash them thoroughly, and then treat with baryta. Collect the precipitate on a filter, wash it with boiling water, and act upon it with hot alcohol;—this dissolves out from the barytic soap the non-saponified grease, which can be separated by evaporation, and then weighed. Tested by litmus paper, it should give no acid reaction.

The quantity of alkali contained in a soap is easily determined by means of the alkalimeter. Ten grammes, in thin slices, are taken, for instance, and dissolved in 150 grammes of boiling water, and this solution is saturated with a normal liquor containing in a quart of water 100 grammes of sulphuric acid, specific gravity, 1.848. The volume of this liquor required for complete saturation, will indicate the corresponding weight of sulphuric acid, which is itself nearly equivalent to an equal weight of dry carbonate of soda. The quantity of pure potash or soda may be thus deduced.

There is no difficulty in ascertaining in the same assay the quantity of the fatty substance. For this purpose ten grammes of pure white wax, free from water, are added to the liquid, after saturation with sulphuric acid, and the whole heated to complete liquefaction; it is then allowed to cool, and when it has become solid, the cake of wax and fatty matter which have united, is removed and washed, dried and weighed; the augmentation in weight beyond the ten grammes employed, will give the amount of the fatty matter. The liquid decanted from the solidified wax, may afterwards be tested, to ascertain the purity of the base. The solution of the sulphate may also be evaporated, and by an examination of its crystalline form, or by means of chloride of platinum, which produces with potash a yellow precipitate, (the double chloride of platinum and potassium,) it may be ascertained whether the base be potash or soda, or a mixture of the two. As to the nature of the fatty substance, it is ascertained with more or less certainty, by saturating the solution of the soap, with tartaric acid, collecting the fat acids and taking their point of fusion. It is possible at least, by this, to prove the identity of the fatty component, for instance, whether the soap is made from oil or tallow, &c. The odor developed by the fatty acids at the moment of the decomposition of the soap by stronger acids, will often indicate the nature of the fatty substance employed in its fabrication, or that at least of which the odor may prevail. A disagreeable smell indicates that the soap has been made of inferior grease. (*Dumas.*)*

It is not to be supposed that a savonnier, ignorant entirely of the rules of chemical analysis, in its bearing upon his trade, will be able to make these latter tests;

* A very handsome Venetian soap, lately examined by Dr. Riegel, was found to contain a considerable quantity of animal gelatin. Its analysis gave, water 33, soda soap 26, gelatin 41.

still there are some ambitious members of the craft, who are desirous of knowing all and everything concerning the art which yields them their bread, and I have given these cursory hints, as an inducement to a fuller investigation of the matter. A little study and experience in manipulation, will make the savonnier chemically informed and practised in all branches of his profession. This book gives the properties and characteristics of all the materials used, but when any analytic examination or testing is to be made, every other assistance and information will be furnished as to the action of re-agents, by either Rose's or Fresenius' work on analytic chemistry.

CHAPTER XXXII.

EARTHY, "MARINE," METALLIC, AND AMMONIACAL SOAPS.

Earthy soaps.—These soaps differ essentially from the alkaline, in possessing no detergent property. They are easily obtained by treating a solution of soap with a solution of earthy salt—an interchange of bases takes place, and an insoluble soap is formed.

Soap of lime.—For instance, this soap is prepared by treating a solution of ordinary soap with lime water or a solution of muriate of lime. Decomposition ensues; the oleic, margaric, and stearic acids set free from the alkali, take to the lime and form an insoluble salt, whilst the alkali alone, (when lime water is used,) or as a muriate, (when muriate of lime is employed,) remains in solution. These soaps are not of any practical use or application. The soaps of baryta and strontia are similarly obtained from the salts of these metals.

Marine soap.—The property possessed by sea and other hard waters, decomponent of soaps, and thus of impairing or destroying their detersive action, renders necessary a means of obviating these inconveniences, and for the purpose there is prepared an article called "marine soap," because of its particular applicability for use on ship board. The composition of sea water is nearly uniform throughout the globe, saving in a few instances, where it is influenced by accidental causes, for example, the water of the English channel contains nine times as much lime as that of the Mediterranean, whilst this latter, in its turn, has twice as much magnesia and sulphuric acid as the former. The general constituency of sea water is three and a-half per cent. of solid matter, composed chiefly of common salt, (chloride of sodium,) muriate of magnesia, sulphate of magnesia, and sulphate of lime, with traces of several other salts. The water of the Dead Sea, as an exception, contains about one-fourth of its weight of solid matter. The excess of lime in the channel water is owing to its flowing over a bed of chalk.

When an attempt is made to use soap in hard or sea water, a curdling appearance is soon evident, and the soda of the soap, by reason of its stronger affinity therefor, has united with the hydrochloric acid of the earthy salts in the water, the bases of which being thus set free, partially unite with the greasy acids eliminated from the soap, and form insoluble earthy soaps, the remainder or uncombined portion of the earth's precipitating, whilst the residual saponic acids congeal and swim upon the surface of the water. If the water contain carbonate of lime, it can be ridded of it by a little caustic potassa; if sulphate of lime or magnesia salt is present, then a little pearlash ley will effect their removal, and render the water soft.

But as it would be inconvenient to be dosing water in this way, a soap, properly prepared, is substituted. It is made of cocoa fat, this from experience being declared the best grease for the purpose, the resulting soap of which, after being made as caustic as possible during its coction, is then ladled off into cooling frames, and treated by thorough mixing with a quantity of dry carb. soda, finely pulverized.

Metallic soaps.—The metallic soaps are known in pharmacy under the name of plasters, but as their examination is not within our province, we refer the reader to Wood and Bache's Dispensatory, or Pereira's Materia Medica.

Ammoniacal soap.—This soap is only used in medicine as a rubefacient; if its action is continued for any length of time, it becomes vesicatory. The common *volatile liniment* of the shops is an ammoniacal soap. Aqua ammonia forms with the oils soapy compounds, of which this is a sample. The ammoniacal soaps are but little known, being prepared mostly for pharmaceutical purposes. According to Boullay, a solid ammoniacal soap can be made by passing ammonia, in a gaseous state, through oil or, better, grease. The mode of making these soaps generally, is explained in any book upon pharmacy.

The action of ammonia upon the oils and fats, is distinguished from that of the other alkalies in producing, besides the ammoniacal soaps, properly so called, a peculiar substance to which Boullay has given the name of *margaramide*. The ammoniacal soap, according to this chemist, has always an alkaline reaction, a property which it loses when exposed to 86° F., acid qualities being then acquired instead. In boiling water, the soap is thinned out, but not dissolved, and on cooling, the greater part congeals at the surface, whilst the liquid retains in solution, glycerin, coloring matter, and a supersalt of ammonia, precipitable by the acetate of lead and nitrate of silver. The congealed portion, when solved in boiling alcohol, lets fall a white crystaline perfectly neutral substance, which is the *margaramide*. The alcoholic mother water retains in solution a small quantity of margaric and oleic acids. The margaramide is a margarate of ammonia, less one equivalent of water. It is unalterable in the air, insoluble in water, but readily so in alcohol and ether. It melts at 140° Fahrenheit, and burns with a smoky flame; is decomposed by boiling concentrated solutions of the fixed alkalies, and is only acted upon by the acids when it is at a certain degree of concentration. This transformation of the oils is particularly rapid, and distinctive in the instance of castor oil.

CHAPTER XXXIII.

SOAPS FROM WAX AND THE VOLATILE OILS.

As regards the action of alkalies upon wax, it has been fully explained in the paragraph appropriated to that substance—the mode of saponification is given below. It has been already mentioned, that the effect produced by the alkalies upon essential oils is partial, and attained to only after a tedious delay of time. Those oils not excepted from this effect, will be spoken of hereafter. There is no chemical change produced in the aromatic oils added to toilet soaps for their perfume; they merely exist in a state of admixture with the paste, the aroma which is emitted by them being due to their spontaneous volatilization.

When treating of wax, we spoke of it as a compound of cerin and myricin, and as saponifiable in a manner with potassa and soda. To effect this saponification, wax is boiled with a solution of caustic potassa; the mixture soon becomes turbid, and the soap separates and swims to the top. This soap, decomposed by the acids, sets free the wax in an altered state; the nature of which change has not as yet been satisfactorily explained. The ancients employed as an enamel or encaustic, a soap composed of twenty parts of wax and one of soda. If ammonia be substituted for potassa or soda, saponaceous emulsion results, which, by cooling of the liquor, collects at the surface, in white flakes.

Starkey's soap.—Starkey, in his experiments upon the

volatilization of salt of tartar, in an attempt to effect that end, combined therewith sundry materials, one of which was the essential oil of turpentine. This last mixture, he observed, resulted in a soapy compound. The soap of Starkey, also called tartarous soap, strictly speaking, is a combination of caustic potassa and soda with the resin of oil of turpentine, and bears the name of its inventor. Although the fixed alkalies are not absolutely without action upon the essential oils, yet their union therewith, even in certain instances, is with difficulty accomplished. For example, if you try to combine any essential oil, particularly that of turpentine with alkali in solution, so as to form a soap, it will be soon seen that the two substances only partially unite, if they unite at all, and then but tardily and imperfectly. Starkey, in composing his soap, had to make use both of time and patience. His method was to put dry alkali in a glass matras, and pour thereon essential oil of turpentine to the height of two or three inches, and to place it aside for spontaneous action. In about five or six months, it can be perceived that a part of the alkali and the oil, which have combined together, form a kind of whitish soapy compound. Separate this soap and leave the rest, so that a new quantity may be formed in the same way.

The tediousness of this process induced an effort to abridge it on the part of Staal, Rouelle, Boullay, Baumè and other chemists. The latter gentleman published a method for making it in a few hours, by triturating potassa upon a slab with oil of turpentine added to it, dropwise. According to Baumè, it is only the resinous part of the turpentine which veritably connects itself with the alkali, and hence the large quantity of the oil required, and more especially, when it is fresh and most ethereal. This explains that the trituration on the slab facilitates the union by evaporating the more volatile parts of the oil.

This soap, however, is used only in medicine, and is of little importance except to the pharmaceutist.

Action of alkalies upon essential oils.—We have said that the alkalies are without influence upon the essential oils, but at the same time qualified or should have qualified the assertion by excepting the oils of cloves, cinnamon and cedar. The first mentioned oil possesses the property of concreting instantly with the alkalies, and forming soaps or savonules.

Bonastre made some very interesting experiments in relation to the influence of alkali upon volatile oils, and particularly on the oils of cloves and Jamaica pepper. We extract from his work as follows.

By soda (in the cold).—If to twenty-four parts of essential oil of cloves, twelve parts of caustic soda ley are added, by stirring a little, the mixture immediately hardens, and becomes opaque. This savonule, when recent, is dry and sometimes even pulverulent, but more frequently in white pearly-looking micaceous plates. It does not attract moisture from the air, especially if the precaution has been taken to press it between bibulous paper, which absorbs its excess of humidity. It is very soluble in alcohol, and if moistened with a little thereof, it sometimes becomes paplike. Exposed to the action of nitric acid, it takes a blood red color. This savonule is but slightly odorous, yet has an acridity almost insupportable.

By potassa.—Potassa, the same as soda, combines with the oil of cloves, and the resulting savonule is as dry and concrete as the preceding, but the union is not so perfect, and it attracts moisture with greater facility from the air; soon becomes dissolved, and the volatile oil recedes in the form of small drops, of a color more brown than before its treatment with alkali. If you operate by heat, after about an hour's boiling, the savonule obtained is entirely brown. Exposed to air it attracts no moisture as does that made by cold; the volatile oil becomes free and appears in oleaginous form—the potassa in small acicular formed crystals. This savonule is very soluble in alcohol, and is of a strong bitter taste.

By ammonia.—Mixed with aqua ammoniæ, oil of cloves becomes granular in appearance. This connection is not so firm as that with soda or potassa. Exposed to air, gaseous ammonia disengages.

By gaseous ammonia.—Eight grammes of oil of cloves, put into a cylinder surrounded by a freezing mixture, after a passage through it for some minutes of a current of very dry ammoniacal gas, is completely solidified. It first takes the aspect of a granulated buttery mass, in which are observable very thin delicate needle-shaped crystals. If the action of the gas is continued, the oil becomes almost as firm as wax. But after a time, and especially by an elevation of temperature or exposure to air, gas disengages, and the volatile oil is left in its oleaginous form. If this compound is kept in a bottle hermetically closed, the crystals are preserved for a long time. They are brilliant, but the color of the oil is singularly changed, and as long as they retain dry ammoniacal gas, do not dissolve.

Later examinations of the subject have thrown such additional light upon it, as enables us to explain the rationale of the union of these certain oils with the alkalies. In relation to the oil of cloves which is extracted from the *Caryophyllus aromaticus*—it is a mixture of a light and heavy oil, the latter of which is separated by the action of concentrated potassa upon the oil, the fatty acid resulting from the decomposition of the compound, styled, by its discoverer, Bonastre, *eugenic acid*, having the power of forming salts with the alkalies. The power possessed by the oil of cinnamon of uniting with the alkalies, is due to the cinnamic acid and the two resins which it contains. The properties of the two latter, as regards their behavior to potash, &c., are explicable upon the same principles as those which elucidate the causes and effects of the union of colophony with soda, &c., in the formation of yellow soaps. The partial saponification of the oil of cedar by alkali must also be attributed to the resin which it contains, for though it has a solid constituent as well as an oily one, we are unable to say, from the information we have, whether either is acid in its nature. The eugenic acid before named, is an oleaginous liquid, whilst the *eugenine*, a yellow pearly crystallized body, also said to be present in the oil of cloves, is, on the authority of Ettling, isomeric with the eugenic acid.

CHAPTER XXXIV.

SAVONS ACIDES, OR OLEO-ACIDULATED SOAPS.

THE intention of making our work perfect in all that concerns the subject matter of which it treats, would fail of completion if we omitted to give some idea of those soaps which bear the title of "*Savons acides.*"

According to Macquer, the alkalies are not the only bodies capable of combining with oils to produce compounds soluble in water and alcohol. The acids having, generally, a very strong causticity, and in particular a decided action upon the oils, it is important to make known, at least, the principal compounds resulting from the union of these two substances, and to reveal the most essential properties of the new formations. There have been several papers written upon the soaps with sulphuric acid base, and among them one by Cornette, and another by Achard, of Berlin, and others by later chemists. The process successful with Achard for making the savons acides, by combining sulphuric acid with the oils, either fluid or concrete, distilled or expressed, consisted in placing two ounces of concentrated colorless sulphuric acid in a glass mortar, and adding thereto, little by little, during constant trituration, three ounces of oil heated nearly to ebullition. In this way are obtained blackened masses, which, when cold, have the consistence of turpentine. These compounds, to be made more perfect and neutral, must be dissolved in about six ounces of boiling distilled water. The excess of acid is thus taken up by the water, and the soapy parts congeal on cooling, and unite into a brown mass of waxy consistence, which sometimes goes to the bottom of the vessel, and at others swims on the surface of the fluid, according to the specific gravity of the oil employed. If the soap still contains too much acid, as is easily distinguishable by its taste, it must be again dissolved in boiling distilled water, and the operation repeated as before until all taste of acidity has disappeared. And in this way is obtained a so called "*savon acide.*" Achard remarks, that the concentrated sulphuric acid operates powerfully upon the oils, and advises that the oil be not added too suddenly, or in too great quantities at a time, for in such case the acid carbonizes the oil, a change easily perceptible by the odors of sulphurous acid which are given off. When the savons acides are made with exactness they harden by age, but when containing an excess of acid, attract moisture and soften in the air. The same process applied to the oils of sweet almonds and olives, to cocoa butter, wax, spermaceti, gave similar soaps. The process requires some attention and careful manipulation. The experiments of Achard present many curious and important phenomena, but it is unnecessary to enter into minute detail. Not only do the alkaline substances and many metallic bodies decompose the savons acides, but what is remarkable, several of the acids also, viz: the nitric, hydrochloric, sulphuric and acetic. However, the effect of the acetic acid is not the same on all the soaps, for there are some that it does not decompose. Many neutral salts also decompose these soaps; some by the greater affinity of their base for the sulphuric acid, others by elective affinity or interchange of bases.

But one circumstance very remarkable is, that no matter in what manner these soaps are decomposed, even though by distillation directly over the fire, the oil which is liberated retains a consistence much firmer than it had naturally, and even sometimes becomes as concrete as wax.

Another general observation upon the decomposition of savons acides by the alkalies, and which is no less important, is that in using these reagents, take care to only employ that proportion exactly needed for the saturation of the acid, for any excess will inevitably combine itself with the liberated oil, even much more easily than by a direct union. Now the mode of accounting for these changes is not at all difficult, for the addition of sulphuric acid, in certain proportions, decomposes the fats by separating their acids from the base with which they are in nature combined, and uniting itself with the glycerin to form a peculiar acid called the *sulpho-glyceric*. It is thus that the *savons acides* have their affinity for alkalies, that property being due to the presence of free fatty acids, and also the decomponent power which the stronger base has upon the sulpho-glyceric acid. When, however, the oil of vitriol exceeds a certain ratio, then its action upon the oil is different from the above recorded, for in this instance it joins with the fatty acids to form compounds, such as sulpholeic and sulpho-margaric acids, (sulphate of glycerin being a simultaneous product,) either of which acids, when diluted with water, is decomposed or changed into one or more new modifications. As to the detergent action of these *savons acides*, that may be referred to the union of their free fatty acids with the grease of the clothes, thus virtually forming a soap of such grease, and consequently rendering it soluble. The chemical relations of these changes in fats by the action of sulphuric and other strong acids are a subject of great interest, and are now occupying the attention of several skillful investigators. Their results may throw a clearer light upon the matter than we have to direct us at the present writing.

CHAPTER XXXV.

TOILET SOAPS.

UNDER this head are comprised all those soaps which are of a finer nature, and in the manufacture of which extra care and cleanliness have been bestowed. They are principally used for the toilet. The varieties are numerous, for instance the white or colored, transparent, flotant and pasty soaps. A thorough treatise upon their mode of preparation, embracing all the recipes for the latest and most fashionable soaps, will be found in a work on "Perfumery," lately published by Messrs. Carey & Hart, Philadelphia.

The toilet soaps are either of potash or soda base, and are made of the purest materials. Those with soda are generally formed of olive oil, palm and sweet almond oil. Of the fats, those most used are lard and suet. As this last fat always communicates a disagreeable odor to the soap, it must first be well purged before being used. The potash toilet soaps are prepared only with fats, and almost exclusively with lard. Care should be taken to get them as neutral as possible. This branch of the art, though it seems simple, nevertheless requires much experience and information as to the best modes of conducting it successfully;—time will furnish the first;—the latter can be had from the book above referred to.

Those toilet soaps made with soda and lard, or with olive, sweet almond or palm oil, are called hard soaps; those with potassa base are styled soft soaps.

As most of the ordinary toilet soaps are made from the better species of white soap, we give below the mode of cleansing and refining, requisite to make it applicable for the purpose.

Purification of soaps.—When it is intended to make a toilet soap, the foundation or body should be as white and of as excellent beauty and quality as can be imparted. To obtain a superior soap from white soap, proceed as now directed. Take soap in thin shavings and melt it over a water bath with rose and orange flower water and salt. To twenty-four pounds of soap take four pints of rose water, four pints orange flower water, and two large handfuls of salt. The next day if it is entirely cooled, cut it up into small bars and dry it in a shady place. Then melt it anew as before in the same quantities of rose and orange flower water and strain it. Cool and dry it again;—this done, the soap is exempt of bad odor. It must be powdered, exposed again for several days to air, but out of the way of dust. It is now ready to receive the perfume it is destined to have, and to be moulded and pressed into forms as may be desired.

Another method of purification is to melt six pounds of best white soap in three pints of water, and when liquid, to strain it through a linen cloth. Replaced in a kettle with a pint of water and a tablespoonful of salt, a brisk fire is kindled thereunder, and by whipping or stirring, the contents are made to foam and froth. The fire must then be put out and the batting continued until the mass is sufficiently inflated. The fire is again kindled, and the kettle kept on until its contents swell and foam, then it is to be withdrawn and emptied into the cooling frames. When the soap has solidified, it is taken out, cut into cakes and pressed.

By either of these plans any of the other toilet soaps can be refined. To color the soaps yellowish brown, use either dead leaves or powdered orange peel, thinned out with water and rubbed into paste, free of granules. After adding the coloring matter, the soap is strained again, put upon the fire, and perfumed, *à la bergamotte*, with two drachms of the oil to every pound. When other scents are added, let them be in proper proportion. The odors are, however, multifarious, and generally designate the soap; for instance, soap perfumed with a mixture of essences, is called soap *au bouquet*, and so on, you can make or alter titles to suit your own taste.

Toilet soap—by admixture of olive oil and suet soaps.—Weigh out 15 pounds handsome olive oil soap and 10 pounds of suet soap. After having reduced the whole to thin shavings with a plane, it is placed in an untinned copper kettle, heated by water bath, and moistened with two and a half pounds water. The quantity of water varies, however, with the state of dryness of the soap, but ought always to be as little as possible. The temperature during the operation, must not exceed 212° Fahrenheit, and the process should be completed as rapidly as can be. In this way the soap is prevented becoming too brittle, a property, which hinders its being moulded and pressed into tablets. Upon these preliminaries depends, mainly, the facture of perfumed toilet soap, known as cake soap. The only requisite now, is to determine the coloration of the soaps, and the proper proportion of perfume. We confine ourselves to a few examples, as the process is similar for all the principal toilet soaps.

Soap à la rose.—To 25 pounds of first quality white soap, either lard, Windsor, almond, or the above soap, add whilst in a melted state, six ounces three drachms of vermilion, and mix it in as thoroughly as can be, and as soon as the mass is poured into the cooling frame, add the aroma in proportions as follows:—

Essence of rose	1 ounce	3 drachms
" cloves		4 "
" cinnamon		4 "
" bergamot	1 "	1 "

We must remark, that in pouring the paste into the frames, it should be passed through a cloth or very fine hair sieve, to separate impurities and undissolved particles. After 24 hours' cooling the soap is ready to be cut into cakes and pressed.

Savon au bouquet.—This soap is obtained as is the preceding, except that you color with three ounces five drachms impalpably fine powdered brown ochre. The perfume is the following mixture:—

Essence bergamot	2 ounces	
" cloves	1 "	5 drachms
" neroli		6½ "
" sassafras	1 "	5 "
" thyme	1 "	5 "

Cinnamon soap, or savon à la cannelle.—Prepare this soap by liquefying 15 pounds suet soap, 10 pounds palm oil soap, and color the paste with eight ounces yellow ochre. The perfume is

Essence of cinnamon	3 ounces	4 drachms
" sassafras		6 "
" bergamot		6 "

The musk soap is colored with brown ochre, and among the essences tincture of musk should predominate.

Orange flower soap, or savon à la fleur d'oranger.—This soap is made of a mixture similar to the foregoing, and among the essences composing its aroma should be that of Portugal. The color imparted is an orange yellow inclining to green. As to the other perfumed soaps, they are prepared in like manner, the difference being in the essence added to them.

Benzoin soap, or savon au benzoin.—Benzoin is a dry, fragile, inflammable resin, of a penetrating agreeable odor, especially when burned. It exudes spontaneously or by incision from a large tree, the *styrax benzoin*, growing in the islands of Java and Sumatra. By sublimation the benzoin gives a white silky crystalline acid, called "flowers of benzoin." This acid is employed as a perfume, and by some is pretended to have the property of effacing freckles and splotches. Benzoin dissolved in alcohol gives a tincture, some drops of which, when thrown in water, renders it milky, thus making a mixture formerly much used as a cosmetic, and called *lait virginal.* Its admixture with soap is only as a perfume. To compound the soap, take of purified white or palm soap, and after having melted it thoroughly, stir in a sufficient quantity of benzoic acid.

Cologne water soap, or savon à l'eau de Cologne.—First prepare the Cologne by solving in ten pints of alcohol, 36° Baumè, the following essences:

Essence of	bergamot	- - - -	4 ounces
"	cedrat	- - - -	1 "
"	lemon	- - - -	1 "
"	lavender	- - - -	2 drachms
"	rosemary	- - - -	2 "
"	cloves	- - - -	1 "
"	thyme	- - - -	1 "
"	neroli	- - - -	1 ounce.

The mixture, after having stood several days and been frequently well shaken during the interval, must be filtered. When the alcohol is good, and very strong, excellent Cologne is made without recourse to distillation. The purified white soap is melted and perfumed with this water in the same manner as by the essences.

Another excellent recipe for Cologne water is the following. To three quarts alcohol 36°, add of

Essence of cedrat	2	drachms
" bergamot	2	"
" lemon	2	"
" lavender	1	"
" rosemary	1	"
" neroli	½	"

After eight days' contact, and frequent shaking during that interval, the mixture is filtered.

Vanilla soap, or savon à la vanille.—The vanilla emits one of the most agreeable and aromatic odors known. It comes as a bean or pod, and the plant upon which this pod grows is the *epidendron vanilla,* a tree of Mexico, Peru, &c. There are three kinds of vanilla: the first, the pod of the *vanilla pompono* or *bora* of the Spaniards. Its length is from five to six inches; breadth from one-half to three-fourths of an inch. It is brown, soft, viscid, almost always open, and of a strong smell, but less agreeable than the *leg.* The second is the *legitime,* or *vanille de lev,* or *vanille leg,* and is most esteemed in France. It is about six inches long, from a quarter to one-third of an inch broad, narrowed at the two ends and curved at the base, somewhat soft and viscid, of a dark reddish color, and of a most delicious flavor, like that of the balsam of Peru. The third species, *vanille simarona,* or *bâtarde,* is a little smaller than the preceding, of a less deep brown color, drier and not so aromatic. It is brought from St. Domingo, and is said to be the product of a wild plant.

The second species is that almost exclusively made use of. By means of alcohol all the resinous odorous part of the vanilla can be extracted. Some spoonfuls of this tincture impart to soaps a most agreeable perfume.

Musked soap (*savon musqué*), *for whitening and soften-*

ing the hands.—Take four ounces marshmallow root, well cleansed and dried in the shade, and reduced to powder; four ounces starch, four ounces wheat bran, two ounces peeled almonds, one and a-half ounces orange seeds, two ounces oil of tartar, and oil of sweet almonds, and half a drachm of musk. All being intimately mixed together, and rubbed down to fineness, add in a half ounce of pulverized orris root, to every ounce of powder. Then take of rose or orange flower water, in which fresh marshmallow root has been macerated for a day or more, and therewith form into a paste, the aforesaid powder. This paste is to be shaped into forms or balls, and dried. Nothing, it is said, is more softening to the hands, and more promotive of their whiteness.

Naples soap.—Into a copper basin, pour of strong ley, twelve pounds, lamb suet three pounds, (previously purified with rose or orange flower water,) oil of ben or olives, one pound. Boil these materials until they have acquired a sufficient consistency to be transvased into another large vessel, shallow and broad. Cover this vessel with glass, and expose it for six days to a strong solar heat, taking care to stir the contents several times daily. At the end of this time, perfume by adding four ounces tincture of ambergris, one ounce tincture of musk, and one and a-half ounces odorous oil. After two months, the soap has acquired its proper consistence.

Kasan soap.—This article, improperly written Kesan and Keesan soap, is peculiar to that portion of the Russian empire, bordering on either side of the Volga, and called Kasan. Its superior properties for shaving use, and the rich and durable lather which it gives, has made it a favorite with purchasers, so that its consumption is such as to make it a profitable item of importation. It is said to be made of the fat of some animal incident to the country of its production, and by long exposure of its mixture with ley, to the heat of the sun on the roofs of houses. It is a rich looking soap, and in color is very little dissimilar from that of lump gamboge. Not knowing the exact mode of its preparation, our supplies are necessarily drawn from abroad.

SAVONNETTES.—These are made of the preceding soaps, which, instead of being shaped into cakes, are moulded in circular forms, and are known by the title of "wash balls."

FLOTANT SOAPS, OR SAVONS LEGERS.—Under this title are designated those soaps, which, when in a state of paste, are batted or inflated with air, by which means its buoyancy becomes such, that the soap floats in the water. It is the most preferable soap, not only on account of whiteness and mildness, but the facility with which it gives a lather. These soaps are colored and perfumed exactly as are the toilet soaps generally. It must be remarked, however, that the olive, palm, and sweet almond oil soaps are the only ones that can be converted into *savons legers*, it being impossible to inflate those soaps made of suet or lard. Their mode of preparation is, in liquefying either of the previously named fine soaps, by melting it, when reduced to shavings, in a water bath kettle, with twelve pounds water to every

Fig. 19.

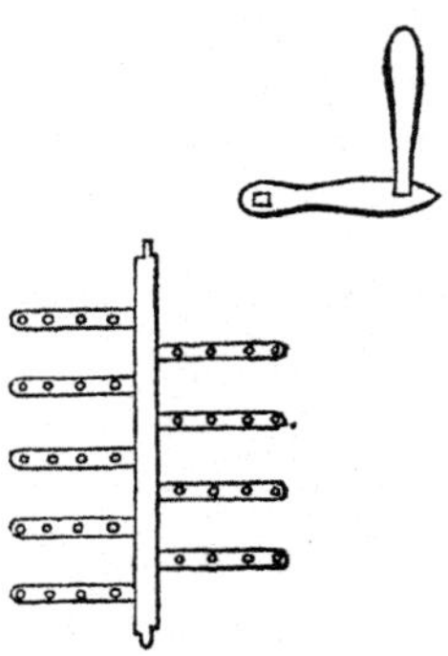

Fig. 20.

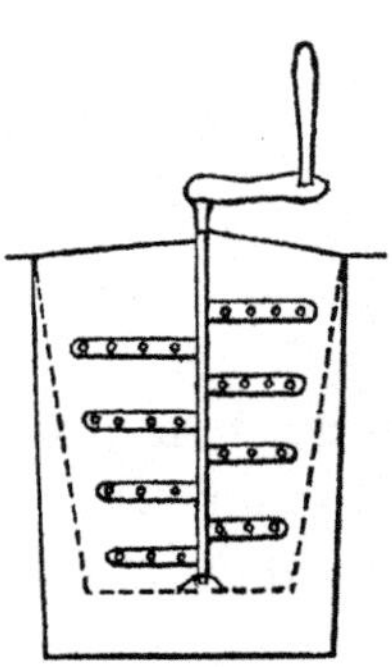

fifty pounds soap. When the mass is in perfect fusion, it is agitated with a twirling fan (Figs. 19, 20), until it froths and foams to the top of the vessel.

Fig. 19 represents the twirling fan detached from the water bath; it is made to revolve on a pivot, and to be taken on or off at will. After the perfume has been added, the contents are emptied out into cooling frames or moulds.

Transparent soap.—This beautiful soap, so much admired and used, is generally shaped into savonnettes or cakes, and when well prepared, assimilates in appearance to handsome transparent candy. The method of facturing it was for a long time unknown. It is made by solving suet soap in alcohol. For the purpose, the soap must be in thin shavings and *perfectly dried* so as to be entirely freed of any water which it may have retained. Thus prepared, fifty pounds are placed in the water bath B (Fig. 21) of the still A. After having also poured

Fig. 21.

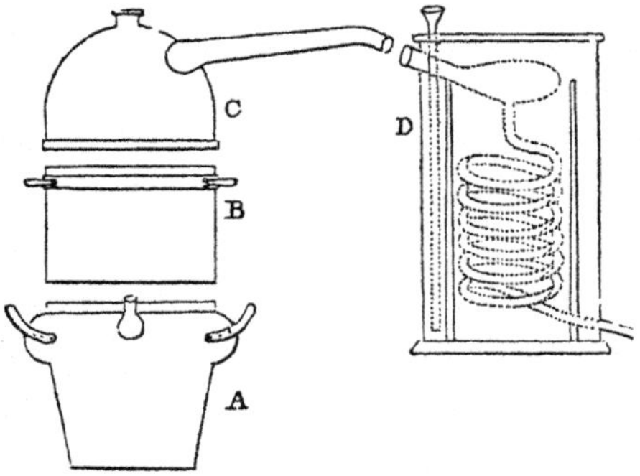

in twenty-eight pints of alcohol 36°, put on its head C, carefully lute it up, and raise the temperature of the water bath just to 212° Fahrenheit. If the heat be too brisk, a certain quantity of alcohol volatilizes, without the solution of the soap being completed. In the first operation only distil off about five pints of alcohol, and after being assured of the perfect liquefaction of the mass, allow a repose of some minutes, and then turn the liquid into the moulds or forms, destined and arranged to receive it. The capacity of these moulds should be one-third larger than the intended size of the form which they contain, because allowance must be made for the shrinkage which this soap undergoes. After having cooled and become sufficiently firm to be taken from the moulds, the soap is then placed in a very airy place, in order that its drying may be facilitated. When it has been cooled in frames, after eight or ten days, according to the season, it is taken out and cut up into bars, and these again subdivided into tablets, which are pressed against dies, bearing on one side some ornamental device, and upon the other, the name and place of business of the manufacturer. These tablets or pressed cakes are then dried in a moderately heated room. When colors are required, use for the transparent soaps, an alcoholic tincture of alkanet for red; a tincture of turmeric for yellow; an orange by a mixture of the two preceding tinctures; green, by an alcoholic tincture of chlorophyle; blue, by an alcoholic solution of indigo carmine. By admixture and modification, the shades can be varied or multiplied. Thus, the conjunction of blue, yellow, and red furnishes a cinnamon color; yellow and blue, differently proportioned, give all the tints of green, &c. &c. The perfume is the same as for the other soaps and in like doses. To the still is affixed a cooling worm, D, for the condensation of the alcohol that runs over. It is placed in a barrel, the water in which is kept constantly cool by fresh additions.

Toilet soft soap.—We generally meet with this kind of soap under the name of "shaving cream." It is admirably adapted for the beard, and for this purpose, has very justly been accorded the palm of superiority. Its preparation is attended with some difficulties. To make it, take fifty pounds of lard, and mix it with seventy-five pounds of caustic potash ley marking 17° Baumè. This mixture is then heated slowly, and when the empatage is completed, and the whole of the ley thoroughly united with the fatty matter, carry the mass to ebullition, but arrest it so soon as watery vapors cease to be given off. This is the signal by which to know the terminus of the operation, and is observed also by the matter becoming so thick that it can hardly be agitated. It is now run into frames and cooled. Generally speaking, these soaps are aromatized, but not colored.

Pearl soft soap.—It is only within a few years that the process for making this elegant soap has been known. The mode differs but little from that above described, the beauty of the soap being due to very careful and exact manipulation, and which we will endeavor to explain clearly and comprehensively. Weigh out first twenty-five pounds of lard, and then twelve pounds of potassa ley, 36° Baumè. Melt the lard in an enameled vessel or stone jar, heated by a sand bath, the temperature of which should be managed. Stir continually with a wooden spatula, and when the grease is half melted and looks milky, pour in half of the ley, still stirring and keeping the temperature constant and uniform. The empatage takes place gradually, yet in about an hour the fat tends to the surface in an oleaginous stratum and the granulations of soap fall to the bottom. It is then that the second portion of the ley must be added;—the empatage is immediately restored, and the granules disappear. After conducting this operation for four hours, the paste becomes so compact that it cannot be conveniently stirred, and must then be lightly batted or beaten. At this time cease the fire under the sand bath, and transfer the vessel into a basin of warm water, so that its contents may cool gradually. The soap itself is now finished, but its pearly aspect is a physical property, to be developed by a strong and constant trituration of the paste, in a marble mortar. All the parts which before seemed separated, unite and form a perfectly homogeneous soapy mass. The perfume almost universally imparted, is essence of bitter almonds, or otto of roses; hence the two titles usually given it, "rose" and "almond" shaving cream. The first is generally of a pink shade, obtained from a little carmine, or better, of tincture of alkanet.

The remarks and instructions as to the toilet soaps would be extended much further, but anything more than has been here given would be infringing another copyright. A full and explicit treatise upon the toilet soaps, their mode of manufacture, and the tools used for the purpose, together with every information appertaining to the art of the Perfumer, and numerous valuable recipes for all the most fashionable French preparations, will be found in a little work on "Perfumery" lately published by Messrs. Carey & Hart, of Philadelphia. That portion of the work which is not original, was culled from the latest and most authentic sources, and its whole contents may be relied upon as correct, and embracing every instruction needed for the successful pursuit of the trade of perfumer.

On another page is a list of the soaps by their fashionable titles. It is not inconsistent with the accustomed morals of the trade, frequently to sell one kind of soap under two or even a half dozen different names. This is a valuable hint to the savonnier, for whenever the overreaching desire of a competitor may induce a high-sounding title for any particular soap, he can be successfully rivaled if the competing products are as euphoniously designated. There is a good deal of *humbug* in this art, as well as in others. I will not say that it is pardonable, but indeed the vagaries of fashion are so whimsical and capricious, that every power of invention would soon be exhausted, in humoring the fancies of that fickle goddess; hence, the necessity of some innocent mode of successfully flattering the public. What better way can there be than by graceful titles, handsome labels, and brilliant envelops? If *savon au bouquet* ceases to be in vogue this year, add another essence to its melange of perfumes, label and wrap it in a new and richer covering, and then advertise it as *savon deana morhofina*, or, by any other designation your wits may invent for it. This is the latest Parisian method of getting up "new and improved" soaps. The plan takes well in this country, to judge from the extent to which it has been, and is yet imitated. But it is not confined to soaps only, for the spirit of enterprise characteristic of Yankee tradesmen has extended its application to all branches of perfumery, so that a plain "pearl powder" of a year since, meets with no purchasers unless sold under the Chinese appellation of *Meen-hum-fooni*. A dentifrice, no matter how excellent, to find a sale, must be handsomely put up in fashionably shaped packages, embellished with gold-bordered labels, centered, by the picture of a row of pearly white teeth, on a gum of healthy redness; whilst olive oil soap, which might remain for a year uncalled for, is a prolific source of revenue, under the designation of *savon de blanc de baleine*. But these are some of the tricks of trade, a knowledge of which is acquired almost intuitively, so it is needless to extend our remarks any further, lest the public might get an insight into the matter, and thus be created a source of inconvenience to the profitable success of the art of soap and perfume making.

LIST OF THE PERFUMED SOAPS.

Savon à l'ambre.
" aux mille fleurs, or au bouquet.
" à la frangipane.
" à la fleur d'oranger.
" au réséda.
" à la tubéreuse.
" à l'acacia.
" à la julienne.
" au pois de senteur.
" au jasmin.
" au narcisse.
" à la jacinthe rose ou bleue.
" à l'hémérocalle.
" au jasmin jonquille.
" au chèvre-fenille.
" au gaïac.
" au vitiver.

Savon au storax.
" au baume de Peru.
" au benzoin.
" au chypre.
" à l'œillet.
" à la menth fine.
" à la menthe Anglaise.
" à la lavande.
" au romarin.
" à l'héliotrope.
" à la violette.
" à l'iris.
" à la jonquille.
" à la rose royale.
" à la palymyrene.
" à la girofleé.
" à la mousseline.
" aux fleurs d'Italie.

Savon de blanc de baleine, (*Spermaceti.*)—This soap, when really made of spermaceti, is superior and emollient, but much that is sold under this title is merely refined white oil or suet soap.

All the preceding titles can be altered to suit the prevalent taste of the times, and altered back again whenever the freak of fashion renders it politic. For instance, if your soap sells slowly, change its form and envelop, and call it *savon de Monterey*, *savon amolissant*, or *savon des houris*. As an aid to the savonnier we annex a list of synonymes, for any of the soaps in the preceding list.

Savon du zephir.
" de Diane.
" des nymphes.
" du baigneur.
" des Circassiennes.
" Oriental.
" Virginal.

Savon à la passiflore.
" à la campanule—lilas.
" aux bluets.
" au zinia.
" au datura.
" au camélia.
" au dahlia.

Savon au bouton de rose.
" au bouquet de violette.
" à la rose blanche.
" au jasmin amande.
" au beurre de galaam.
" d'albâtre.
" perlé.
" à l'étoile d'amour.
" à l'ambroisie.
" des sultanes.
" à la maréchale.
" à la pervenche.

Savon aux liserons roses.
" aux *pensez-à-moi.*
" à la belle de nuit.
" à la balsamine.
" à la grenade.
" à l'hortensia.
" aux chrysanthêmes.
" au myrte du Canada.
" au myrte des amours.
" au géranium.
" à la rose du Bengale, &c. &c.

The shade, label and envelop of the soap should always correspond. On those soaps, named after flowers, it will be in good taste to have the plant represented upon the wrapper, or else on the die with which the soap is compressed. Everything, however, depends upon the taste of the manufacturer; and an ingenuity in these respects, assiduously exercised, will bring a corresponding reward.

The present advanced state of the art has entirely varied the old form of tablets of soaps, so that now, instead of the square cakes, the paste is moulded of oval or any other shape that the fancy of the savonnier may desire. There is, indeed, a great deal of taste exercised also in the adornment of the wrappers, which, independent of the quality of the soap, and the really beautiful designs impressed thereupon by the dies, are frequently specimens of art worthy minute examination and liberal admiration. The effect of a tasty arrangement in the wrapping of soap is quite evident in the preference given to such by purchasers. These remarks embrace all that is necessary to be said in relation to the matter, and with them we conclude the subject.

CHAPTER XXXVI.

AREOMETERS AND THERMOMETERS.

An areometer is a convenient glass instrument for measuring the density or specific gravity of fluids, and thence the strength of alcoholic liquors, these being inversely as their specific gravities. Areometer and hydrometer are synonymous, the first deriving its name from the Greek words αραίος, *rare*, and μετρον, *measure*, and the latter from ὑδωρ, *water*, and μετρον, *measure*, hence the same instrument is frequently denominated both hydrometer and areometer. This apparatus is often referred to throughout the work; for instance, in speaking of alcohol, or alkali, its strength is announced as being of so many degrees, for instance 17° or 36° Baumè, that is, its force or value is of that specific gravity, corresponding with the degree to which the hydrometer sinks in either the alcohol or alkaline solution, and which is easily seen by reference to the following table from Ure, arranged for the convenience of operatives. But for those liquids lighter or rarer than water, viz., alcohol, ethers, &c., the scale is graduated differently than for the heavier or more dense, examples of which are the acids, saline solutions, syrups and the like. There are several kinds of hydrometers, but that called Baumè's is the most used, and to this our remarks are applied. That for the denser liquids, (fig. 22,) is formed of a slender glass tube, A, D, about five or six inches long, and of diameter proportional to the intended sensitiveness of the instrument; in fact the space intervening between each degree is proportional to the ratio of the diameter of the tube, and that of the bulb C, which sometimes is made spherical and at others cylindrical; but regardless of its form, there is alway sattached beneath a globe, D, filled with shot or mercury, and serving as ballast to keep the hydrometer in an upright position in the liquid. This instrument is graduated by plunging it into distilled water of 58° Fahrenheit, and adding mercury to the globe aforesaid, until the instrument sinks to the line *a*, which is its zero point. This zero point, thus ascertained, is to be marked accurately upon the glass or its accompanying paper scale, and the instrument again plunged into 90 parts of distilled water, holding solved ten parts of previously dried chloride of sodium or common salt. The point to which it sinks in this liquid, say *b*, for instance, is then also marked carefully upon the scale and rated as ten compared with its zero point. The interval between these two points is then spaced off into ten equal divisions, according to which the remainder part of the tube is graduated, so that each degree is intended to represent a density corresponding to one per cent. of the salt. These instruments are not critically accurate, but the results which they furnish are sufficiently correct for practical purposes. The hydrometer for the liquids rarer than water is but little different from the preceding, in form, but necessarily has a modified scale, which is graduated as is shown by fig. 23. The instrument should be sufficiently heavy in ballast to sink in a saline solution of 10 parts of dried chloride of sodium in 90 parts distilled water to the bottom of its stem *a*, to be marked as the zero of the scale. Now when it is again placed in distilled water alone, it floats or sinks to a point somewhere about *b*, which is to be the ten degree

Figs. 22 and 23.

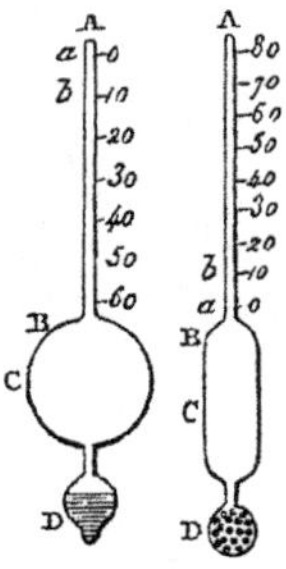

mark. By dividing the rest of the stem into as many ten degree intervals as its length will permit, and spacing each division into ten uniform smaller intervals, you have an areometer rating the number of degrees written upon its scale.

As it would be troublesome, and with many impracticable, to estimate the specific gravities of their liquids in a scientific way, these little instruments are a great convenience, for by taking out a portion of the fluid to be tested, and placing it in a slender graduate, you ascertain its degree Baumè, by noting the point to which a hydrometer sinks therein, and afterwards its specific gravity, by comparing it with its corresponding point in the table. For instance, suppose your hydrometer sinks in alcohol to 35°, then its specific gravity is 0.8538, and this again can be translated into its absolute spirit strength by comparison with any accurately calculated alcohol tables.

Specific Gravity numbers corresponding with Baumè's areometric degrees.

Liquids denser than water.						Less dense than water.			
Degrees.	Specific gravity.	Degrees.	Specific gravity.	Degrees.	Specific gravity.	Degrees.	Specific gravity.	Degrees.	Specific gravity.
0	1.0000	26	1.2063	52	1.5200	10	1.0000	36	0.8488
1	1.0066	27	1.2160	53	1.5353	11	0.9932	37	0.8439
2	1.0133	28	1.2258	54	1.5510	12	0.9865	38	0.8391
3	1.0201	29	1·2358	55	1.5671	13	0.9799	39	0.8343
4	1.0270	30	1.2459	56	1.5833	14	0.9733	40	0.8295
5	1.0340	31	1.2562	57	1.6000	15	0.9669	41	0.8249
6	1.0411	32	1.2667	58	1.6170	16	0.9605	42	0.8202
7	1.0483	33	1.2773	59	1.6344	17	0.9542	43	0.8156
8	1.0556	34	1.2881	60	1.6522	18	0.9480	44	0.8111
9	1.0630	35	1.2992	61	1.6705	19	0.9420	45	0.8066
10	1.0704	36	1.3103	62	1.6889	20	0.9359	46	0.8022
11	1.0780	37	1.3217	63	1.7079	21	0.9300	47	0.7978
12	1.0857	38	1.3333	64	1.7273	22	0.9241	48	0.7935
13	1.0935	39	1.3451	65	1.7471	23	0.9183	49	0.7892
14	1.1014	40	1.3571	66	1.7674	24	0.9125	50	0.7840
15	1.1095	41	1.3694	67	1.7882	25	0.9068	51	0.7807
16	1.1176	42	1.3818	68	1.8095	26	0.9012	52	0.7766
17	1.1259	43	1.3945	69	1.8313	27	0.8957	53	0.7725
18	1.1343	44	1.4074	70	1.8537	28	0.8902	54	0.7684
19	1.1428	45	1.4206	71	1.8765	29	0.8848	55	0.7643
20	1.1515	46	1.4339	72	1.9000	30	0.8795	56	0.7604
21	1.1603	47	1.4476	73	1.9241	31	0.8742	57	0.7656
22	1.1692	48	1.4615	74	1.9487	32	0.8690	58	0.7526
23	1.1783	49	1.4758	75	1.9740	33	0.8639	59	0.7487
24	1.1875	50	1.4902	76	2.0000	34	0.8588	60	0.7449
25	1.1968	51	1.4951			35	0.8538	61	0.7411

These hydrometers are made sometimes with varied scales, suitable to the purpose for which they are to be applied; thus, that which is intended for testing the strength of alkalies, is called an alkalimeter; that for acids, an acidimeter; that for alcohol, an areometer, &c. All of them as well as the thermometers about to be described, are accurately manufactured by Stephen Heintz, a skillful glassblower, living in Kensington, Philadelphia.

Thermometers.—The thermometer is an instrument

made of glass exclusively, when intended for practical purposes. Fig. 24 shows one made by Heintz, with the

Fig. 24.

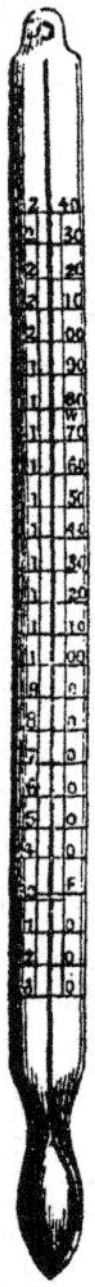

scale of Fahrenheit, graduated on the glass, so that, when having been dipped in liquids, it can easily be wiped clean. It derives its name from two Greek words, θερμος, *warm*, and μετρον, *measure*, and is, as its title indicates, a measurer of the variation of temperature in bodies. The principle upon which it is constructed, "is the change of volume which takes place in bodies, when their temperature undergoes an alteration, or, in other words, upon their expansion." As it is necessary, in the construction of thermometers, that the material used to measure the change of temperature, shall be of uniform expansion, and with a very distant interval between its freezing and boiling point, as fulfilling these requisites better than any other body, metallic mercury is generally used. There are several different thermometrical scales all constructed upon the same principle, but varying in their graduation; the boiling and freezing points of each, though corresponding in fact, being represented by different numbers. The Fahrenheit scale is most used in this country, that of Celsius, called the Centigrade, in France and the continent generally, except Spain and Germany, where Reaumur's scale is preferred. The relation between the three scales is shown by Fig. 25. In the Fahrenheit

Fig. 25.

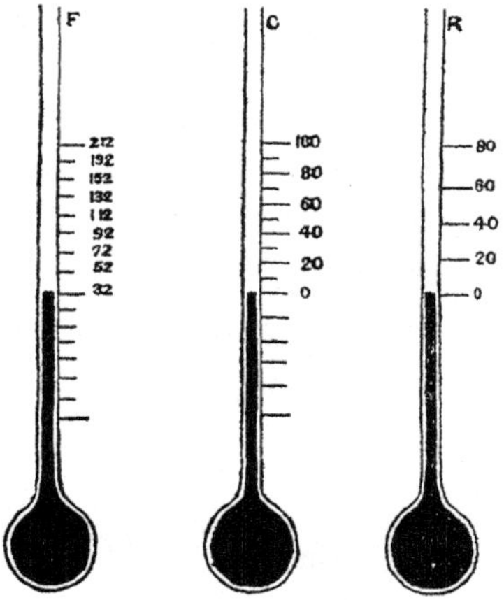

thermometer, the interval between the freezing and boiling points of water, is divided into 180 degrees. The freezing point is placed at 32°, and hence the boiling point at 32+180=212. Reaumur divides the distance between the two extreme points of water into 80°, and Celsius spaces his thermometer (the Centigrade), into 100 equal intervals, the zero point, as in Reaumur's, being placed at freezing. As the decimal divisions of the latter scale render calculation more easy, it is for this reason most preferable for use in practical operations; but as it frequently happens that the manufacturer has no choice in the kind, but is compelled to take such as can be conveniently obtained, we annex a table showing the comparative value of each.

In the graduation of the scale it is only necessary to have two fixed determinate temperatures, and for these the boiling and freezing points of water are universally chosen. The scales can be extended beyond either of these points, by continuing the graduation. Those degrees below zero or 0° have the minus (—) prefixed to distinguish them from those above; thus, 55° F., means fifty-five degrees above zero, Fahrenheit's scale, and —9° C., nine degrees below zero, Centigrade scale. The thermometers for general use, very seldom, however, extend either way beyond the boiling and freezing points of water, but for manufacturer's use, they are graduated sometimes to 400° or 600°; the figure represents one ranging as high as 240°.

The following table shows at a glance, the corresponding value of each scale.

TABLE OF THE CORRESPONDING DEGREES OF FAHRENHEIT'S, REAUMUR'S AND THE CENTIGRADE THERMOMETERS.

Fahr.	Reau.	Centig.	Fahr.	Reau.	Centig.	Fahr.	Reau.	Centig.	Fahr.	Reau.	Centig.
212	80.0	100.0	148	51.5	64.4	84	23.1	28.0	20	5.3	6.6
211	79.5	99.4	147	51.1	63.8	83	22.6	28.3	19	5.7	7.2
210	79.1	98.8	146	50.6	63.3	82	22.2	27.7	18	6.2	7.7
209	78.6	98.3	145	50.2	62.7	81	21.7	27.2	17	6.6	8.3
208	78.2	97.7	144	49.7	62.2	80	21.3	26.6	16	7.1	8.8
207	77.7	97.2	143	49.3	61.6	79	20.8	26.1	15	7.5	9.4
206	77.3	96.6	142	48.8	61.1	78	20.4	25.5	14	8.0	10.0
205	76.8	96.1	141	48.4	60.5	77	20.0	25.0	13	8.4	10.5
204	76.4	95.5	140	48.0	60.0	76	19.5	24.4	12	8.8	11.1
203	76.0	95.1	139	47.5	59.4	75	19.1	23.8	11	9.3	11.6
202	75.5	94.4	138	41.1	58.8	74	18.6	23.3	10	9.7	12.2
201	75.1	93.8	137	46.6	58.3	73	18.2	22.7	9	10.2	12.7
200	74.6	93.3	136	46.2	57.7	72	17.7	22.2	8	10.6	13.3
199	74.2	92.7	135	45.7	57.2	71	17.3	21.6	7	11.1	13.8
198	73.7	92.2	134	45.3	56.6	70	16.8	21.1	6	11.5	14.4
197	73.3	91.6	133	44.8	56.1	69	16.4	20.5	5	12.0	15.0
196	72.8	91.1	132	44.4	55.5	68	16.0	20.0	4	12.4	15.5
195	72.4	90.5	131	44.0	55.0	67	15.5	19.4	3	12.8	16.1
194	72.0	90.0	130	43.5	54.4	66	15.1	18.8	2	13.3	16.6
193	71.5	89.4	129	43.1	53.8	65	14.6	18.3	1	13.7	17.2
192	71.1	88.8	128	42.6	53.3	64	14.2	17.7	0	14.2	17.7
191	70.6	88.3	127	42.2	52.7	63	13.7	17.2	1	14.6	18.3
190	70.2	87.7	126	41.7	52.2	62	13.3	16.6	2	15.1	18.8
189	69.7	87.2	125	41.3	51.6	61	12.8	16.1	3	15.5	19.4
188	69.3	86·6	124	40.8	51.1	60	12.4	15.5	4	16.0	20.0
187	68.8	86.1	123	40.4	50.5	59	12.0	15.0	5	16.4	20.5
186	68.4	85.5	122	40.0	50.0	58	11.5	14.4	6	16.8	21.1
185	68.0	85.0	121	39.5	49.4	57	11.1	13.8	7	17.3	21.6
184	67.5	84.4	120	39.1	48.8	56	10.6	13.3	8	17.7	22.2
183	67.1	83.8	119	38.6	48.3	55	10.2	12.7	9	18.2	22.7
182	66.6	83.3	118	38.2	47.7	54	9.7	12.2	10	18.6	23.3
181	66.2	82.7	117	37.7	47.2	53	9.3	11.6	11	19.1	23.8
180	65.7	82.2	116	37.3	46.6	52	8.8	11.1	12	19.5	24.4
179	65.3	81.6	115	36.8	46.1	51	8.4	10.5	13	20.0	25.0
178	64.8	81.1	114	36.4	45.5	50	8.0	10.0	14	20.4	25.5
177	64.4	80.5	113	36.0	45.0	49	7.5	9.4	15	20.8	26.1
176	64.0	80.0	112	35.5	44.4	48	7.1	8.8	16	21.3	26.6
175	63.5	79.4	111	35.1	43.8	47	6.6	8.3	17	21.9	27.2
174	63.1	78.8	110	34.6	43.3	46	6.2	7.7	18	22.2	27.7
173	62·6	78.3	109	34.2	42.7	45	5.7	7.2	19	22.6	28.3
172	62.2	77.7	108	33.7	42.2	44	5.3	6.6	20	23.1	28.8
171	61.7	77.2	107	33.3	41.6	43	4.8	6.1	21	23.5	29.4
170	61.3	76.6	106	32.8	41.1	42	4.4	5.5	22	24.0	30.0
169	60.8	76.1	105	32.4	40.5	41	4.0	5.0	23	24.4	30.5
168	60.4	75.5	104	32.0	40.0	40	3.5	4.4	24	24.8	31.1
167	60.0	75.0	103	31.5	39.4	39	3.1	3.8	25	25.3	31.6
166	59.5	74.4	102	31.1	38.8	38	2.6	3.3	26	25.7	32.2
165	59.1	73.8	101	30.6	38.3	37	2.2	2.7	27	26.2	32.7
164	58.6	73.3	100	30.2	37.7	36	1.7	2.2	28	26.6	33.3
163	58.2	72.7	99	29.7	37.2	35	1.3	1.6	28	27.1	33.8
162	57.7	72.2	98	29.3	36.6	34	0.8	1.1	30	27.5	24.4
161	57.3	71.6	97	28.8	36.1	33	0.4	0.5	31	28.4	35.0
160	56.8	71.1	96	28.4	35.5	32	0.0	0.0	32	28.0	35.5
159	56.4	70.5	95	28.0	35.0	31	0.4	0.5	33	28.8	36.1
158	56.0	70.0	94	27.5	34.4	30	0.8	1.1	34	29.3	36.6
157	55.5	69.4	93	27.1	33.8	29	1.3	1.6	35	29.7	37.2
156	55.1	68.8	92	26.6	33.3	28	1.7	2.2	36	30.2	37.7
155	54.6	68.3	91	26.2	32.7	27	2.2	2.7	37	30.6	38.3
154	54.2	67.7	90	25.7	32.2	26	2.6	3.3	38	31.1	38.8
153	53.7	67.2	89	25.3	31.6	25	3.1	3.8	39	31.5	39.4
152	53.3	66.6	88	24.8	31.1	24	3.5	4.4	40	32.0	40.0
151	52.8	66.1	87	24.4	30.5	23	4.0	5.0			
150	52.4	65.5	86	24.0	30.0	22	4.4	5.5			
149	52.0	65.0	85	23.5	29.4	21	4.8	6.1			

CHAPTER XXXVII.

WEIGHTS AND MEASURES.

THE convenience of the decimal divisions has induced in this work the so frequent use of the French denominations of weight, that it becomes necessary to give the reader some clues by which to know or estimate their values. The *unit* of the French weight is the *gramme*, which is the weight of the hundredth part of a cubic metre of distilled water, at the temperature of melting ice. Below are some comparative tables which will serve to give every explanation.

		Grammes.		Troy grains.
Milligramme	=	.001	=	.01543
Centigramme	=	.01	=	.15434
Decigramme	=	.1	=	1.5434

Troy weight.

English weights.			French weights.	
1 grain	=	$\frac{1}{24}$ dwt.	= 0.06477	gramme
1 pennyweight	=	$\frac{1}{20}$th of an ounce	= 1.55456	gramme
1 ounce	=	$\frac{1}{12}$th of a lb. Troy	=31.09130	grammes
1 pound imperial			= 0.3730956	kilogramme.

Avoirdupois weight.

English.				French.	
1 drachm	=	$\frac{1}{16}$th of an ounce	=	1.7712	gramme
1 ounce	=	$\frac{1}{16}$th of a pound	=	28.3384	grammes
1 pound or 1 pound imperial			=	0.4534148	kilogramme
1 cwt.	=	112 pounds	=	50.7824600	kilogrammes
1 ton	=	20 cwts.	=	1015.6490000	kilogrammes.

		Grammes.		Troy grains.
Gramme	=	1	=	15.434
Decigramme	=	10	=	154.34
Hectogramme	=	100	=	1543.4
Kilogramme	=	1000	=	15434
Myriagramme	=	10000	=	154340

Or:

French. English.

1 gramme=15.438 grains troy=0.643 dwts.=0.03216 oz. Troy
1 kilogramme=2.68027 lbs.=2 lbs. 8 oz. 3 dwts. 6 grs. Troy wt.
1 kilogramme=2.20548 lbs.=2 lbs. 3 oz. 4⅘ drs. Avoirdupois
1 myriagramme=22.0485 lbs. Avoirdupois
1 quintal=1 cwt. 3 qrs. 25 lbs.

As regards *measures*, the French system, which, by reason of its decimal subdivisions is the most convenient standard, has for its unit of superficial measure, "the *are*, a surface of ten mètres each way, or 100 square mètres. The unit of measures of capacity is the *litre*, a vessel containing the cube of a tenth part of the mètre, and equivalent to 0.220097 parts of the British imperial gallon. The standard temperature is 32° F. All the divisions and multiples of the units are decimal."

The measures of length are,

Myriamètre	=	10000	mètres
Kilomètre	=	1000	"
Hectomètre	=	100	"
Mètre	=	1	"
Décimètre	=	0.1	"
Centimètre	=	0.01	"
Millimètre	=	0.001	

The measures of surface are,

Hectare	=	1000	sq. mètres
Are	=	100	"
Centiare	=	1	"

The measures of capacity are,

Kilolitre	=	1000 litres
Hectolitre	=	100 "
Decalitre	=	10 "
Litre	=	1 "
Decilitre	=	0.1
Centilitre	=	0.01

The unit of solid measure is the *stere*, or cube of the mètre, equal to 35.31658 English cubic feet.

Measures of length. Long measure.

English.		French.
1 inch or $\frac{1}{36}$ of a yard - -	=	2.539954 centimètres
1 foot equal=$\frac{1}{3}$ of a yard=12 inches	=	3.0479449 decimètres
1 yard=3 feet - - -	=	0.91438348 mètre
1 fathom=2 yards - - -	=	1.82876696 "
1 pole or perch=$5\frac{1}{2}$ yards -	=	5.02911000 mètres
1 furlong=220 yards - -	=	201.16437000 "
1 mile, or 1760 yards - -	=	1609.31490000 "

French.		English.		
1 millimètre	=	0.039370		
1 centimètre	=	0.393708		
1 décimètre	=	3.937079		
1 mètre	=	39.37079	=	1.093633 yards
1 décamètre	=	393.7079	=	10.936630 "
1 hectomètre	=	3937.079	=	109.366300 "
1 kilomètre	=	39370.79	=	4 furlongs, 213.633 yards
1 myriamètre	=	393707.9	=	6 milés, 1 furlong, 156.288 yards.

French. English.

1 toise=6.3945 feet=2.1315 yards=76.735 inches

1 aune or ell=3.893 feet=46.79 English inches.

Square measure.

English.		French.
1 square yard - - - - -	=	0.836097 mètre carré
1 rod or pole=$30\frac{1}{4}$ sq. yards -	=	25.291939 mètres carrés
1 rood=1210 sq. yards - -	=	10.116775 ares
1 acre=4840 sq. yards=40.4671 ares	=	0.404671 hectare.

French.	English.
1 mètre carré=1 centiare - -	=1.196033 sq. yard
1 are=3.95 English poles - -	=0.98845 rood
1 hectare=2 acres, 1 rood, 5 perches,	=2.473614 acres.

Measures for liquids.

English.		French.	
1 pint or $\frac{1}{8}$th of a gallon	=	0.567932	litre
1 quart or $\frac{1}{4}$th of a gallon	=	1.135864	litres
1 imperial gallon	=	4.5434579	"

Dry measure.

English.				French.	
1 peck	= 2 gallons		=	9.0869159	litres
1 bushel	= 8	"	=	36.347664	"
1 sack	= 3 bushels		=	1.09043	hectolitre
1 quarter	= 8	"	=	2.907813	hectolitres
1 chaldron	=12 sacks		=	13.08516	"

French.	English.	
1 litre=1.760773 pints=.8803865 qts.=	.2200966	gallons
1 decilitre - - - -	= 2.2009667	"
1 hectolitre - - - -	=22.0096670	"

As a greater convenience for common purposes, the French denominations were, in 1812, arranged as follows:

1 toise or 6 feet=2 mètres	= 6.5618334	English	feet
1 foot or 12 inches=$\frac{1}{3}$ mètre	= 1.0936389	"	foot
1 inch or 12 lines -	= 1.0936389	"	inch
1 line - - -	= 0.0911365	"	"

1 aune or ell=1$\frac{1}{5}$ mètre	= 3.937	English	feet :—Or
1 aune - -	=47.244	"	inches

1 bushel=$\frac{1}{8}$ hectolitre=762.85 cubic inches.

1 old Paris foot	= 1.066	English	foot
1 old Paris inch	=1.066	"	inch
1 old line	=0.0888	"	"

The metre, the square metre, and the cubic metre are the radical standards of the three measures; for there are only three, as solidity and capacity, differently named and used, are the same in reality. From these radical denominations, namely, the *metre* of the lineal measure, the *are* of the superficial measure, the *litre* of measured capacity, and the *stere* of solid or cubic admeasurement, the larger ones are procured by multiplying by ten and the lower ones by dividing by the same. Thus,

Deca,	prefixed,	means	10	times
Hecto	"	"	100	"
Kilo	"	"	1000	"
Myria	"	"	10,000	"

The number is understood to multiply the surface of the solid and not its side; thus, one decare is ten ares, not a square of ten times the side of an are, and so of the others.

The denominations below the radical ones are expressed by a sort of Latin prefix: thus,

Deci is one-tenth.
Centi is one-hundredth.
Milli is one-thousandth.

It may be some little trouble to get in the way of using these tables, but when this experience is once accomplished, they will be found very convenient in calculating proportional quantities; it is for these reasons that they have been preferred by the author. In reciting the French recipes, those who dislike the change can, by reference to the tables, see the value of the new standards side by side with that of the old denominations, so that if their introduction is not an improvement upon established usage, it is free from any just charge of being an innovation, for we present both standards, with the choice of either.

PART II.

CHAPTER XXXVIII.

CANDLES.

THE invention of gas and its introduction into many of the larger cities of the civilized world, have greatly restricted the manufacture of candles, which, anterior to this discovery, were the most prevalent means of illumination. A candle is, as well known, nothing more than a fusible solid cylinder with a wick of cotton in its centre to aid its combustion, but as the melting point of bodies applicable to this purpose is variable, so also do the candles differ in quality according as their material is more or less fusible. The artificial light which a burning candle emits is dependent upon the combustion of the materials of which it is made. This being melted by the radiated heat of the flame, and drawn up whilst fluid to the heated part of the wick, in consecutive portions, by the power of capillarity, serves as food, which is immediately consumed to make way for successive portions. A candle may in truth be called "a portable gas apparatus," because it carries its own nourishment or combustible, and melts it in quantity as wanted, hence the necessity of apportioning a proper wick. The more fusible fats require a large wick, because it must be of sufficient size to suck up and consume all the melted fat which accumulates in the cup formed at the top of the candle, otherwise there will be a running or guttering. This shows the superiority of the harder fats for candle purposes, because, being less fusible, they require a thinner wick, and consequently, whilst the intensity of light is unimpaired, there is a less consumption of combustible matter. Candles with unsuitable wicks are defective. If the wick is too large they give a flaring and a smoky flame, whilst a deficiency in its size and number of threads, causes the candle to burn dimly and gutter. Candles are in burning quality proportional to the fusibility of their combustible matter. The variety of materials of which candles are manufacfactured, will hereafter be commented upon, each separately, and the requisite directions for their purification and improvement, thoroughly explained, in detail, under their respective heads.

"Simple as is the contrivance of candles, they do not appear to have been generally known to the ancients, who continued to make use of the lamp only. We read, however, of a species of candles sometimes used among the Romans, made of strings of papyrus, or rushes dipped in pitch and surrounded with wax. Candles of wax and tallow, were, according to Pliny, likewise occasionally employed in religious offices. In the twelfth century, those of wax were generally seen in churches, and as refinement increased, they came gradually into use among the nobility and wealthy all over Europe, as did those of tallow among the middle classes."

The great advantage in the cleanliness and portability of good candles, has retained for them a very extensive consumption, more especially in those places where the limited population has as yet prevented the introduction of gas. We cannot tell the quantity manufactured per annum, but as the trade in sperm and stearic candles, belongs almost exclusively to the United States, which, in that of tallow, also has a lion's share, it can be judged how extended and important is this branch of art, not only to the interests of individual manufacturers, but to commerce generally, when there is scarcely a ship that leaves for a southern port, but is partially freighted with candles.

The excise duty which existed upon candles in England, until 1831, was then taken off, so that the impediment to a competition on their part is in a slight degree removed if they had the power of practising it; but their home consumption is so large, and their deficit in tallow so great, that it has to be supplied by this and other countries. Notwithstanding the improvements in lamps, and the invention of burning fluids, there is no fear of the consumption of candles depreciating lower than what it is at present, for the regions of the West are filling up so rapidly, as to afford a market for every pound that is now made. Besides, there are certain circumstances, under which there is no substitute for them except gas; for instance, in the printing offices, working shops, and other situations where the operative requires a light immediately neighboring to him it would be too expensive to furnish a lamp for each. Moreover the tidy housewife will gladly banish from her parlor or sitting room a greasy troublesome lamp, when she can use a graceful looking, handsome candle at as economical a rate. To sustain the successful career of the art depends then upon the skill and enterprise of the chandler, at whose option, a successful competition with oil may be continued, provided his ingenuity, in the production of a candle, can keep pace with the times in respect to improvement and economy.

As it may sometimes be desired to contrast the relative value of different candles as illuminating agents, the determination can readily be arrived at by the well known method of the comparison of shadows. A more accurate means is by a *photometer* or measurer of light, but the skill required for its management renders it inefficient in other hands than those of experienced manipulators, so we confine our notice to the most ready and available method. To ascertain which candle gives the greatest intensity of light, erect in a darkened room at six or eight inches distant from the flame, a board of thirty-two centimetres length by thirteen millimetres breadth, and pierced with two holes, being careful that the flame of the candle shall be immediately opposite the centre of the hole. Placed behind, at equal distances from the plank, and in a direct line with the holes are two screen frames tightly and smoothly stretched over with white paper. Upon these screens, standing upright and moving on castors is received the light: they are then gradually pushed back until the shadows projected upon them are of equal intensity. The candle which throws the most distant shadow is the better in the ratio of the square of the distances.

Other minutiæ in respect to the subject will be mentioned and described in their appropriate places.

CHAPTER XXXIX.

ILLUMINATION.

ARTIFICIAL illumination, the economical production and diffusion of which are matters of general and individual interest, is obtained for domestic purposes from such materials as will by combustion afford a permanently luminous flame. The term combustion, as usually applied, is a rapid combination of a combustible body with oxygen, with an attendant disengagement of heat and light, resulting in the destruction of the burning body and the formation of new compounds. We cannot positively state more in explanation of the causes which lead to this dis engagement of heat and light, because the conflict of the various plausible theories leaves a doubt upon the subject. Certain it is, however, that chemical combination is always a source of heat and light; but whether these phenomena of combustion are explicable upon certain changes in the amount of latent heat in bodies entering into combination, or, by Berzelius's theory of its origin, "from the oppositely electrical substances neutralizing one another, in the same manner as the electric equilibrium is restored, during the discharge of a Leyden jar," we will leave to the decision of philosophers, as it is not in the author's power to settle this mooted point.

Oxidation is a species of combustion, but in this process the union with oxygen is not sufficiently rapid to produce a *perceptible* evolution of heat and flame. A body having so great an affinity for oxygen, that its union therewith will be attended by an emission of heat, is called a *combustible*, and the oxygen or other gas in which it is burned, is styled a *supporter* of combustion. The evolution of heat is apparently proportional to the amount of oxygen consumed, but varies with the nature of the material. A candle burns much faster in an atmosphere of oxygen than in that of common air, but the increase of heat and light is in a corresponding ratio. That species of oxidation taking place with those bodies, which burn with flame, at a heat less than that of redness, is called *low combustion*. Thus, when tallow or oil is thrown upon a heated iron plate, not visibly red-hot, it melts and emits a pale lambent flame, only perceptible in the dark. The same occurrence is evident, when a portion of oil is heated in an iron capsule until it catches fire, and the flame then extinguished; the hot oil still continues in a state of low combustion, and is ready for thorough ignition immediately upon the application of an atmosphere of oxygen, and so also a candle recently blown out, although no point of the wick remains red, is, upon the same reasoning, sometimes rekindled in oxygen. Here we have the explanation of the apparently unaccountable causes of frequent fires in store houses for woolen and other rags more or less imbued with oil. The large surface exposed to the air so facilitates this low combustion, that the oil becomes converted into resin, and the rapid absorption of oxygen producing this phenomenon, generates sufficient heat to cause a spontaneous ignition of the combustible. This affinity of combustible bodies for oxygen is developed only at high temperatures, and hence in most cases it must receive a *start*, if the expression may be used, or, in other words, to commence the combustion, the combustible should be heated to a certain point, but as this degree is far below that emitted during the combustion, it follows that when once ignited, it continues burning until its entire destruction. But it is not meant to convey by the word destruction, the idea of the total loss of the material, or the perfect annihilation of any particle thereof, for this would be contrary to fact; nothing is lost, there being merely a change of state or resolution of the solid or liquid bodies into gaseous matter and vapor.

The modes by which artificial light is at present produced, are three, viz., gas, lamps, and candles. The principles upon which their illuminating power is based, are identical in each of them; the variation in fatty material and the modification of the burning apparatus, are foreign to the theory of combustion, their part and object being only to facilitate its progress, and thus influence the economy of the light. Granting, then, that all the ordinary substances from which we obtain light may be represented by gas and oil, let us proceed to explain the rationale of their combustion, which furnishes us the flame that bears so importantly upon the domestic economy; for without it, to say nothing of the pleasure and convenience it affords, there would be a loss of that time now vouchsafed to us by these means, and which would otherwise form an incalculable deficit in the life period of mankind.

In the language of Prof. Faraday, let us suppose we have at hand for the purpose of obtaining light, both oil and olefiant gas. Although wax, tallow, or anything of that sort, be used for the purpose, yet they all fall under one or other of these denominations, inasmuch as they are like the vapor of oil. These are the two substances to which we generally refer the light. The oil contains carbon, hydrogen and oxygen, and the gas only carbon and hydrogen. The composition of oil is given by Dr. Ure, in 100 parts, as follows,

Carbon - - - - -	78.0
Hydrogen - - - - -	11.5
Oxygen - - - - -	10.5

The olefiant gas is that made from coal, and burned in our streets and stores as a means of illumination. Its composition, per cent., is

Carbon - - - - - - -	70.
Hydrogen - - - - - -	30.

Let us suppose that we have these two things represented, one by a vessel containing hydrogen, and the other by a piece of charcoal existing separately, and in their elementary states.

The constituents of the two bodies are given now, but the oxygen present in one of them will be passed by, inasmuch as it merges into the other bodies when burning in the air. Consider the hydrogen, which is a very light gaseous body, as burning; the carbon is ready at hand, and the combination of the two will be explained directly. The charcoal can be burned in the oxygen of the air, or pure oxygen prepared for the purpose. The oxygen of the air is identical with the gaseous oxygen, and these two instances illustrate all the modes in which is burned every one of the combustible substances, either in the ordinary way, or by extrordinary means, for the production of light. Whether we burn the oil or the gas, the same thing happens in all these cases, the same final result takes place as in this particular instance. For example, there are the carbon and hydrogen, combined in the form of gas, and its combustion partakes in its character of the two compounded—viz., of the flame of the hydrogen, and the luminosity of the carbon, combined in one act.

Now when these two bodies are burned, as they are in either the oil or gas lamps, the hydrogen produces water, which is a compound of hydrogen and oxygen, and carbonic acid, constituted of carbon and oxygen, as is proved by receiving the products of the combustion of a lamp into a glass jar, and testing them. But what are the changes in the atmosphere, which, under all these circumstances, is supporting the combustion of the light, and what is it that is thrown into that atmosphere from the oil or gas which is burnt? The oil, by reason of the hydrogen it contains, produces water by the union with that portion of oxygen which it takes from the atmosphere, a fact very easily determinable by holding a glass jar over a jet of ignited hydrogen. The light takes the oxygen from the atmosphere, and forms with it water which condenses in vapor upon the sides of the vessel. The carbon by the same cause is converted into carbonic acid, a vicious gas, destructive of life and insupportable of combustion, and hence the dimness of light, after a lengthy burning in close apartments, is partially owing to the presence of this body, and partly to a deficiency of oxygen, by reason of the absence of all inlet for fresh air, the source which furnishes it.

The committee of the English House of Commons, having proposed the inquiry as to the relative vitiation of air by the breathing of men and the burning of candles, received from Mr. Ure, one of the chemists to which it was addressed, the following reply.

"Wax contains 81.75 parts of carbon in 100, which generate by combustion 300 parts of carbonic acid gas. Now, since 125 grains of wax constitute the average consumption of a candle per hour, these will generate 375 grains of carbonic; equivalent in volume to 800 cubic inches of gas. According to the most exact experiments on respiration, a man of ordinary size discharges from his lungs 1.632 cubic inches of carbonic acid gas per hour, which is very nearly double the quantity produced from the wax candle. Hence the combustion of two such candles vitiates the air much the same as the breathing of one man. A tallow candle, three or four in the pound, generates nearly the same quantity of carbonic acid as the wax candle; for though tallow contains but 79 per cent. of carbon instead of 81.75, it yet consumes so much faster as thereby to compensate fully for this difference. When a tallow candle of six to the pound, is not snuffed, it loses in intensity in thirty minutes, eighty hundredths, and in thirty-nine minutes, eighty-six hundredths, in which dim state it remains stationary, yet still consuming nearly the same proportion of tallow. A wax candle attains to its greatest intensity of light when its wick has reached the greatest length, and begins to bend out of the flame. The reason of this difference is, that only the lower part of the wick in the tallow candle is charged with the fat so as to emit luminiferous vapor, while the upper part remains dry; whereas in the wax candle, the combustible substance being less fusible and volatile, allows a greater length of the wick to be charged by capillary attraction, and of course to emit a longer train of light."

The following table expresses, according to Péclet, the illuminating powers of different candles, and their consumption of material in an hour; the light emitted by a *carcel* argand lamp, consuming 42 grammes (=42+15½ grains) in an hour, being called 100:—

	Intensity of light.	Consumption per hour.
Tallow candles, 6 in pound	10.66	8.51
Stearin or pressed tallow, 8 in pound	8.74	7.51
" " 5 "	7.50	7.42
Wax candles, 5 in pound	13.61	8.71
Spermaceti candles, 5 in pound	14.40	8.92
Stearic acid or adamantine, 5 in pound	14.40	9.33

The subjoined table shows the economical ratios of the candles, where the second column gives the quantity of material in grammes, which is requisite to produce as much light as the Carcel lamp.

	Quantity of material.	Price per kilogramme.	Cost of light per hour.
Tallow candles, 6 per pound	70.35	1 f. 40 c.	9.8 c.
" 8 "	85.92	1 f. 40 c.	12.0 c.
Pressed tallow, 5 "	98.93	2 f. 40 c.	23.7 c.
Wax candle, 5 per pound	64.04	7 f. 60 c.	48.6 c.
Spermaceti candles, 5 per pound	61.94	7 f. 60 c.	47.8 c.
Stearic candle, 5 per pound	65.24	6 f.	37.1 c.

These results can be compared with those given above; thus, a kilogramme or 1000 grammes = 15.440 grains = 2¼ pounds avoirdupois.

A series of experiments, with a design of ascertaining the relative illuminating power of the different kinds of lights, was not long since undertaken by Mr. Ure, who has furnished the following tabular view of the cost per hour of an illumination equal to that of the mechanical lamp of Carcel, reckoned 100, or that of eleven wax candles, three to the pound.

TABLE OF COST PER HOUR OF ONE HUNDRED OF ILLUMINATION.

	Pence per lb.	Pence.		Pence.
1. Parker's hot oil lamp with Southern whale oil -	3¼	0.4875	or about	½
2. Mechanical, or Carcel lamp, with sperm oil -	11	1.2804	"	1¼
3. Parker's hot oil lamp, with sperm oil - -	11	0.902	"	1
4. " " " common olive oil -	7½	0.900	"	1
5. " " " cocoa-nut olein or oil	5¾	1.031	"	1
6. French lamp in general use, with sperm oil -	11	1.7072	"	1¾
7. Wax candles - - - - - - - -	30	5.892	"	6
8. Spermaceti candles - - - - - - -	24	5.352	"	5¼
9. Stearic acid " - - - - - - -	16	4.224	"	4¼
10. Palmer's spreading wick candles - - -	10	2.800	"	2¾
11. Tallow (mould) candles - - - - - -	9	2.520	"	2½
12. Cocoa-nut stearin of Price & Co., - - -	10	4.41	"	4½

All kinds of artificial light are designated by the term terrestrial light, in contradistinction to solar light, or that proceeding from the sun. The chemical agency of each is analogous, though the terrestrial light is much the feebler of the two. The source of this latter kind of light is the combustion of inflammable matter, the excessive heat, given out during the process of which, renders the burning substance luminous. This luminosity increases proportional to the intensity of the heat, and augments from a dull red to one of shining brilliancy, and to this incandescence of their material in a gaseous state, are to be attributed the different kinds of flame, for instance, of candles and gas light, serving for the purposes of illumination. In conclusion, therefore, the illuminating power of the coal gas and of oils and candle materials, is due to their constituency of carbon and hydrogen, assisted by the oxygen of the atmosphere, which converts the first into carbonic acid, and the latter into water.

CHAPTER XL.

THE PHILOSOPHY OF FLAME.

It has already been written, that the temperatures at which bodies enter into rapid combustion, are very various, but when this kindling or *accendible* point, so called, is attained by inflammable materials, they take fire and burn with a flame. All flames are not equally luminous, for instance, that of pure hydrogen gas is scarcely visible in the day time, but still is so hot as to impart a white heat to a piece of metal brought in contact with it. The luminosity of flame, therefore, does not depend upon its temperature, but is due to incandescent solid matter, ignited by the inflamed hydrogen of the combustible bodies; and this solid matter, to which the light of all the common flames is owing, is finely divided carbon. Davy called a luminous flame, *solid matter heated to whiteness*, and as an illustration of this fact we have the "Drummond light," which is produced by the projection of ignited hydrogen gas in contact with oxygen upon lime. The instance of the combustion of metallic zinc illustrates the incandescence of flame as due to the solid particles of the oxide of that metal diffused throughout the flame, and serving as radiating points for the light and heat. The flame of gas is then owing to the particles of carbon which compose it, being rendered incandescent by the ignited hydrogen, and in that state burned and passed off as invisible carbonic acid gas. In all instances, therefore, of a bright light during combustion, there is evidence of one of the combustible bodies being solid, for to its incandescence the flame is owing.

In the flame of a candle a portion of the carbon is in a solid form, that of smoke, without which the light given out would be very limited, for it is this portion of heated carbon that radiates the light. When, however, the supply of air is insufficient, the imperfect combustion of this smoke causes a disagreeable increase of unconsumed particles, which, escaping in their heated state, cool the flame by an abstraction of its heat, and thus create a loss of the burning material, and an unpleasant atmosphere in the apartment where it is burned.

To provide against this inconvenience in the case of oils, the Argand lamp was constructed, so as to perfect the combustion of the excessive carbonaceous matter by allowing a current of air through the centre of the flame, because, in order to a clear and brilliant flame, there must be a complete consumption of the carbon, or, in other words, it must have oxygen enough to convert its carbon into carbonic acid gas. The larger the flame, the greater the quantity of light, and its brightness is increased by every circumstance promotive of the rapidity and intensity of combustion, for instance, as in the Argand burner, by a stream of air, or, as in the Bude or Gurney light, by a current of pure oxygen gas. The light is not only thus augmented by the accelerated combustion, but by the carbon being more highly ignited by the increased temperature of the flame.

"In an ordinary flame, such as that of a spirit lamp or candle, the gaseous matter is in free contact with the air only at the exterior of the flame, where alone perfect combustion takes place. In the interior, combustible gases still unburned, exist. That such is the structure of the flame may be easily seen by depressing it upon a sheet of wire gauze of small mesh (through which the flame is unable to pass), so as to view it in section. The combustion is then seen to be limited to the margin of the flame, because it occurs only at the point where the two substances which enter into union are in contact. But if the structure is examined a little more minutely, it is seen to consist of three parts which are represented

Fig. 26.

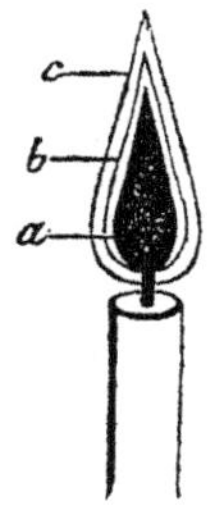

in Fig. 26. 1. The shaded interior represents the unburned combustible, and may be styled the cone of vaporized combustible; 2, around this is the brilliant part of the flame, or the flame, strictly so called; 3, on the very exterior is another portion, perceptible, with attention in the flame of a tallow candle,—this is the sphere of complete combustion."

"In *b*, where the supply of air is insufficient for complete combustion, it is the hydrogen principally, which burns, the carbon being liberated in solid particles, which are heated while hot, by the combustion of the gas. The sphere *b* is the luminous part of the flame, in truth, for the light depends entirely upon the deposition of carbon arising from the consecutive combustion of the two elements of the vapor. The carbon does not burn entirely until it reaches the exterior of the flame."

But to maintain a steady continuance of the flame and a uniform light, there must be consecutive supplies of the combustible material, which serves as its fuel, and in the instance of candles, the fat or fuel which must be in a liquid state, is conveyed to the flame by means of a wick, where it is volatilized and raised to its accendible point. This fat is rendered fluid in that part of the candle immediately below the lighted end of the wick, by the radiant heat of the flame. As soon as that portion taken up to the point of combustion is consumed, another consecutive portion melts and arises to take its place, and so the candle, when once lighted, continues to burn until it is spent. The dim light given out on first being inflamed, is owing to the deficiency of food, but as the size of the flame increases, more fat is melted, and the wick is then better, and more perfectly supplied with fuel. If the wick is either too large or too small, the burning quality of the candle will be impaired. The continued and uniform flow of melted fat is supplied to the wick by the force of capillarity.

The smoke arising from the burning of badly made candles, consists of unburned carbon in a state of very minute division, and is occasioned by a deficiency of oxygen to the centre of the flame. The water produced by the burning candle is dissipated throughout the atmosphere in the form of heated vapor;—the carbonic acid simultaneously generated, making its escape in a gaseous form. In close apartments, surrounded by a cold atmosphere, the water is condensed upon the window glasses or the walls, and where any number of lights are burned, is in such quantity, as to moisten a cloth rubbed over a limited space, whilst the carbonic acid, being destructive of life, vitiates the atmosphere, proportionally as it accumulates;—hence the necessity of a thorough ventilation of those lighted apartments that are occupied by crowded assemblies. These remarks conclude the subject, and with those preceding, upon illumination, will serve to furnish a correct idea of the principles upon which light is yielded by the burning of oil and candles.

CHAPTER XLI.

RAW MATERIAL FOR CANDLES.

THE grease of animals, wax, palm oil, and butter of cocoa, are almost the only fats used in the composition of candles, but for this purpose, more especially the first named, require a previous preparation or purification. The consistence of fats, their color and odor, vary with the animal whence they are taken. They are solid and inodorous in the ruminant; soft and strong smelling in the carnivorous, and generally fluid in the cetaceous. The fat of young animals is white and abundant;—less so in those which are older. The fat which is secreted in the neighborhood of the kidneys, and under the skin, is firmer than that about the movable viscera. The chandler whose desire is to produce white, solid, sonorous candles, must use some caution in the selection of stock, and choose the most consistent, but not too brittle tallow;—that by preference procured from the ruminants. Beef and mutton suet are therefore the better articles. The first denomination comprises also the suet of the bull and cow, not forgetting, however, that beef suet is softer than that of the two former animals; under the latter title, the fat of the ram, ewe, he and she goat is included;—the fat of the last two being the more solid of all. Experience has proved that the nature of the food has much influence upon the quality of the fats, but this difference is unimportant, when the fat of the ruminants is selected as above directed. It is an established fact, however, that animals killed in winter give a better fat for candles than those yield which are slaughtered in summer. The fat, as it comes from the animals, is enveloped in membranes, and enclosed in cellular tissues, and in that state is called "rough tallow or suet," in which form there is much more facility in judging its source and quality, than after it has been melted. As soon as the butcher separates the fat from the carcass, it is strung upon sticks and hung up to dry; but it must not be left in this condition too long a time, for, in warm weather especially, the blood which it always retains, putrefies, and the suet contracts a bad odor. It is then very important to melt the suets as quickly as possible after the animal has been killed. For this purpose, exclusively, there are large tallow refineries erected, and to these depots, or to the extensive chandler, who pursues the trade through all its different operations, and renders for himself, the rough tallow is carried and sold. The refiners dispose of the product to the smaller manufacturers, and to shipping merchants for export. The most improved modes of rendering the tallow are given in detail in other parts of the book.

In large cities, the chief source of the rough tallow is the butcher, and the only way to amass any amount of stock is to do as is generally done by the refiners, gather it from the stalls of each on market days. In New York, the butchers have an association of their own, and possess a large foundery exclusively for the rendering of tallow; it is here they send their rough suet, and being then credited with the quantity, receive an equivalent share of profit corresponding with the market price of the refined tallow. At the West, whence the most of the tallow comes, there is a refining room to every packing house, and the daily product of rough suet is rendered therein, and packed in barrels and hogsheads to take the same course of the meat whence it was extracted, viz., either to an eastern or foreign market. Those using the most care in its manufacture, generally receive a preference from purchasers, and hence the high reputation of "Wilson's" brand, of which mention will be made hereafter. There is an immense trade done in tallow, but as we have no means of accurately determining its amount, we forbear giving any approximate statistics. The fact of its being a cash article, and always of ready sale, is an evidence that there is a healthy demand for the whole product of the country. All particulars as regards its quality, or other circumstances connected with it, will be found in the following pages.

THE MODES OF RENDERING THE FATS.—We have said that the refineries furnish the most of the rendered tallow, but we further mentioned that those manufacturers whose capital is sufficient, prefer saving the profit they would have to pay others, and refine their own tallow and suet. To make our work complete, here follow the plans, most approved, for the purpose. There is an advantage too, in a factor's preparing his own stock, and for this reason, that his experience will enable a better selection of rough material, and more care and exactness in its purification, both matters of importance, when it is considered, that there are often much loss and inconvenience to the chandler by the bad quality of tallow. The fat as it comes from the animal is warm, but when suspended in a proper drying room, it cools and congeals, the blood and membranes become dry, and the suet can then be extracted more readily. Care should be taken not to leave it too long, nor must the temperature of the room be elevated, lest putrefaction of the blood ensues, and the tallow becomes spoiled. The danger of these defects happening, is much diminished when the fat is dry. It is in this state carried to the melting room, but before proceeding further, we will give you a description of both that and the drying-room, which, for convenience, should adjoin each other.

The drying-room is an apartment more or less in size, and proportional to that of the manufactory. It is filled with long poles, stretched across its breadth, and suspended at each of their ends, to the ceiling, by two cords. The walls are pierced on either side with holes for ventilation, and free circulation of air. It is important that the drying should complete quickly, in order to avoid the putrefaction before alluded to. It is upon these rods that the rough tallow is hung as soon as it is taken

Fig. 27.

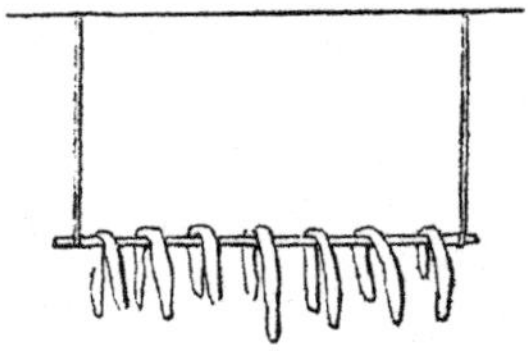

Fig. 28.

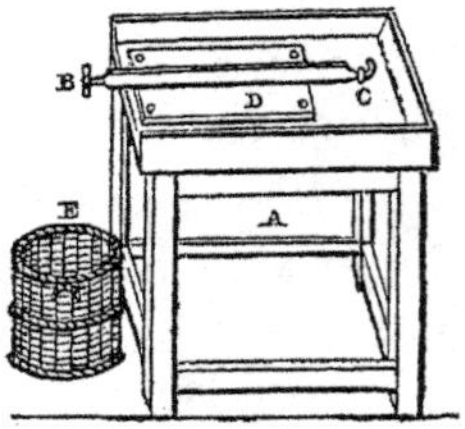

from the animal. Fig. 27 shows the arrangement. A corner of the drying-room should be appropriated to the hachoir, for, by this arrangement, independent of the convenience it affords, the workman is enabled to reject the undried portions, and hang them up again for further desiccation, without much loss of time. The hachoir is a strong table A (Fig. 28), upon which is fastened, by one of its ends, a chopping-knife B, propelled vertically, but in a circular direction around the centre C. This knife is more convenient and expeditious than a cleaver, which is sometimes used, and by awkward management becomes dangerous in unskillful hands. It is advantageous to fix upon the table where the edge of the knife constantly comes down, a piece of beech wood (plank D) of an inch thickness, and fastened by screws, so that when it becomes worn out by repeated hacking, it can be replaced by a new piece. In this way the top of the table is saved, and moreover, as it becomes thin by repeated planing, can be renewed for a less price than the top of the table. It is this economy in small matters that leads the proprietor to prosperous fortune, and it is only by a strict adherence to that principle, that success is to be attained in any branch of manufacture. As a labor-saving substitute for this hachoir, I would recommend the circular combination of knives, used in the sperm candle factories, for cutting up the pressed cakes of the crude material. Its motion is rotary, and one workman, it seems to me, could accomplish as much with it in an hour, as by the other in a day. With this, or the first named machine, the workman cuts the rough fat into small pieces, and brushes them off into a basket E, placed under the table for their reception. This basket, when full, is carried to the melting caldron, which is placed in a brick-work furnace arranged against one of the walls of the melting room, and neighboring to a window. In a moderate sized factory, this caldron is about three feet in diameter, and two feet two inches deep. Its borders incline inwardly, that the tallow which, in melting, may scatter and spatter upwards against the sides, shall fall back again into the caldron. Fig. 29 shows the

Figs. 29, 30.

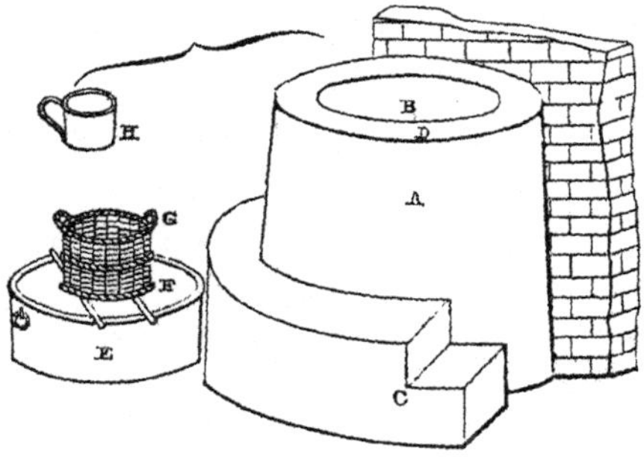

whole arrangement. At the base of the furnace are some steps C, which give facility to the workman in stirring or taking out the contents of the kettle. At A the furnace, and at B the caldron are seen. The hearth is not visible in the figure. Its entrance is behind the wall, against which the furnace is built. This precaution is necessary to prevent inconvenience of smoke, and to save the workman from being incommoded by the heat proceeding from the fire doors. The caldron is of copper, egg-shaped at the bottom, and so built in, that the heat only operates upon its base, and not upon its sides. There is always a bath of melted suet on the bottom, and this protects the metal from injury by fire, but the precaution alluded to is necessary, else the pieces of unmelted suet, coming in contact with the sides, would become scorched and acquire a brownish tint, of which the whole melting would partake, and thus the suet get so colored, that no mode of bleaching could restore its primitive purity and whiteness. As soon as the workman commences to charge the caldron with minced fat, the fire should be kindled under it, and moderately kept up. The workman now mounts the steps, and with his long wooden paddle, stirs around the contents until all is melted; this done, the bottom of the kettle is then covered with fluid fat, and the danger of scorching is diminished;—constant stirring, however, is necessary. New portions of rough suet are then added, and the same manipulations repeated again and again until the kettle becomes two-thirds full. Take care that the suet is perfectly melted before drawing it off, then with a long handled coppel dip it out and empty it into large copper vessels E, Fig. 30. Upon each of these is placed a frame F supporting a tallow-sieve G, through which the melted suet is strained. The sieve is most generally a fine willow basket, of cylindrical form, and so closely plaited as to prevent the passage of the membranes and other rough impurities; but a better substitute would be a copper box pierced with innumerable small holes, like a strainer or cullender, or still better, a brass wire sieve;—in this way much less grease is lost, there being none of the fluid tallow absorbed, as is the case with willow twigs. The frame F is merely four strips of wood jointed together, and should be strong enough to support the weight of the sieve when it is full. There are some workmen who place the sieve in the caldron and dip out the clear fluid which rises through its meshes, but this is not the proper way. As soon as a vessel is filled, it is taken into the foundery, and an empty one substituted. Cover each as it is filled, and allow sufficient repose that any foreign matters which it contains may, by their greater specific gravity, subside. As soon as this occurs, and before the tallow has congealed, the supernatant clear liquid is ladled off with the pail H into the small wooden pails. I I I (Fig. 31), ranged in a row upon the floor of the

Fig. 31.

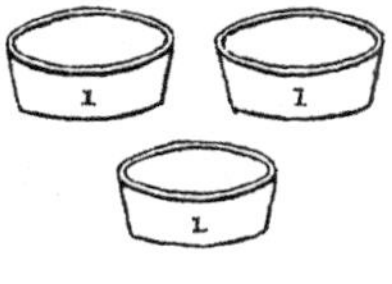

foundery, and at some distance from the furnace. When the tallow is to be exported, it should be run into casks, which, after their contents have thoroughly cooled, are to be tightly headed up and coopered. This then is *rendered* tallow. The form of the small cooling vessels is that of a flattened truncated cone to afford facility in ejecting the block of cooled tallow, by merely upsetting them after the contents have congealed. The dregs remaining at the bottom of the pan E, technically called *boulèe*, consist of the impurities contained in the grease.

To extract any remnant of fat that may still be adherent to the *boulée*, it is thrown into boiling water. The grease melts and rises to the surface, and can be ladled off, whilst the dirt, &c., falls to the bottom of the caldron. The *cracklings* or membranous matters which are left upon the sieve, contain even still more of fatty matter, which is only separable by strong pressure. The remaining marc after pressure, is used as food for dogs, but more generally, by reason of its azotized constitution, for the manufacture of prussiate of potassa and Prussian blue. I have seen lard cracklings resulting from the old methods of curing lard, which contained enough per centage of greasy matter not only to pay for the expense of its extraction, but their original cost additional.

The kind of press used for the purpose of refining the cracklings, is shown by Fig. 32. It is formed of two

Fig. 32.

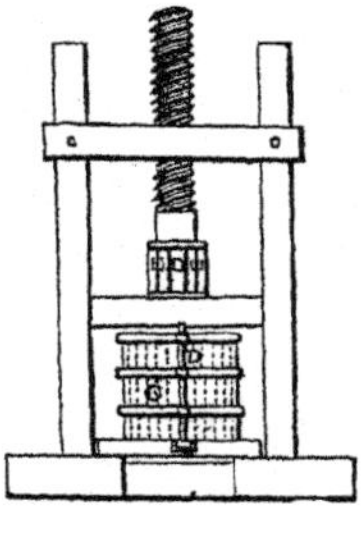

Fig. 33.

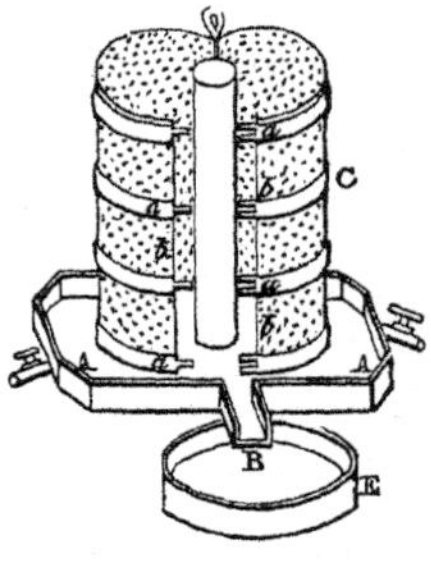

strong upright stanchions, and two proportionably strong cross pieces, firmly jointed in the side beams. The upper cross piece carries a box through which works an ordinary press screw, in the usual manner. Upon the lower cross piece, is placed a wooden trough A (Fig. 33) at least two inches deep, and to the front of which is adapted a gutter B for the conveyance of the liquid fat, which assembles in the trough, to a vessel E placed at and beneath its mouth. Upon and within the trough is placed a wrought iron boiler plate cylinder. This cylinder is formed of two semi-cylinders joined together. Throughout its height, it is divisioned off alternately into equal parts by zones or belts. The zones *a a* are a good inch broad;—the partition, *b*, &c., four or five inches in width. The top as well as the lower zone is narrow. All the wider divisions are cullendered throughout their circumference with innumerable small holes, through which the liquid suet is to flow when pressure is applied. All the narrow zones are secured by a strong wrought iron ring, formed of two pieces working on a hinge adjusted at the back. Upon the front is a movable broach D, which bolts them together, and makes the cylinder compact, so that it can resist the pressure applied. When the marc is exhausted, by draining out the broach D, the circumference of the cylinder is loosened or extended, so that its contents can be removed without difficulty.

This explanation finished, we proceed to give directions as to the *modus operandi.*

Whilst the marc or cracklings are yet warm, they must be taken from the sieve and placed in this cylinder and pressed out by the power of the screw, until nothing more of fluid fat will exude, even with the force of two men upon the lever or a tourniquet which can be arranged by the side of the press. The tallow runs into the gutter B and through a sieve, which should be properly placed for the purpose, into the vessel E and thence emptied after it has settled, into the conical tubs I, Fig. 31. The residual marc of exhausted cracklings is, as said above, easily emptied out by loosening and removing the pin D, but it is better not to disturb it or relax the pressure until it has entirely cooled. It is very doubtful whether by this operation, all the grease is separated; hence some other plan is requisite, and consists as follows. The shape of the cylinder is unaltered, but its diameter must be increased four inches. In the centre and upon the trough, is fastened a cylindrical wrought iron steam tube, reaching in length, just about an inch above the top of the cylinder. The tube tightly fastened and luted to the trough is hermetically closed, and connects with a pipe leading from a steam boiler, which furnishes vapor as may be wanted. A tube fitted with a cock, placed at the bottom, leads off the vapor as soon as it condenses into water. There should also be a stop-cock in the steam conduit, so that the steam may be shut off when necessary to do so. By these means the heat requisite to maintain fluidity being uniformly supplied, the screw pushes out every portion that can be extracted.

The different species of suet should be rendered separately; do not add any salt, for it decrepitates during the combustion of the candles, and causes them to run. The tallow keeps well enough without being salted, as the disagreeable odor it acquires by age will befall it nevertheless. Many processes have of late years been proposed for the purification or "rendering," as it is technically called, of crude tallow, and in consecutive order we make note of such as are worthy that attention. Much industry has been bestowed upon those points of the process, which required amelioration, with a view to producing that effect, as well as bettering the product, and removing much of the inconvenience to the workmen by reason of the disagreeable and unhealthy vapors escaping during the operations of rendering. The complaints of neighbors in those localities in Paris, where many of these rendering founderies are situated, some years since became so clamorous, that the proprietors, to protect themselves against the penalty of sustaining a nuisance, had recourse to professional advice as to ways and means of abating the evil complained of. Several chemists were consulted, and among the modes developed by their ingenuity, one was to conduct the offensive vapors arising from the rendering caldron, over a bed of burning coals. The vapors being rich in carbon and hydrogen, became themselves such good fuel, that the coals required no renewals. D'Arcet proposed melting the rough minced tallow in an open caldron along with water and sulphuric acid. It is true that this decreases the fetidity of the vapors, but experience has proved that the bad odor is still undestroyed. Acting upon the hints obtained, a series of experiments was entered into with the design of either condensing the unwholesome and diffusive vapors which are given off in the rendering operation, or else neutralizing their odor, whilst at the same time the resulting product of suet might at least be unimpaired, if not improved in quality. The first essay was to melt the suet in a close vessel, over a water bath, and under a certain pressure. For this purpose was used a vessel of one and a-half gallons capacity, and carrying a pressure of 100 pounds. Three litres of water, and one and a-half kilogramme of unminced rough suet having been put in, the whole was boiled over a furnace for forty minutes. At the end of that time, the apparatus being opened, it was seen that the suet had not entirely melted, many portions, though assuredly much reduced in bulk, still retaining their original state. It having been concluded, after thought in the matter, that these phenomena were owing to the insufficiently reduced volume of the suet, before it was put in, as well also to a too brief ebullition, the essay was repeated and the two conditions supplied. After an hour's boiling, the melting was complete, and the fluid tallow, pressed through a cloth with the hand, gave a residue, weighed when it was cold, of ninety-six grammes. By pressure, this marc would have been still less. The suet thus obtained was of good quality; but notwithstanding its prompt fusion, very disagreeable vapors were continually being emitted through the valve of the vessel, though in very small quantity compared to what would be thrown off in the usual way of rendering tallow in an open vessel.

Another essay, after the opinions and advice of D'Arcet, consisted in placing in the same marmite, after it had been well cleansed, 1½ kilogramme of rough suet, 750 grammes of water, and 124 grammes of sulphuric acid, and heating it up as in the preceding experiment; —but in twenty minutes the foaming of the mixture, a portion of which forced its way through the valve, soon condemned this method as impracticable. Still, in taking to pieces the apparatus, the suet was found to be entirely melted, and when it had cooled, was whiter and firmer than that obtained by the aforementioned method. The expectation that no inconvenience would be experienced by unpleasant vapors, in using this plan, was not realized, though it must be confessed that the odor given off was changed in its nature and much less deleterious than that arising during the other experiment. This experiment performed in an open vessel, gave the same results both as to the odor and quantity of suet. The melting took place in thirty minutes, and the residual cracklings or marc did not appear to have retained any particular greasiness. The many advantages accruing from this mode of rendering the tallow, both as regards its improved quality, the economy of time, the absence of the necessity of pressure, &c., suggested the policy of an arrangement for burning the fetid vapors as they were emitted, and thus render this process still more satisfactory. To evade the necessity of stirring the contents to prevent their scorching, and also the frequent opening of the vessel for this purpose, it had placed within, at about an inch from the bottom, a cullendered copper diaphragm, so that the materials were saved the danger of a too close contact with an over-heated metallic surface. Into the marmite thus arranged, was placed the same quantity as before, of rough suet, water and acid, and the whole hermetically closed;—the vessel being, however, fitted with a tube so crooked that its mouth of exit lead into the hearth of the furnace, and served as a conduit pipe for the vapors. This experiment was entirely successful; the odor was almost nullified; the vapors rushing forth freely through the pipe after having traversed the fire, retained only the smell analogous to that given off by water when thrown upon hot iron. The suet drawn out after half an hour's ebullition was completely melted, and on cooling became white, firm, and sonorous. The proportion of crackling remaining in the strainer cloth did not exceed ninety-six grammes. The suet not having influenced the litmus paper, the absence of all acid was conclusive.

The preceding experiment having induced the idea of a possibility of condensing the vapors instead of burning them, and thus evading the above-mentioned contingencies of accident, the following essay was made as a test.

Into an alembic furnished with a worm, were introduced seven and a-half kilogrammes of rough tallow or suet, with an equal quantity of non-acidulated water. No odor was perceptible during the whole operation, and the water resulting from the condensation of the vapors which passed over into the receiver, was clear and retentive of scarcely any smell of grease. An hour's ebullition melted the suet but imperfectly, and the necessity of acid being thus inferred, the requisite quantity was added;—a half hour's boiling then sufficed to complete the liquefaction. In this experiment the cracklings weighed but 224 grammes. The tallow of these last processes, wherein sulphuric acid was used, upon examination by competent chandlers, proved of superior quality, and furnished candles firmer, and whiter than those from the ordinarily prepared tallow. These candles, burned with a slightly diminished flame, perhaps, but without running, or any disagreeable smoke, and moreover, endured longer than candles from tallow prepared without sulphuric acid.

Thibault repeated these essays on a large scale, by melting one hundred kilogrammes of suet with the proportions of water and sulphuric acid above indicated, and obtained ninety-two of melted suet and eight of wastage; whilst a like quantity of the same suet, rendered simultaneously by the ordinary process, experienced fifteen per cent. of wastage. In these two essays, conducted in an open vessel, the odor emitted, though strongly disagreeable, did not present the peculiar characteristic of that which escapes during the process usually practiced. The plan, as advised by Darcet, is then advantageous in producing tallow economically, and of good quality. To this is added another gain; there is no necessity for a press; the acid in the cracklings so acts upon the woof of the cellular tissue, that its rupture is rendered easy, and the greasy particles which they retain, are readily separable, by boiling in water.

The principal object in all these experiments having been, then, to devise ways and means of obviating the disagreeable effects of the fetid fumes, it is decisive that the process for burning the vapors is incomplete thus far, that, during their combustion, large volumes of thick smoke fly off to incommode the neighborhood, not to mention the danger of a conflagration from the rushing over of the materials into the fire, by reason of the vessel becoming too highly heated. This danger, as well as the objection above named, is obviated by the other plan of condensing the vapors by means of a refrigerant. It is very doubtful whether the proportion of acid prescribed

Fig. 34.

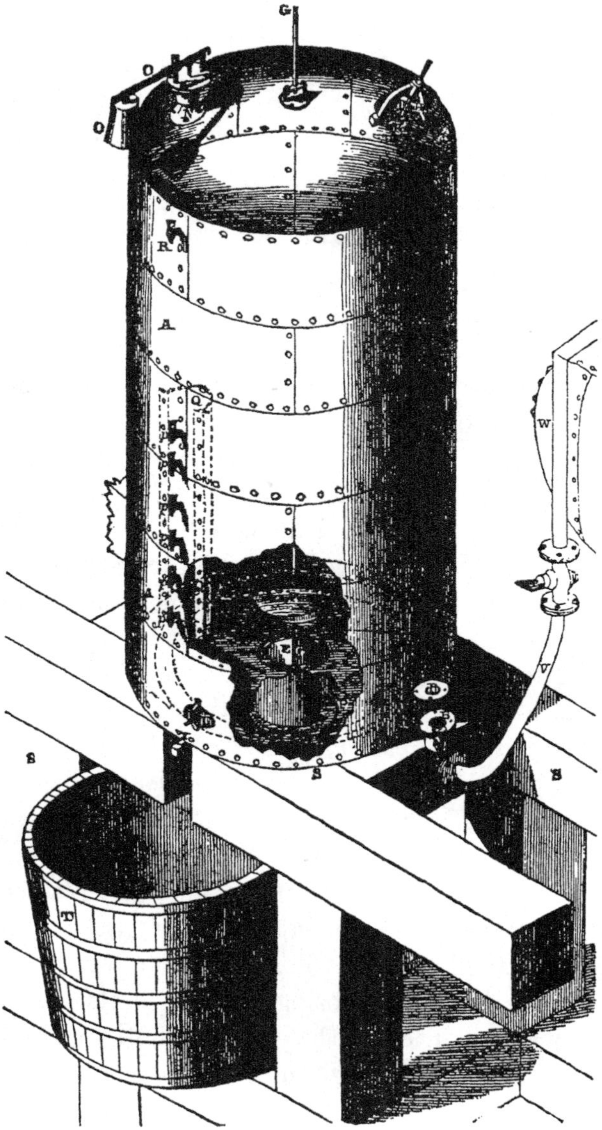

need be strictly adhered to; half that quantity would perhaps produce as favorable results.

In this country the most generally practiced method of rendering tallow is in the open pan, by which, as before said, many disadvantages both to health, comfort, and convenience, are incurred. It has only been a few years since a company in New York were mulcted in excessive penalties for damages claimed to have been sustained by the residents adjoining their "tallow refinery." All these inconveniences can be, in a great measure, prevented by the use of an apparatus, which I saw and examined at Cincinnati, during the last summer. It is the invention of the senior partner of the house of E. Wilson & Co., of that city, at whose extensive establishment there are three in operation, each of about 1200 to 1500 gallons capacity. I have obtained a copy of the drawing (Fig. 34), filed with his specification in the Patent Office, and will endeavor to say as much as possible in description of the arrangement, without conflicting with the rights or interests of the patentee. Each of these rendering caldrons is composed of boiler iron plates, tightly riveted together in the form of an enclosed cylinder, in length about two and one-half times greater than the diameter, and is furnished with a diaphragm or false bottom. The drawing itself is very explicit, and the mode of working these machines, and the use and application of their various appointments, will be mentioned in reciting the process as practically carried through in the laboratory of the inventor. It is as follows: The false bottom being arranged in its place, and the discharging hole closed up, the steam-tight iron tank or cylinder is filled through the man-hole with the rough tallow or lard material, to within about two and a-half feet of the top. This done, the man plate K is securely fitted into the man hole H, and steam let on from an ordinary steam boiler, through the foot valve, into the perforated pipe C within the tank. Set the weight on the valve at the requisite pressure, and during the steaming, frequently and carefully essay as to the state of the contents of the tank by opening the try cock R; if the quantity of condensed steam in the tank is too great, it will be indicated by the ejection of the fatty contents in a spurt. In such case it is then requisite to immediately open the regulating cock X and draw off the condensed steam, through it, into the receiving tub T, until the fatty matter ceases to run from the try cock aforesaid. After ten or fifteen hours' continued ebullition, the steam is stopped off, and that excess already in and uncondensed, allowed to escape through the try cock and safety valve. After sufficient repose, the fatty matter separates entirely from water and foreign admixture, and forms the upper stratum. It is drawn off through the cocks P P in the side of the tank, into coolers of ordinary construction. The tank, being emptied of its lard or tallow contents, the cover F is raised by means of the rod G, from the discharging hole E, and the residual matters at the bottom, let out into the tub T. If, on inspection, the contents of this tub have retained anything of fat, it must be again returned to the tank, when they are being filled for a fresh operation. Mr. Wilson told me he had ascertained, by actual experiment, that to produce the best result, the steam pressure should not be less than fifty pounds to the inch, though the weight used by him was seventy-five pounds; and that he intended to carry it to one hundred pounds, as by that means the rapidity of the process would be augmented. I should, however, advise against so high a pressure in the preparation of tallow; it may do well enough for lard; but if these closed tanks are made to operate as digesters, the effect produced by the decomposition of bones and other matters, which, in the wholesale way of preparing fats at the West, are generally thrown in indiscriminately with the rough suet, would be to deteriorate its quality. No, the better way is to take a little more time, and be sure of a better result. The process is sufficiently economical as it is, for whilst, by a pressure of between fifty and seventy pounds, the bones, &c., are made to yield all their oleaginous or fatty matter, there is no influence occasioned which will convert them into a contaminating constituent. In making lard, from the whole carcass of the hog, excepting the hams and shoulders, a yield is always obtained by the use of this apparatus, full twelve per cent. greater than by any of the other methods; whilst, in rendering tallow, the gain exceeds the product furnished by the ordinary plans at least six per cent. To say nothing of the economy both of time and labor, (fifty per cent. of each,) the material obtained is so much superior, that it always commands, if not the preference, at least a slight advance of price, in the market. The marc or residuum, thrown out into the tub T, being rich in azotized and phosphated matter, when admixed with bog or street earth, and gypsum, makes manure equaling the best guano. Some slight objections might be urged against this mode of operating, but, according to Mr. W., experience proves them to be untenable.

The inconveniences spoken of in the first part of this chapter as to stench and fetid vapors, are mostly obviated by the peculiar management of the tanks; and the product of tallow is so pure and free of foreign admixture, that it is at once ready for conversion into candles. Wilson & Co. have three refineries, one at Louisville, Ky., and another at Madison, Indiana; but their main scene of operations is at Cincinnati. It is to be regretted that their tallow goes most exclusively abroad, being shipped as soon as it is refined, upon European account; for otherwise,

Wilson's brand would become a favorite guarantee with our tallow chandlers. I should recommend this apparatus to all companies or individuals purposing to erect new rendering establishments, or to improve old ones, and for full and explicit particulars reference is made to the inventor at Cincinnati, Ohio.

Some enterprising operators have already adopted the arrangement, and furnish an evidence of its advantage in the superior and excellent products obtained. An investigation of the matter, during a tour of observation through the lard and tallow-packing country, has induced a preference for this mode of rendering fats. I am aware of, and have examined, the many foreign inventions for preparing and refining fats under pressures, but all of them are more or less defective;—there being no better proof of their inefficiency, than that none of them have been continued in operation. One suggestion may very properly be made in regard to the cooling of the tallow refined by Wilson's process, and that is, to prevent any partial granulation, and to ensure more solidity and consistence, it should be run off into small capacities, say open barrels, and allowed to congeal as rapidly as possible.

If all the tallow coming to the market was prepared with as much care as is Wilson's, and those brands made in a similar way, the chandler would not be compelled to any preparatory process previous to converting it into candles. But so much of the commercial article is defective both as to purity and color, that it is considered necessary to introduce here some method by which it can be converted into a suitable state for making first quality candles. To accomplish this, Mr. Dunn proposed to subject the melted tallow to streams of atmospheric air forced below the surface thereof, so that the agitation caused by its entrance, might facilitate the dissipation of the impure odors, and at the same time, bleach the fatty matter under process. To carry out this plan, the tallow or other fatty matter to be purified, must be kept heated by steam pipes or other means, from 170° to 230° F., in any convenient vessel, over which should be placed a hood leading into the chimney, and serving to conduct off any disagreeable vapors. By means of perforated pipes, placed at the bottom of the vessel, streams of air are injected from a forcing apparatus through the melted fat, for a length of time varying with the kind of grease under treatment. Thus, in treating different qualities of tallow, the impurities of smell are quickly dissipated, and the bleaching perfected in ten or fifteen hours, whilst, when operating upon palm oil, fifteen to twenty-five hours are required to perfectly destroy its color. The period of action must, however, be determined by the amount of material under process, the nature of its coloring matter, and the extent of whiteness required. This process might be advantageously modified by adding a small portion of pearlash, and after its completion, allowing sufficient repose to ensure a perfect separation of the oil from the substratum of saturated water.

The most improved and successful process, however, for bleaching and cleansing rendered tallow, is that devised by Mr. Watson, an English chemist, the agent used being the oxygen evolved from the manganate of potassa, or chameleon mineral, in its act of decomposition. The rationale of the bleaching operation is, as has been before stated, in the fact of the coloring matter, insoluble in itself, becoming so changed and altered by the absorption of oxygen, that it can be and is abstracted by the assistance of water or other menstrua. This process is equally applicable to other fats and oils, and to the dirty stearin residuum after the pressure of inferior lard for its oil.

The tallow or fatty matter is melted either in a leaden vat made without being soldered, or a strong deal wood cistern, and therein commingled with an aqueous solution of one-twentieth part of its weight of chameleon mineral, chemically known as manganate of potassa.

To this mixture is gradually added and intimately incorporated so much sulphuric acid diluted with four or five times its bulk of water, as will impart to the liquor which separates, after a good stirring and a few minutes repose, an acid taste. The temperature of the mixture is then raised to and kept at about 150° F., or from that to 212° F. After an hour's agitation, the heat is discontinued, and a rest allowed, so that the oleaginous matter may rise to the surface, and be drawn off from the lower deposit of acid liquor and oxide of manganese. Thus is bleached the tallow or greasy matter. An increase of chameleon is sometimes necessary to produce a whiteness of extra brilliancy. The quantity of water used for its solution is not material;—it can be as much as twenty times the weight of the chameleon. Tests of the state of efficacy of the process may be made during the operation by taking out small samples and cooling it. Its degree of whiteness can then be observed, and the process regulated accordingly. When steam is used for melting the greasy matter, the conduit pipe should be fitted with a stop-cock so that it may be shut off or let on at pleasure.

The mineral used in this process is called chameleon, because, like the animal whence its title is derived, it is remarkable for the changes of color which, in solution, it spontaneously assumes. It exists in two states, as a manganate and oxy-manganate of potassa, the latter salt having for its acid a more highly oxidized manganese than the former. When it is prepared by strongly igniting peroxide of manganese with caustic potassa or carbonate or nitrate of potassa, the extra atom of oxygen required for the conversion of the peroxide into manganic acid, is derived, in the two former instances, from the air, but when the latter is used, the nitre itself supplies the oxygen, and an increase of product is obtained when the process is conducted in open vessels.

The manganate or real chameleon mineral is prepared by calcining together at a red heat in a covered crucible, a mixture of one part of the black peroxide of manganese, with three parts of fused potassa. The mass is of a green color when cold. It is to be dissolved in cold water and the solution allowed to clear by settling, and not by filtering, for fear of its becoming decomposed.

Dr. Gregory's process is to intimately mix four parts of finely powdered peroxide of manganese with three and a half parts of chlorate of potassa, and adding thereto five parts of hydrate of potassa (caustic potash of the shops), dissolved in a small quantity of water. The mixture is evaporated to dryness, powdered and afterwards ignited in a crucible at a low red heat, insufficient for fusion. Digested in a small quantity of cold water, this affords a deep green solution of the alkaline manganate.

The oxy-manganate of potassa is made by fusing with a strong heat, a mixture of equal parts of peroxide of manganese and hydrate of potassa, or one part of peroxide and two parts of nitre. This mass is to be dissolved in water, and if the solution be green, it should be reddened by the cautious addition of a few drops of nitric acid. The clarified liquor is to be evaporated to the point of crystallization.

Both the manganate and oxy-manganate of potassa are readily decomposed by organic matter, the least portion of which, when added to a solution of either salt, reducing its acid to a peroxide state. Hence, its bleaching power, resulting from the ready disengagement of oxygen, and the consequent destruction of many vegetable and animal colors.

The change of color which a solution of chameleon mineral spontaneously undergoes, a change that has originated its name, is thus explicable. The first green color of the solution is due to its being wholly of manganate of potassa; the consecutive red color to its rapid partial conversion into per-manganate, and the purple tints to a mixture of these two degrees of oxidized manganate of potassa. The manganic acid has three atoms of oxygen to one of manganese; the per-manganic, seven atoms of oxygen to two atoms of metal. Manganic acid cannot be obtained by itself, and upon its proneness to decomposition in an uncombined state depends its bleaching power, because, as soon as it is set free from its connection with the base potassa, it becomes an hydrated peroxide, and evolves oxygen, that which exerts all the influence in whitening oils. This salt is particularly efficient in bleaching those stearins of dark color which pertinaciously resist the action of other blanching agents; but its expense is a consideration which prevents its general application to the cheaper qualities of fats and oils.

CHAPTER XLII.

OF WICKS.

THE object to be attained in making a candle is to form a cylindrical tube of combustible material, which, upon being ignited, will throw out a sufficiently steady and uniform flame, with as little consumption of material as possible, and without any emission of smoke or disagreeable odor. For this end there is placed in the centre of the candle, throughout the whole length, a spongy cord or bundle of soft spun cotton threads, serving as its axis, and technically styled its *wick*. It is this wick, which, when lit at the top protruding beyond the candle, becomes imbued with the melted tallow just below the point of combustion, and draws it up by capillary attraction in small consecutive portions to be burned, and thus serves as a means for the constant presentation of requisite aliment to the flame. The material of the wick is a matter of importance; it should be of such filamentous substance as possesses the property of easy imbibition of the melted liquid fat, so that, by the union of its threads, a bunch of capillary tubes may be formed, to suck up gradually, as food for the burning wick, the fluid fat which collects in the cup at the top of the candle being consumed. It is only requisite to observe the behavior towards fire of the animal filamentary matters, such as hair, bristles, wool, silk, &c., to know that neither of them can be used for making wicks. A greased thread of wool or silk crisps up immediately upon ignition, and is carbonized. Among the vegetable substances, there are many which present all the qualities desirable for good wicks, and among them that most preferable and generally used, is cotton. That the wicks may be suitable, the cotton threads should be finely and uniformly spun, white, well cleansed, and without smell, for the least unevenness occasions, during the combustion of the candle, a coal, which, becoming detached from the wick, falls into the melted suet and thus overflows it and produces a running;—hence candles are said to gutter or run. The finer the cotton, the greater number of capillary tubes formed by the threads, the more perfect is the suction of the melted grease and the more considerable the white portion of the wick below the point of combustion.

Some years ago it was proposed to make wicks of the wood of the fir tree, but the light which they gave was not at all comparable with that from cotton wick.

We have recommended the preference of fine white cotton thread for candle wicks, but for the coarser kinds of candles, it is not intended that this direction should be strictly adhered to. For the commoner dipped candles, it will answer to take such thread as is used for weaving coarse calico cloth, the most important consideration being that it should be uniform, free of knots, dirt, and imperfections, and perfectly cylindrical throughout its whole length. Candles made from the best and whitest tallow, sometimes run, and for no other reason than the want of care taken in selecting the wicks. We repeat again that the imperfections of cotton, as above cited, should condemn its use for wicks, else the little knots along the threads, during the burning of the candle, are reduced to a coal sooner than the wick, and falling off into the melted suet, by the excessive heat carried with them, causes more of the tallow to liquefy than can be consumed, and hence the candle runs. When a candle which has already been partially burned, is again relit, it is clearly seen that the flame at first descends just even with the tallow at the foot of the wick, and seems as though it would become extinguished; and it is further observable that it is not revived until the heat has melted a certain quantity of tallow, which, in its liquid state, assembles in a kind of small cup at the top of the candle. The flame then rises, and the wick remains white, even so much as three or four lines above the liquefied tallow beneath it. The combustion goes on then in the top part of the wick, its aliment of melted tallow being conveyed to it by the capillary power of the bunch of cotton yarns forming the wick. There are some, not versed in the theory of the matter, who aver that the tallow is drawn up by the act of combustion, but this is not so, it being a physical cause that produces the effect.

Experience has proved that when a small glass tube, open at both ends, is plunged into a liquid, that liquid immediately rises in the tube, higher in proportion above the level of the same liquid in the vessel into which the tube is dipped, as the bore of the tube is decreased in size, so that this effect is most conspicuous when the width of the hole of the glass pipe is so small as to resemble a hair. To this action has been given the name of capillarity, and to the small tubes which are used, capillary tubes, from the Latin word *capillus*, signifying a hair. The rise of the oil in the wick of a candle or a lamp, is due to this action. "Capillary action is, in short, only an instance of the operation of that species of attraction which is exerted among the elementary particles of matter within very small or insensible distances, and which is called *molecular attraction*."

It thus follows, that a bunch of cotton threads forming a wick, is nothing more than an assemblage of cylinders, so arranged together, that their intervening interstices may each operate as a tube. It is not necessary that the tube shall be absolutely capillary, in order that capillarity may manifest itself, for this property is evident in pipes of even two and a half lines diameter. It is an established fact, that the smaller the bore, the higher the liquor rises, nevertheless it is not to be inferred from this, that the rise of the liquid in capillary tubes is in direct proportion to their diameter, or that the exact height to which it will ascend in a tube of given diameter can be calculated from a knowledge of the height to which it rises in another standard tube more or less large, and which served to make the experiment. It requires, for that, to know the cause which produces this effect, and as yet it has not been satisfactorily explained. Whatever it may be, one thing we know, the effect exists, and it is this only which has a bearing on our subject. This law of capillarity applies to all bodies sufficiently porous for the admission of liquids, such being in fact a mere assemblage of capillary tubes. For instance, if a piece of sponge is placed in contact with water, at one of its ends the fluid immediately passes through and entirely moistens it. It is also by this effect of capillarity, that the sap of the plants mounts from the root to the tip of the branches, and, as before said, the liquid grease is drawn up through the wick to its extremity, and there decomposed by combustion; its elements producing the flame which illuminates. We will not here enter into a detail of the theory of combustion; what has been written is only to show the falsity of the idea that it is the act of combustion which draws up the liquid fat through the wicks,—it being entirely foreign to the effect.

After these remarks, it is evident that when the wick is formed of large threads of cotton, the interstices between each being larger, the grease is less elevated than if the threads were finer; but this is not the only consideration that should induce the preference of fine spun cotton for the wicks. As a matter of evidence, take for example two candles of equal dimensions, and made of the same tallow, each with a wick of like diameter, but in the first candle A, formed of ten threads, and in the second candle B, of 100 threads; by way of being better understood and to facilitate the calculation, let us suppose, that the wick of the candle A presents ten capillary tubes, and that of the candle B 100 capillary tubes; these tubes of both candles will be sufficiently small to effect the ascent of the melted tallow up to the end of the wick. Supposing then the two candles to have been lit simultaneously, now mark the difference.

The melted tallow rises abundantly in the tubes of the wick of the candle A, which are very large, but in a greater quantity than can be consumed by the flame, but the heat at the focus is so strengthened, that it volatilizes the excess of fat and converts it into a dense smoke accompanied by a disagreeable acrid vapor; and the waste of tallow is considerable. The wick of the candle B, on the contrary, consisting of capillary tubes ten times smaller, nevertheless occasions the ascent of the melted tallow, in the same proportion, but as the threads by their greater fineness become better and more uniformly imbibed, the combustion is consequently perfect, without smoke or vapor, and with less consumption of fat. It is then not to be wondered at, that, notwithstanding the two candles were lit simultaneously, B should outlast or burn longer than A. By a repetition of this experiment it will be also observed that the candle with the finer wick required but very little snuffing, burns well, sheds a beautiful light, and does not run.

The spun cotton is purchased in skeins, and wound off into balls or clews by women.

In 1842, Mr. Wm. Kempton, in England, obtained a patent for the application to tallow candles of plaited wicks combined with bobbin, braid, cord, or yarn in such a manner as to produce an advantageous effect. It is well known that the use of plaited wicks in the manufacture of candles, has been restricted to those which are made of pressed tallow, or of the solid fatty acids which melt at comparatively high temperatures, and where the additional heat on one side of the candle from the inclination of the wick in that direction creates no uneven consumption of the material. On the contrary, where plaited wicks have been used for those candles, made of more fusible material, the candles have guttered on that side towards which the wicks bend. Now Mr. Kempton's invention, although admirably adapted for those candles generally strung with plaited wicks, is yet more particularly applicable, he says, to those of tallow, as they can be made with his plaited wicks, so as to burn uniformly, and without requiring to be snuffed.

"If the cord, braid, bobbin, string, or yarn, is to be combined in the plaited wicks, I cause it to follow the course of one of the strands. In making a plaited wick, the cord, braid, bobbin, or other string may be used and incorporated therewith, in which case I cause two of the strands, or the three strands, each to contain a cord, braid, bobbin, or string, of a substance or manufacture different from the yarns used in compiling the plaited wick, the object being to restrain the too great tendency of plaited wicks to turn out in one direction, particularly when common tallow is used for making candles with plaited wicks; and in thus naming cord, braid, bobbin, or string, I do so in order to state that the object of the invention being to restrain the bending of plaited wicks by the application of fibrous materials in a manufactured state, and in addition to the ordinary materials of plaited wicks used in making candles. And when the cord used is to be affixed to the plaited wick after it is made, I bind tightly around the ordinary plait and the additional cord, at intervals, a fine cotton or other thread, and thus combining the plait and the cord together. Care must be taken that whatever be the description of the cord or bobbin, affixed to, or applied in a wick, that it shall not be of such a strength and relative dimensions in respect to the plaited wick employed as to prevent that plaited wick turning sufficiently out of the flame, in order that the burned end may be dissipated. In affixing a cord, bobbin, &c., to a plaited wick for a candle, care is to be observed that it is on that side of the wick where the strands of which the plait is composed, run upwards and from the centre outwards to the edges of the plait, for it will be found, on examining a plaited wick which is burning in a candle, that the wick at all times bends on that side where its component strands run in a direction upwards from the sides of the plait into the centre thereof. This will be better understood by reference to the figures.

Fig. 35.

Fig. 36.

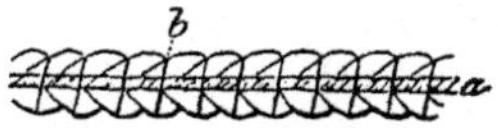

Fig. 35 exhibits part of a plaited wick on a magnified scale, in which the strands run upwards and outwards from the centre to the sides, *a* being the small hard twisted cord which is attached to the plaited wick by a binding thread *b*, which I believe is the best mode of attaching the cord to the plait. Fig. 36 shows a view of part of a plaited wick on a magnified scale, taken on the reverse side to that shown in Fig. 35, and it will be seen that the strands of which the plait is composed, proceed from the side upwards, and to the centre of the plait, and it is this side on which the plait will bend when being consumed in a candle. These plaited wicks are of ordinary construction, but not so tightly plaited. They are composed of three strands of soft cotton yarn, with several yarns in each strand, which are simply laid side by side, and in that state the three strands are plaited, and the twisted cotton cord which I am using, therewith consists of three strands of cotton yarn, each strand consisting of three yarns of cotton, the strands being twisted separately at first and then afterwards the three together; and I find that taking any length of wick and the same length of cotton cord, that the plaited wicks weigh eight and a half parts and the cord three parts. I am thus particular in giving such exact quantities, because much attention is required in seeing that the twisted cord or other preparation of fibres into a cord or yarn should not be too strong for the wick with which it is used, otherwise the wick will be prevented turning out sufficiently, and it will be desirable, when commencing with any cord, to try a few candles first, to ascertain whether the combined wick is too strong, or too weak, and when properly adjusted, the manufacturer may constantly continue using the same materials, and the workman after a little practice will quickly judge of the proper using of my invention, and plaited wicks will require less controlling by the cord used; and when the matter of which the candles are made melts at higher temperatures, the wicks and cords above described, are for common tallow candles.

"It has already been stated, that when the ordinary plaited wicks are burnt in common candles, the excess of heat arising from the preponderance of flame on one side of the candle causes the tallow or other substance to melt rapidly on that side and gutter or run down. Now, by the addition of the twisted cord, or other preparation of fibres into a cord, when the same is incorporated in the plaited wick, it will prevent the wick bending too much out of the flame, and when it is affixed to one side of a plaited wick, it causes the flame to descend lower on the side of the wick opposite to that which turns outwards; in addition to its preventing the wick bending so much outwards, thus counterbalances the excess of heat on the other side, and the candle burns evenly, steadily, and brilliantly. In some cases I find it advantageous to use two plaited wicks in the same candle, each wick having a cord or thread as above described, the cords or threads of the two wicks being towards each other when in a candle, by which arrangement it will be evident that the two wicks will turn out from the flame in opposite directions, and for this purpose, I prefer to use hempen yarn or twine."

In making mould candles, the combined wicks are to be inserted in the moulds in the usual manner, taking care, however, that the wicks are right end upwards, as above described and explained. When making "dips," the wicks are plunged successively in the tallow, as usual, keeping the cotton straight at the commencement of the dipping. The above specification is so lamely written as to require some attention in order to comprehend its meaning. As it embodies an improvement which may be practically advantageous, it is here introduced as a companion to another late contrivance by Mr. Wm. Tindall, which consists in a peculiar mode of applying plaited wicks to candles by means of a flat metal rod, wire, or instrument; the object being to make the candles with double wicks, which will turn from the centre whilst burning, and be consumed without requiring any snuffing. It is well known that the lighted end of an ordinary plaited wick bends over upon that side in which the strands run upwards from the outer edges to the centre; whilst on the opposite side of the wick the same strands will be found to run upwards from the centre to the outer edges; the patentee takes advantage of this circumstance, and arranges the wicks accordingly upon the instrument employed for inserting them in the candles. The ordinary plaited wicks may be used, and the instrument may be made in any manner that will answer the purpose. The form used by the patentee is represented in the margin; it is a thin narrow rod, of steel, brass, or other metal, rather longer than the candle intended to be made, having two flat sides, with a handle at one end, and a notch at the other. The following is the mode of employing the instrument:—the plaited wick is first placed against one of its sides, commencing near the handle *a*, in such a manner, that the strands on that side of the wick which comes next the instrument, shall run upwards towards the notch *b*, from the centre of the wick to the outer edges; the wick is carried up to the notch and then twisted, or turned half round; which twisting or turning enables the operator to conduct the wick from the notch along the opposite side of the instrument, with the strands on that side of the wick which comes next the instrument running upwards from the centre to the outer edges, in the same manner as the strands on the other side. Both ends of the wick being then fastened, by

Fig. 37.

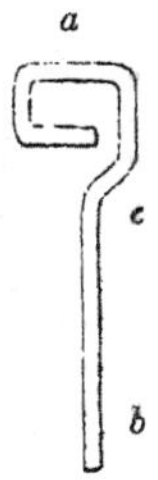

thread or other means, to the instrument at *c*, it is ready to be placed in the candle mould.

Fig. 38.

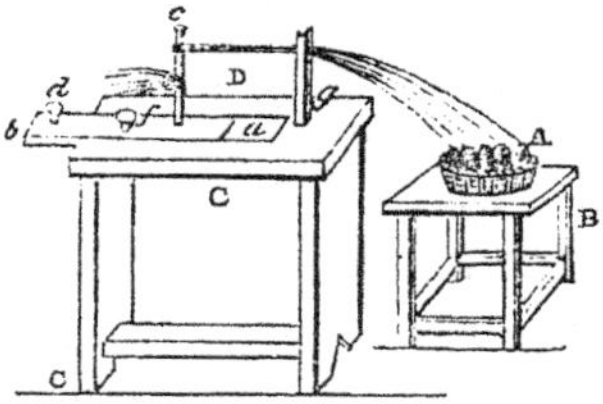

The coupoir, or wick cutting machine.—Fig. 38 represents this apparatus, which consists of a strong table C C, the board forming the top of which is surmounted by a second plank, with a longitudinal opening *a*, along its centre, serving to let glide in by a groove the shelf *b*, carrying the iron broach *c*, vertically fastened on. A wooden knob *d*, placed at the end of this shelf, answers as a handle to draw it in and out. The key *f* opposite, is to fasten the shelf, and consequently the iron broach, at any convenient point, proportional to the length to be given to the wick. At the end of the table, on the left of the workman, is firmly fixed a knife, *g*, the edge of which is turned towards D.

At the side of the table is placed a stand B, and upon it are set the baskets holding the balls or clews of thread. The workman, generally a woman, occupies a stool before the bench, and adjusts the lengths that the wicks are intended to have, by regulating the distance between the knife *g* and the broach *c* accordingly, and with the screw *f*, which renders the latter as well as the slide, immovable. The wicks are then cut in proper manner, and of uniform length.

The workman should take care that all the ends of the cotton are even, else the wick is defective, and, technically speaking, *barlongue*. He places the wick between the palms of his two hands, without removing it from the broach, and by gliding one hand over the other, twists it slightly so as to prevent the threads from separating, and to shape around the broach a kind of loop which is termed the *collet* of the wick. Afterwards, and without raising the wick off the broach, he throws it upon the end of the table at his right. In this operation, the workman must hold on steadily to the ends of cotton which remain in his right hand after he has cut the wick.

When he or she has thus prepared a number of wicks sufficient to fill a broach or rod, (*baguette*,) they are laid flat by the side of each other, and then inspected; but without being taken off the rod. If any thread is seen detached, or detaching itself from the bunch, it must be picked out, and all the ends of the wicks being assembled together, the workman clips them evenly, and with the pressure of one hand near the collet, forms an indentation around which a fastening is placed, and the bunch is then thrown towards the end of the table. Each of such bunches is technically designated a *brochee*.

These packets or brochees are not all composed of the same number of wicks; they vary according to their size, or what is the same thing, according to the number of candles to the pound. Each consists of twelve where the candles are to be four to the pound; of fourteen for fives; fifteen for sixes, and sixteen for eights, and so on.

The word *brochee* is only applicable to the wicks serving for dipped candles, or candles *à la baguette*, as the French call them. In preparing the wicks for moulded candles, the broach is filled as full as it can be without regard to the number, and then emptied of its load upon sundry small baguettes. It is from these baguettes filled with wicks that the workman threads the moulds, taking one at a time and substituting another when that is emptied.

As to the wicks for the dipped candles, when the broach is full, its burden is distributed upon smaller wooden rods, smooth throughout their length, and sharp pointed at one end so as to facilitate their entrance through the whole of the collet of the wicks. These wooden rods are always of two and a half feet length.

Another cutting machine, possessing all the advantages of the before-mentioned, and many others additional, is that manufactured after an English pattern by Mr. Brock Watson, Philadelphia. It cuts, spreads, and twists the wick in one operation, and so rapidly, that a dextrous workman can prepare for 1000 pounds candles, per hour. The machine is neither costly nor complicated, and the mode of working it both simple and easy, as its management can be so regulated as to occupy only the intervals between the moment of charging the melting kettle with tallow, and the time it is ready for dipping use. In this way the wages of one man are saved, for, by the old plan,

Fig. 39.

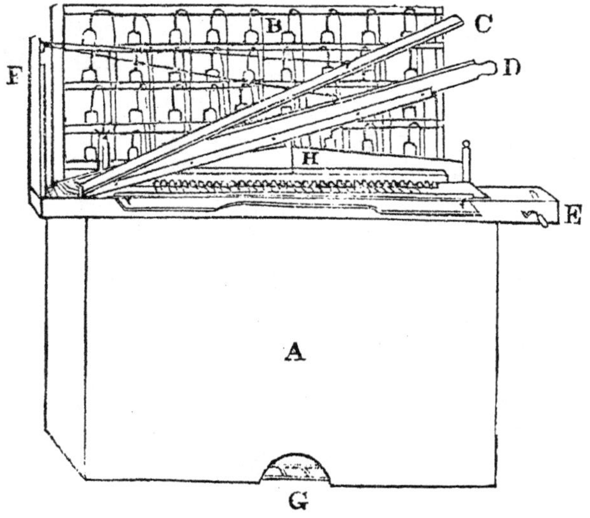

a separate hand is required exclusively for cutting and preparing the wicks. Being very compact, and readily taken apart, it is also convenient for transportation, and when put up, occupies no more space than an ordinary counting-room desk.

Fig. 39 represents the apparatus. A is the body of the machine, serving as the bed of the upper works, in the interior of which are the pulleys that regulate the movement of the carriage B, worked by the treadle G. The carriage B, a kind of frame-work running on wheels, contains a range of boxes, placed shelfwise, as receptacles for the balls of wick, one of whose ends runs through a notched reed below H, and comes forward upon the twisting board E having fastened to its back edge a knife serving as the under blade of the movable clipper D. This, when drawn down vertically, severs the wicks evenly. The twisting box E C consists of two boards hinged, and moving on rollers. A turn of the crank near the end communicates that motion which twists the wicks after they have been cut by the knife E, and this knife having effected its purpose, is immediately drawn upwards again by a counterpoise F. At the front of the twisting box D, is a sliding board so fixed, that it can be graduated to regulate the required length of the wicks.

Motion being communicated to the machine, the wick yarn is then cut, spread, and arranged on the baguette simultaneously, and in complete readiness for dipping;—as soon as the workman removes it, another rod rolls into its place. When either the wick or rods give out, a fresh supply must of course be substituted.

The size of the wicks should vary not only with the dimensions of the candles, but also according to the greater or less facility with which the tallow or combustible material melts.

After what has been said relative to the different qualities and properties of fats and tallows, this remark is important. The materials used as the body of candles have not all the same melting point, consequently the wick should always diminish in size ratably as the suet or other candle material is more difficult and slow of fusion. For instance, a candle of *pure* stearin, with a small fine bougie wick, will give a beautiful white flame and as much light as an ordinary candle of the same size with a wick twice as large; and with this additional advantage, that whilst the common candle loses an inch of its height, that of stearin in the same time is scarcely consumed the length of its conical tip.

The application of wicks to candles is regulated, 1st, by the degree of heat at which the tallowy material melts; 2d, by the size of the threads which compose the wicks; 3d, by the thickness of the candle. Repeated experiments are necessary to ensure complete success.

Of the kinds of wicks applicable to moulds and other species of candles, we will speak when treating of each separately.

CHAPTER XLIII.

OF THE MANUFACTURE OF CANDLES.

THE chandler forms from tallow two kinds of candles, the one called dipped or candles *à la baguette*, because they are made by stringing the wicks on baguettes or rods, and dipping them repeatedly in the melted fat. The other species bears the name of mould candles, being cast or made in moulds. We have already said that the mutton suet alone is the best for good candles, but as it is very firm and rather brittle, especially in cold weather, it is considered preferable by *some* to combine it with equal parts of beef tallow. The suet of the ram is still more solid and consistent. The important points to be observed in manufacturing candles, are, 1st, to have the tallow perfectly free from all impurities; 2d, to render it as hard as possible so as to prevent any greasiness to the touch, or *running*, during combustion, and 3d, that it shall retain a permanent whiteness of color. A number of essays have been made to accomplish these different ends, but we shall introduce here only such as are known to be practicable and perfect.

Much of the tallow, as it comes into the market, is not sufficiently pure to be formed into good candles, and consequently must be remelted and treated anew, and mixed in proper proportion with mutton suet. After equal parts (these being the best equivalents), of mutton suet and tallow have been weighed out, they are passed over to the workman, who cuts them into small pieces upon a table technically styled the *chopping-board*. This is a strong table, the top of which is rectangular or a lengthened square, ledged around with boards of six inches height, an open space of a foot breadth, being left in the front to facilitate the work. In this space is fixed a large sharp blade, arranged like a chopping-knife, and similar to that described at page 328. It is very important that the suet and tallow be divided into small pieces, for otherwise they dispose badly in the caldron; and as they melt only superficially, if they are too large, the liquefaction progresses very slowly, and the liquid becomes scorched, and acquires a more or less brownish color, which deteriorates its quality and that of the candles. The smaller the lumps are, the shorter their exposure to heat, and consequently the danger of scorching is lessened. The pieces as they are cut should be brushed off from the table into a basket placed under it so as to receive them. This basket, when full, is carried to the caldron, and its contents emptied in it. Many chandlers, especially those who operate in a small way, have their kettles movable and placed upon a tripod, and build the fire beneath them. This is a bad mode of operating;—the smoke arising, blackens the suet, and the ashes carried up by the current of rarefied air, soils and deteriorates it. There cannot be too much care and precaution in the manipulations of melting. Experienced chandlers have a copper caldron mounted in a brick furnace. This caldron is of the same form as that represented by Fig. 29, and the furnace door is constructed on the outside of the building, so as to avoid smoke and inconvenience to the workmen from the heat of the fire. Near the bottom of the caldron, and at the height of some two or more inches, that is, above the space it is presumed will be occupied by the deposit of impurities, is affixed a pipe or tube, fitted with a stop-cock, through which the clear melted upper stratum of fat can be drawn off without disturbance of the residual lees.

The caldron being charged with the tallow and suet, mixed half and half of each, and well minced up, a small quantity of rain or river water is poured in, and the fire in the furnace then kindled. The addition of water at this stage is called by the workman the *filet*, and varies in proportion, some chandlers using more, and others, less. This water is intended to prevent the suet from any danger of being scorched by too close contact with the bottom of the caldron which is directly exposed to the fire; and also to facilitate the complete rectification of the melted tallow. The foreign matters which, by careless rendering, might be retained by the suet, would, in precipitating to the bottom, by virtue of their greater specific gravity, be still, on cooling, admixed with a certain portion of grease. Now this is obviated by the addition of water, which, forming a substratum, holds back the dirt and foreign matters, and leaves the tallow above much cleaner and purer.

There are, however, some objections to this method, and a different arrangement is now substituted. The modus operandi is the fusion of the tallow by steam in an apparatus of the following construction (Fig. 40).

Fig. 40.

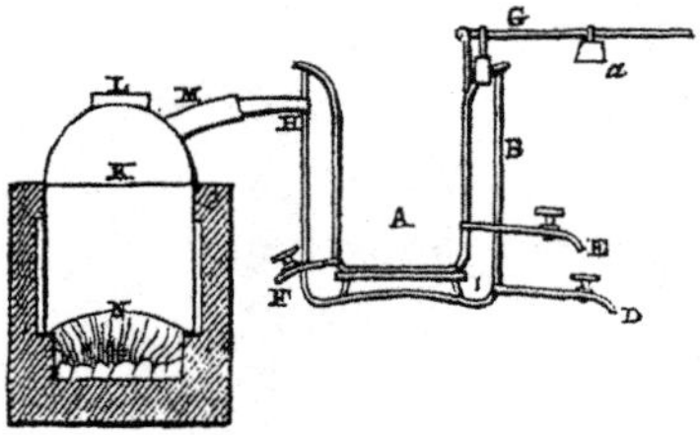

A is a copper caldron, enveloped by a second caldron, or jacket, of copper or iron, of three to four inches diameter and three inches depth, greater than the inner kettle. The bottom of the jacket is exteriorly concave, and the two kettles are hermetically fitted together at their mouths. A tripod I which rests upon the bottom of the jacket, serves as a support for the weight of the inner kettle. At a point the most convenient near the bottom of the kettle, is placed a tube D, fitted at its outlet with a cock, through which is evacuated the condensed water. At the side of the jacket near the top, is a tube H, the use of which is apparent. The interior kettle carries two tubes, each fitted with a stop-cock. That one, E, about two inches from the bottom, serves as an exit pipe for the clear rectified tallow floating upon the substratum of drossy deposit;—the tube F which is even with the bottom, being a conduit for the emission of the settled water, and along with it the precipitated dirt and impurities.

By the side of this double caldron or water bath is constructed a furnace in which is placed a steam boiler K, which is heated by the fire N. This caldron, the mouth of which is at L, has a tube M, adjusted to the tube H, the connection forming the pipe through which the steam is conveyed into the jacket. Upon the top of the kettle is placed a safety valve, and the pressure regulated by the weight *a* on the lever G. Another safety valve is also fitted to the mouth L of the steam boiler K. The ash pan of the furnace is seen at O, its opening being placed on the other side of the wall to save inconvenience from dust, &c. In lieu of this steam boiler, it will do equally well to connect a pipe with the large boiler that feeds the engine of the establishment. In small factories, however, where the work is exclusively by manual labor, I would recommend, in preference to the boiler described above, one of the portable steam generators invented by Messrs. Greer &

Amer, of Philadelphia. They are economical as to fuel, and being portable, and occupying but little room, are easily manageable. The conduit pipe can be fitted on with a coupling screw, and thus rendered movable at will. The enveloped kettle serves well as a water bath, for other purposes, when the stop-cock of the conduit is so managed as to prevent more steam running through into the jacket than is sufficient to heat it to 212° F.

It has been a mooted question whether the addition of water in melting the suet is injurious or not. The experiments of others, competent to decide in the matter, authorize me to say, that no detriment results from its presence; provided a sufficient repose is allowed for the complete separation and deposition; and care is taken to draw off the superstratum of melted tallow free from any of the water upon which it rests. Consequently, the caldron A should, previously to being charged with tallow, always receive water enough to cover its whole bottom to the height of an inch, guarding, however, against its level being closer than within six or seven millimetres of the stem of the tube E. By this arrangement, all probability of water running through the cock E is removed, and if the settling is allowed six hours, the tallow then drawn off is very pure, as the entire filth is left with the water which remains at the bottom of the caldron.

It must be recollected that after each operation, the caldron should be cleansed of all residual filth, and rinsed out through the cock F. In case there should arise to the surface of the tallow during its fusion, brought up by occasional stirring with a wooden paddle, any dirt or impurities lighter than the melted suet, they must be skimmed off with a cullendered ladle. Be careful not to throw cold water upon the suet while it is hot, for otherwise a commotion will ensue, ejecting the liquid fat in all directions. The suet should never be allowed to boil;—the heat must not exceed 212° F. It is folly to raise it higher, that degree being more than sufficient to give the requisite fluidity to the tallow. When the refined suet is ready to be drawn off, after having stood long enough to drop its impurities, it should be strained through a very fine hair sieve into a clean iron bound wooden tub. This tub is placed by the side of the caldron, and on a movable stool of height sufficient to allow an easy passage under the cock fitted near the bottom of the vessels that are to receive its rectified contents.

There are but very few chandlers who take any of the aforementioned precautions in preparing the tallow for dipped candles;—the consequence is, a slovenly, imperfect article, alike disreputable to the manufacturer, and prejudicial to the profits of the trade which he follows. A more enlightened spirit would soon suggest the bad policy of such negligence and indifference, and it is well for the consumer that he has that protection to his interests, in a brisk competition among the manufacturers, which he might fail to receive, if the palm of excellence was an optional matter with the operative.

CHAPTER XLIV.

DIPPED CANDLES.

DIPPED candles are, from their low price, the most used perhaps of any other kind, especially by the poorer classes of the community, and in the workshops. It is doubtful, whether there is any economy in their preference, for the slovenly manner in which they are generally prepared, renders them both disagreeable and ultimately expensive. They are called dipped, because a certain number of wicks, strung along equi-distant from each other upon a wooden baguette or rod, are covered with successive coatings of suet by being plunged or dipped into a bath of melted tallow. The bath vessel, called an *abime,* is a trough (Fig. 41), of walnut wood, firmly

Fig. 41.

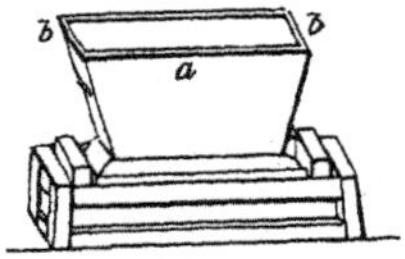

and tightly jointed. Its form is shown by the figure. The two larger sides *a*, incline towards each other at the bottom and meet in a plank of three inches breadth. These two sides are rectangular, two feet in height, and three feet long. The two smaller ends *b*, are vertically placed, and tightly fitted to the side pieces. A wooden handle upon each of these ends serves as a convenience for the transportation of the bath from one place to another. This trough is fixed upon a foundation of wood. The ledges of this foundation are slightly inclined, so that the suet which falls upon them can run off into vessels placed beneath to receive it. When the trough is not being used, it should be closed with a cover, made especially for the purpose, so as to exclude all dust and dirt.

The workman commences operations by filling the trough with hot tallow, and thus forms the bath into which the wicks are dipped, as follows. He takes ten or twelve baguettes, each strung with wicks as beforesaid, and dips them simultaneously into the liquid suet, and lets them soak thoroughly. It is necessary that the tallow should be hot, so that the imbibition may be more complete. But it is not only on this account that the suet should be hot for the first dip; the dry wicks sink more easily into the well heated and consequently more fluid tallow, and distend more readily and perfectly, than if the suet is too cool and dense. The wick which is flabby on entering the liquid fat, does not afterwards fully straighten out, and hence causes the candle to be defective. The first dipping having been made, the batch of baguettes are placed upon the ledge of the trough, and one after the other lifted up and examined. After having separated such of the wicks as may have matted together, the workman carries the whole batch of rods to the *égouttoir*, arranged aside, and represented by Fig. 42.

Fig. 42.

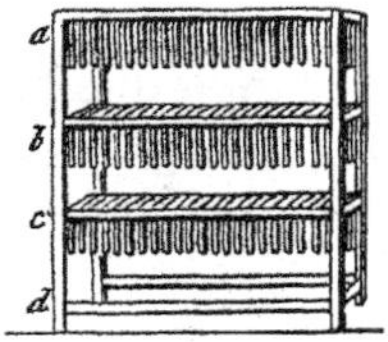

This *égouttoir* of wood is formed of four upright and four cross pieces, well and solidly jointed together. The baguettes are placed upon the cross pieces *a b c*, and the uncongealed tallow dripping from the wicks falls into the basin *d*, beneath. Whilst the workman is arranging the wicks upon another batch of rods and dipping them, those first dipped have cooled so far, that but very little tallow drops from them. In the same manner he goes on until he has soaked all the wicks prepared to be dipped. This quantity should not exceed in number more than can be managed between the time of soaking the first batch, and the period when the suet of that batch has cooled sufficiently for a repetition of the dip. If the suet be too hot, the layer which attaches to the wick will be too thin, and the candle become, in the workman's language, *tavelée*. If, on the contrary, the suet is too much cooled, the layer is clotted, the candle is rough all over, and spotted. The proper degree of heat is reached, when the tallow commences to congeal on the sides of the trough where it forms a thin jagged pellicle. If, during the operation, the tallow has cooled too much, the temperature must be equalized by stirring in an addition of hot tallow. By a renewal of this means from time to time, the bath is kept at an uniform degree of heat and liquidity. The stirring paddle serves also to scrape the bottom and the corners of the trough, and when clotted with cold suet, it is relieved with a copper trowel or iron spatula. The same trowels serve also to gather together the tallow which collects upon the ledges of the foundation.

Some chandlers so arrange this dripping bath, in such case made of metal, as to keep it continually warmed by a bed of coals placed beneath. This mode is not to be condemned, provided it can be so managed as to prevent a too great fluidity of its tallowy contents, and consequently an hindrance to the consecutive dips following the first plunge. When all the wicks have been dipped with the precautions indicated, and they are well cooled, the workman takes the baguettes two at once, and sometimes three in order to hasten the work. Keeping each separate and apart from the other, by placing his fingers between, he turns the palm of his hand upwards, lays the first baguette under his thumb, it resting upon the palm of the hand; and passes the second under the indicator or next finger, it resting upon the third finger. If three baguettes are handled at a time, the third is passed under the index and between it and the middle finger, so that this latter and the indicator are above, and the ring finger and the little finger are beneath.

The fatigue consequent upon this mode of working, rendering desirable some invention by which the workman could proceed with less inconvenience to himself, the following instrument was contrived with a view to meet that purpose. *a b* (Fig. 43) shows an end view

Fig. 43.

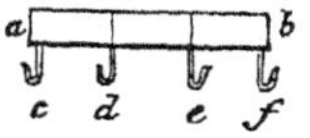

of it. It consists of a strong bar of deal wood, twenty-eight inches in length, (if the baguettes are thirty inches,) and two inches breadth, when but three baguettes are to be worked simultaneously, and of three inches breadth for four. The bar is of one inch thickness throughout its length. On the edges are two iron hooks, *c f*, and in the centre of its breadth are two others, *d e*, fixed tightly in the wood, equi-distant from each other about an inch. The workman shakes each baguette to separate any of the wicks which may have adhered together, dips them into the melted tallow, and upon lifting them out, gives each rod an expert movement so as to disentangle the wicks which have touched together. This dipping operation is a delicate one, and requires experience and dexterity, for if two or more wicks mat together during the cooling of the tallow, there is more time lost than their worth in making them follow the proper direction; and, moreover, such candles are badly shaped. This dip being completed, the baguettes are placed upon the lower tier of the *égouttoir*, in order that the drops of tallow which trickle during the congelation shall not fall upon any other candles more advanced to completion, and thus damage them, particularly if they are nearly finished. As soon as the suet is at the proper temperature, as before explained, then the other dips are proceeded with. The second dip, technically termed the *retournure*, is more easily made than the first, because the wicks, having acquired a consistence, sink much better into the tallow. After dipping the wicks once, they are lifted entirely out of the tallow, allowed to drain a moment, and then dipped their whole length anew.

The tallow congregates at the lower end, and forms an ill-shaped butt; to prevent this, the candles are plunged two or three times consecutively, to a third or a fourth of their length, in order to melt off the excess of suet and render them more truly cylindrical. This operation is called *ravaler*, and requires care and expertness, for if it happens that the tallow is not sufficiently hot to melt off the excessive suet during the ravale, the action of the bath must be continued until the size of the butts becomes uniform with the upper part of the candle.

It is impossible to prescribe the exact number of dips that should be given to the candles, as it depends upon the size they are to have, and on the temperature of the tallow. The dip before the last is called by the workmen, *mettre pres*, and the last, *achever*. Experience will teach as to the moment when the candle is finished, though the better way is from time to time to weigh samples. At the last dip, the workman forms the collet of the candle by plunging it deeper into the tallow; and by this means imbuing the tip of the wick which has not been previously soaked, there is formed a small cone through which protrude two small ends of the wick, to facilitate ignition.

It is hardly necessary to repeat here, that the workman should, after every dip, carry the baguettes strung with candles to the *égouttoir*, and place them in tiers, the more elevated as he approaches the terminus of the operation.

When all the layers are very thin, the base of the candle forms in a point which often exceeds the length of the wick an inch. In such case the candles should be

Fig. 44.

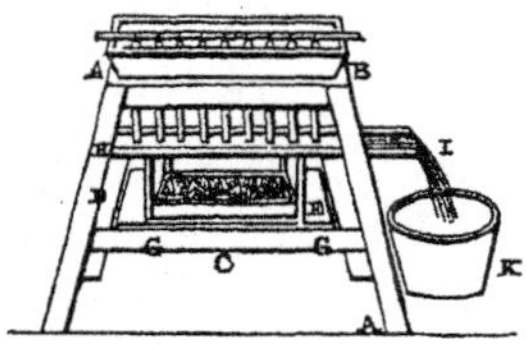

pared off so as to be made of uniform length, and for this purpose, a particular apparatus called a scraper or *rogne-cul*, is required. It is represented by Fig. 44. In the top part of a strong legged, tenoned morticed, and tightly jointed table, A, is adjusted a kind of bottomless hopper B. Between the legs of the table, and upon the lower cross pieces C rests an iron plate D with raised edges, supporting a small square sheet iron furnace E, which, when filled with ignited coals, furnishes the requisite heat. The iron plate D sustains by four small columns, G G, a copper platten H I, a little inclined towards I. This platten has a raised edge of six lines on every side except I, which terminates in a spout for the safe conveyance, without waste, of the suet into the vessel K.

The hopper B, is intended solely to protect the body of the candle from the effect of the fire, but it can be made to serve another purpose, and that is, to prevent the paring of the candles more than to a definite extent. To effect this, the machine must be so constructed, that it can be let down so as to have, at will, a distance from the top to the surface of the copper platten H I, proportional to the lengths the candles are to receive. This is easily done by piercing the two smaller ends of the hopper with holes at different heights; and two holes in each of the upper cross pieces of the foot. It is adjusted at the proper point with pegs.

This generally terminates the operations for dipped candles, there being naught else to do, except string them in pound packages for the market. The dipped candles, as is well known, are very uncouth and unequal in their appearance, and but seldom perfectly cylindrical and neat looking. To obviate these disadvantages, a method has been planned, by which they are made to assume exteriorly, a similarity to mould candles. The process is to pass the candles through a kind of draw-plate, and is practised as follows. A box wood board, twelve inches long, two inches broad, and four or five lines thick, is bored its whole length with ten holes, distant from each other six lines. Fig. 45 shows this

Fig. 45.

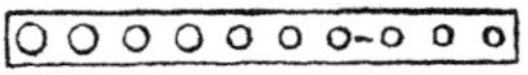

draw-plate. The first hole is perfectly round, and in diameter equal to that of a five mould candle, but all the holes are not of the same size; they increase upwards, for the third is half a millimetre more in diameter than the second, and so on to the tenth, which, consequently, has about four and a half millimetres more than the first. The holes being thus made, are chamferred with care upon one face, leaving them nearly sharp on the other face. These edges being well smoothed, the plate is complete. There should be one for each sized candle, the dimensions of the holes varying accordingly.

When the candles are well dried, they are caught one at a time by the loop of the wick with a small iron catch, and drawn through the large hole of the plate conveniently and firmly affixed to a table. In this way, some excess of tallow is pared off. By repeating the manipulation through all the holes consecutively, the candles are reduced to a uniform size, and become more agreeable to the sight. A little experience enables a rapid dexterity on the part of the workman, so that much more can be accomplished in a day, than would be supposed, by the reader. Besides the advantage accruing to the chandler from a more extended sale, induced by such an improvement, there is a further recompense for the trifling additional expense incurred, in the feeling of satisfaction, that his enterprise is not to be restricted where an amelioration pleasing to the eye, and improvement of the quality, though costly at first, must eventually bring its own reward.

For those tradesmen who work at benches, such as shoemakers or tailors, there is a kind of candle made by the union of three dipped candles, after the third plunge. Thus joined, they are dipped together as one candle, which, when finished, presents three wick heads. Each being ignited, the whole throws out an economical light.

The comparative tardiness of this mode of dipping candles by hand, suggested the necessity of some improvement, by which the quantity as well as quality of the product might be enhanced. To this end many contrivances have been devised, the most prominent and successful of which we are about to comment upon.

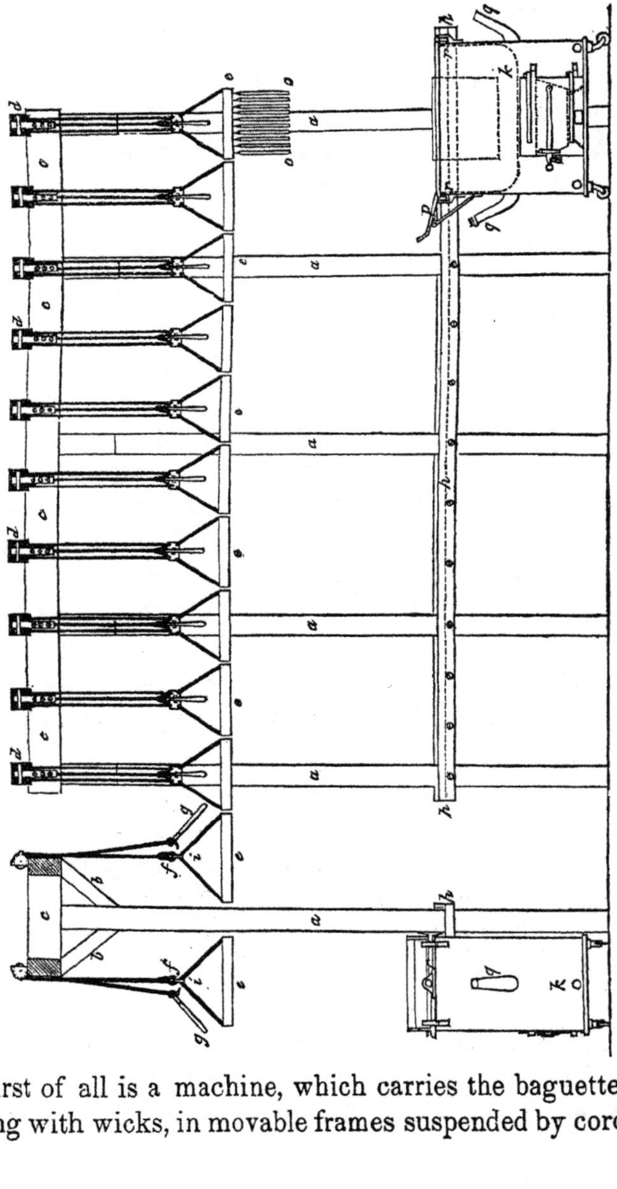

Fig. 46.

Fig. 47.

First of all is a machine, which carries the baguettes, strung with wicks, in movable frames suspended by cords.

over two basins placed one in the other, the larger of which contains the tallow heated to 85° or 95° Fahrenheit, by a small furnace, whilst the suet in the small basin has but a temperature of 54° to 59° Fahrenheit. A workman raises and depresses each frame alternately with the hand, and thus dips and re-dips the candles until they have attained their proper size.

Figs. 46 and 47 exhibit a front and profile of this machine. Fig. 48 presents a part of the apparatus in plan.

Fig. 48.

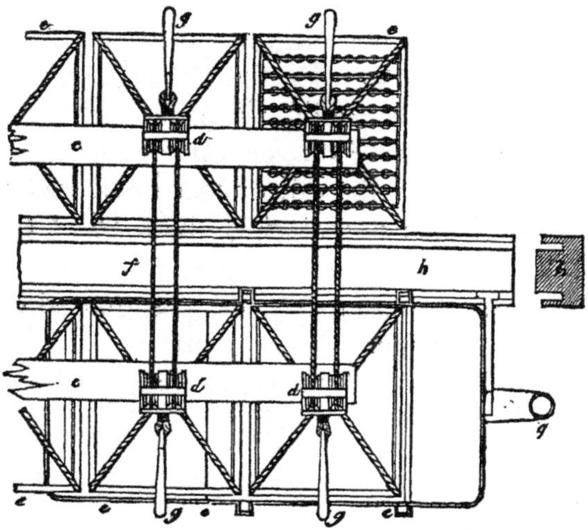

The frame work consists of five oak wood beams, *a*, each bearing, near its top, two abutments, *b*, connected together by the long cross pieces of an horizontal frame *c*, throughout the length of which are five small traverses, and into each of which is tenoned and morticed one of the beams or posts *a*, of the frame.

The iron caps *d d* are fixed upon the sides of the long cross pieces of the frame *c*, and each has two brass pullies. The fastening or connection of one of these caps is seen both by front view and profile, and in extenso, in Figs. 49, 50. Each pulley receives a cord *f*, at the end

Fig. 49.

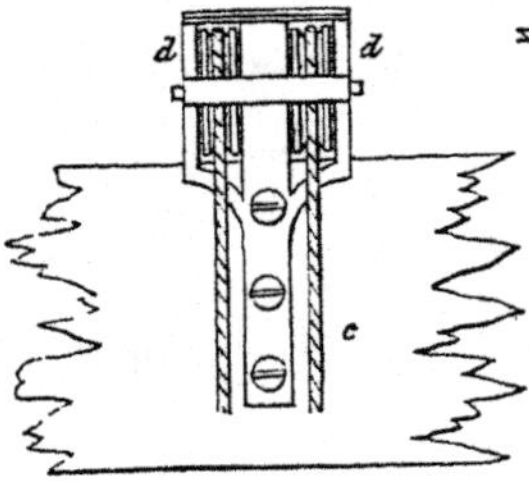

Fig. 50.

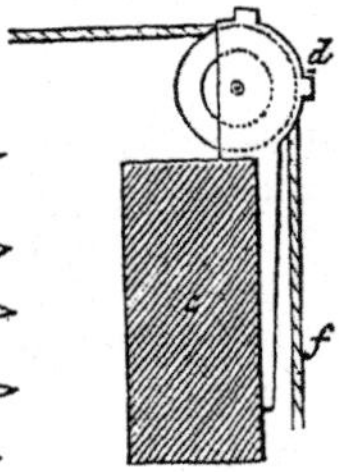

of which is suspended a small wooden rectangular frame, double grooved interiorly, for the reception of the baguettes which carry the candle wicks.

The wooden handles of the cords *g g* have each a hook fastened in a hole in the centre of a small rectangular iron plate *i*, to which are attached the cords of the pulleys, and the frame *e*.

The long traverse or cross piece *h*, is grooved on each side to facilitate the movement of the furnace, seen in vertical cut at Fig. 51. This furnace runs on four cast-

Fig. 51.

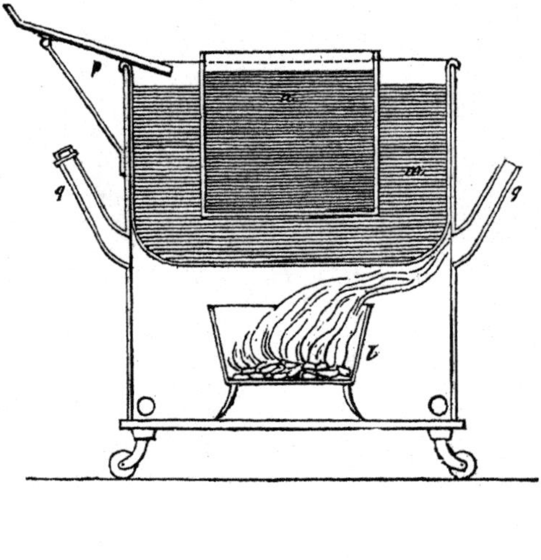

ers, so that its position may be readily changed, and it has two doorways for the service of the small furnace *l*, which holds the coals for maintaining the fusion of the tallow *m*, contained in the larger tinned copper basin. This tallow, thus kept at a temperature of 85° to 95° F., imparts sufficient heat (50° to 60° Fahrenheit) to the contents of the smaller or inner basin, to ensure the adherence of the tallow to the candles *o* (Fig. 47), each time they are dipped therein. Upon the ledge of the larger basin is an iron *égouttoir p*, on which such portions of tallow fall as have not congealed, when a frame of candles is being brought into the smaller basin. The suet which drops, is, by the declivity of the *égouttoir*, directly conveyed into the large basin.

The furnace is fed by two currents of air, admitted through two draught ways, *q*, which are opened and closed at will. The crotchets *r*, in the groove of the cross piece *h*, serve to conduct the furnace from one frame of candles to another.

The mode of operating with this machine is to fill the two basins, placed over the furnace, with suet and tallow mixed in proper proportion, and heated up to the degree before indicated. Place the furnace under one of the frames *e*, to which are suspended the wicks destined to form the core of the candles, and then with the hand lower these wicks into the small basin, and dip and re-dip them successively, until the candles have attained the proper size. This point reached, lift up the frame, which remains suspended as soon as the hook of the handle *g* is fastened in the centre of the piece *i*. When the candles of a frame are finished, and whilst they are cooling and hardening, change the position of the furnace, and convey it beneath the frame placed in front of that previously operated upon, and then recommence the same operation.

This machine is said to be very effective, turning out the candles rapidly, and of good appearance, but there is another, for which are claimed superior advantages; a front view is shown at Fig. 52.

Fig. 52.

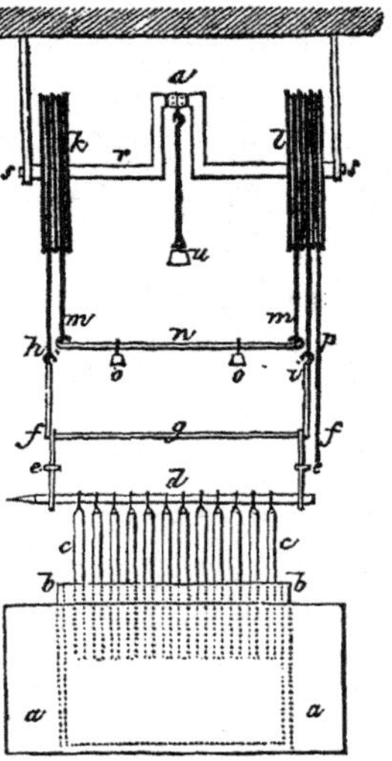

Fig. 54.

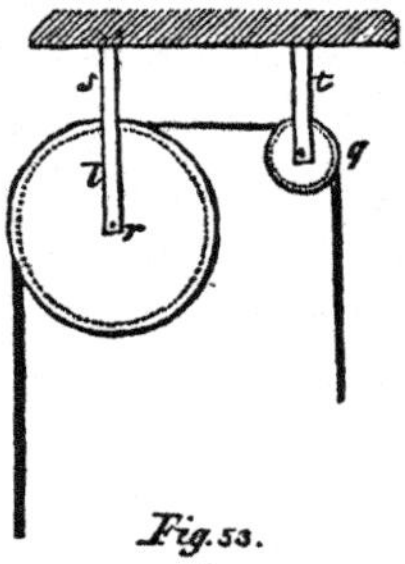

Fig. 53.

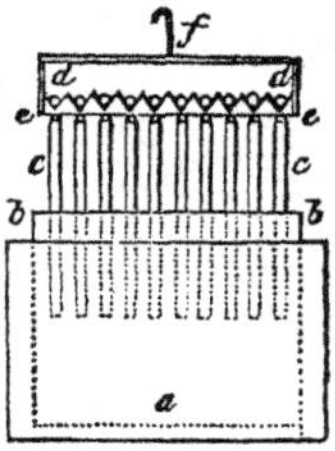

a is a longitudinal view of the caldron; *b*, the chest placed in the caldron, and containing the candles; *c*, the candles; *d*, the round wooden baguettes which carry the candles; they are ten in number, and placed equidistant apart upon the same horizontal plane, as shown at Fig. 53; *e*, two supports, seen in profile at Fig. 53, and both notched to receive and firmly hold the ends of the baguettes, *d*. Each of these supports is fitted with a hook *f*, catching at both ends of an iron bar *g*, suspended by means of two cords *h* and *i*.

The cord *h* is attached at its upper extremity to the circumference of a double grooved pulley *k*, and works in one of the gorges. The cord *i* is also fixed at its upper end in the middle groove of a three gorged pulley *l*, a face view of which is given at Fig. 54. There are two other cords *m*, the upper ends of which are attached, one to the second groove of the pulley *k*, and the other to the interior groove of the pulley *l*. The two lower ends of these cords carry an iron bar *n*, on the length of which are hung the counter weights *o*.

The guide cord *p*. passes in the gorge outside of the pulley *l*, and in the gorge of another pulley *q*, (Fig. 54); *r* is the horizontal beam upon which are fixed the two pulleys *k* and *l*, their trunnions or spindles being received in the two supports *s*, firmly fastened to the ceiling.

The pulley *q* is suspended by a cap also fastened to the ceiling. The beam *r* forms in the centre of its length, a crank or double elbow, to which is suspended a counterbalance *u*, having a weight at its lower end. The movable ring is mounted upon the tip of the crank, and carries a hook to receive the catch of the balance *u*.

To set this machine in operation, the workman draws the cord *p*, at each end alternately; this motion turns the pulleys *k* and *l*, and consequently raises or lowers the supports *e*, which dip the candles into the chest *b*, and draw them out successively. The balance *u* facilitates the movement, by hurrying on the crank after it has passed the vertex, the bar *n*, forming a counter weight, and serving to establish the equilibrium.

With this machine, candles can be made at all seasons, and when practice has rendered its management familiar and easy, its use is a great saving of time and labor, for one workman can turn out twelve hundred pounds per day, whilst by the old method, not a third that quantity could be made in the same length of time. It acts upon the same principle as the preceding apparatus, but being simpler, is particularly adapted for smaller factories. It will throw off one hundred and twenty candles at a time, that is, supposing every frame to carry ten baguettes of twelve candles each. In working by hand, but three baguettes can be managed at once; hence there is a saving by this machine of three to one.

Fig. 55.

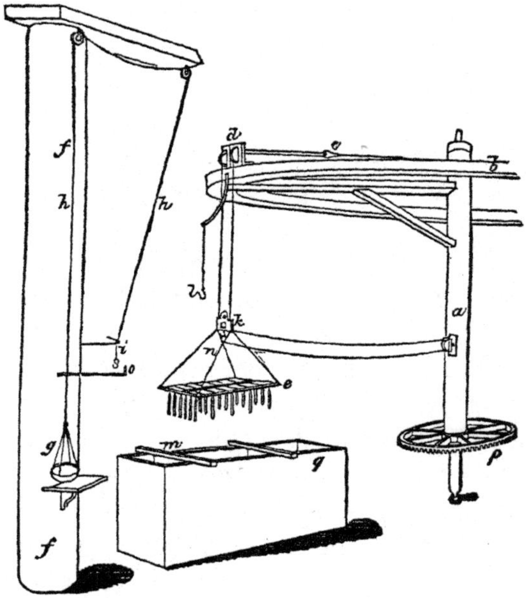

Fig. 55 gives a perspective view of a part of another machine constructed for the accomplishment of the same ends as the two preceding apparatuses. The upright beam *a*, revolving upon a pivot, has near its upper end a large horizontal wooden zone *b*, divided off into twenty equal parts, by the cords *c*, passing upon the pulleys *d*, and carrying at their ends the suspended frames *e*, each bearing twenty baguettes of candles. The equilibrium is maintained by a weight affixed to cords and pulleys, but as this is insufficient when the candles have acquired additional heaviness by successive dippings, there is another balance *g*, suspended by a cord *h*. This cord runs upon two pulleys, and has at its end opposite to the balance, a hook *i*, that is hitched, at will, in one of the holes of the square iron *k*, to which the frame is hung. This arrangement of a second counter weight gives a facility of re-establishing the equilibrium which otherwise would be destroyed, proportional as the candles acquired weight by successive dippings.

This operation terminated, it is necessary to secure the hook of which we have spoken, within a hook *l*, suspended above the frame, for without that, the candles which have been dipped, being heavier than those which counterbalance them, would otherwise descend to the floor. The frame suspended to this new hook, should be drawn towards the pivot, that the workman may fill the dock *m*, with suet. This is done by pulling a cord, one end of which is attached to the frame carrying the candles, and the other passing over one of the pulleys, whose casings are screwed to the pivot. The frame is hooked on to a catch beneath the square iron *k*, by means of a ring *n*.

After having drawn this cord forwards, the workman hitches it by its ring to the hook *o*, in the pillar behind him.

Fig. 56.

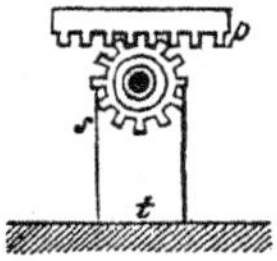

Fig. 57.

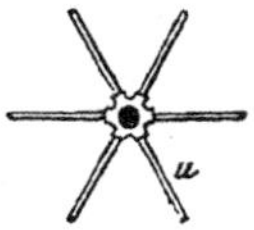

Fig. 58.

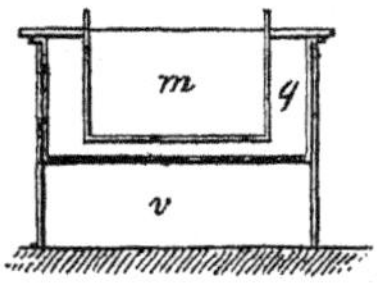

A cog wheel of iron, copper or wood (Fig. 56), is placed near the lower end of the beam *a*, and aided by a second smaller wheel *s*, in which it catches or works, serves to give motion to the beam *a*. It is set in motion by slightly pressing one hand upon the catch fitted to the beam carrying the second smaller wheel *s*, and pushing the other against one of the six branches of the winch A, (Fig. 57,) which is placed upon the end of the shaft of the small pulley *s*.

This shaft or beam is so arranged by the side of the basin *q*, containing the suet, that the workman can turn it without leaving his place. The horizontal beam or shaft which carries the small wheel *s*, has a pivot at each end. These pivots are supported by two pieces of iron or wood (Figs. 56, 59), firmly fastened to the ceiling, with both ends equidistant therefrom, so that the beam shall be evenly horizontal. At one end of the shaft the smaller wheel *s*, is so fixed that it works into the cog *p*, as is clearly exhibited by Fig. 56. At the other end of this same beam are, 1st, the spring catch, which, as particularly arranged, is seen at *x*, Fig. 59; and 2d, the winch with the six arms, Fig. 57, which is movable, and can be taken off when the machine is not in use. These two pieces are firmly fastened to the square of the shaft by means of screw nuts. The winch enables the turning of the shaft by the hand, and it is placed near the workman, about four inches from the place where the frame holding the baguettes of candles descends.

Fig. 59.

Fig. 60.

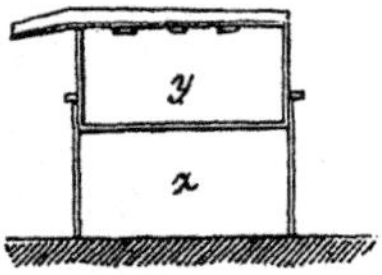

Fig. 58 is a vertical cut of the dock *m*, in which the candles are dipped; *q* the basin, and *v* the furnace which heats this basin.

The iron rognoir seen at *y*, Fig. 60, is a part of this apparatus heated by steam, by means of the furnace *z*, instead of by direct fire, as is usually done. The advantage thus gained, is in the removal of all danger of the tallow becoming scorched or burned, and in the acceleration of the work, three-fourths. This iron should be placed at the side and very near to the wheel with the catch.

With the above machine, a single workman in cold seasons, can easily put through 800 baguettes of candles.

To make our book as complete as possible, we shall describe another apparatus for dipping candles, which, by its real merits and advantages, claims the attention of chandlers. A front and side view of it are seen in Figs. 61 and 62. It consists of a wagon-like apparatus, *a*,

Fig. 61.

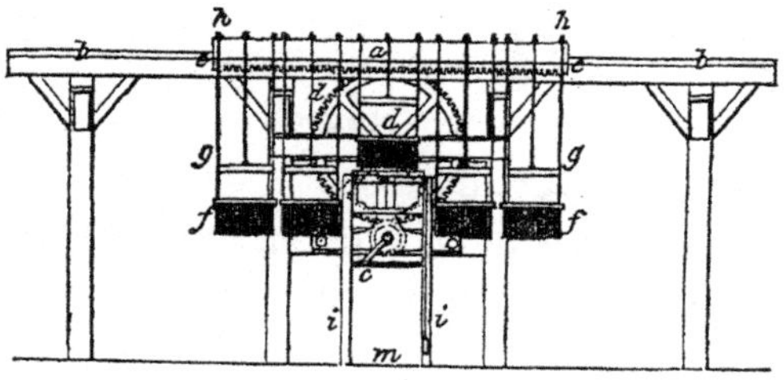

Fig. 62.

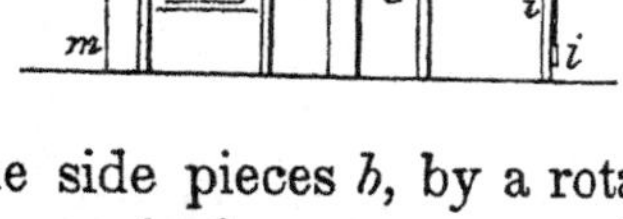

sliding upon the side pieces *b*, by a rotary motion communicated with a winch *c*, to a cog wheel *d*, working and catching in the teeth of an horizontal rack *e*, setting against the bottom of the wagon. *f*, ten wooden frames each bearing 20 baguettes, from which severally hang fifteen wicks ready to be dipped. These frames are suspended by cords *g*, passing upon the pullies *h*. The conductors *i*, formed of four wooden stanchions, and of a movable frame *k*, which has a constant tendency to rise by reason of the counter weight, *l*.

The vessel or wooden chest, *m*.

The mode of working the machine is as follows. After having arranged the wicks upon the frames, the workman, by means of the winch *c*, directs the first frame over the conductors and fixes it upon the movable frame with small nails; then by pressing on the frame, he causes it to descend vertically in the chest *m*, and having thus effected the dip, allows it to remount, and thus continues until the terminus of the operation.

The cord placed in the middle of the frame, serves to give a control in the management of its course.

With this machine a workman can in cool seasons, dip about 10,000 candles daily.

In addition to the foregoing contrivances, Ure, in his Dictionary of Arts, gives a drawing of another, very similar to one which has been already described. It is that which "has been long used in Edinburgh, and is

Fig. 63.

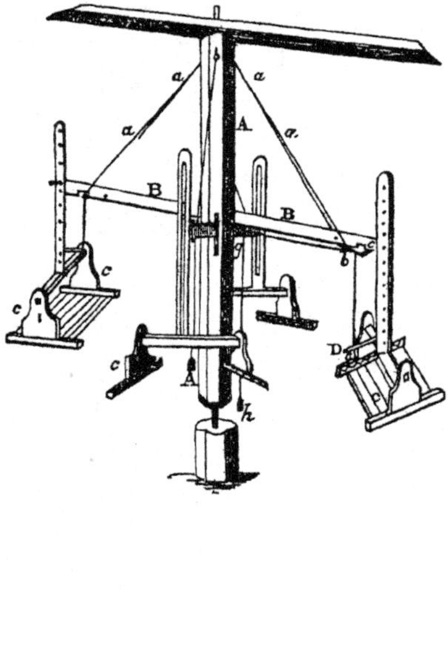

composed of a strong upright post A A (Fig. 63), with turning iron pivots at its two ends. Near its middle six mortices are cut at small distances apart, into each of which is inserted a long bar of wood, B B, which moves vertically upon an iron pin, also passing through the middle of the shaft. The whole presents the appearance of a large horizontal wheel with twelve arms. A complete view of two of them only is given in the figure. From the extremity of each arm is suspended a frame or port, as the workman calls it, containing six rods or baguettes, on each of which are hung eighteen wicks, making the whole number upon the wheel, twelve hundred and ninety-six. The machine, though apparently heavy, turns round by the smallest effort of the workman, and each part, as it comes in succession over the dipping mould is gently pressed downwards, by which means the wicks are regularly immersed in melted tallow. As the arms of the lever are all of the same length, and as each is loaded with nearly the same weight, it is obvious that they will all naturally assume an horizontal position. In order, however, to prevent any oscillation of the machine in turning round, the levers are kept in a horizontal position by means of small chains *a a*, one end of which is fixed to the top of the upright shaft, and the other terminates in a small square piece of wood *b*, which exactly fills the notch *c*, in the lever. As one end of the levers must be depressed at each dip, the square piece of wood is thrown out of the notch by the workman pressing down the handle D, which communicates with the small lever *e*, inserted in a groove in the bar B. In order that the square piece of wood, fixed in one extremity of the chain, may recover its position upon the workman's raising the port, a small cord is attached to it, which passes over a pulley inserted in a groove near *c*, and communicates with another pulley and weight, which draws it forward to the notch. In this way, the operation of dipping may be conducted by a single workman with perfect ease, regularity and even dispatch. No time is lost, and no unnecessary labor expended in removing the ports after each dip; and besides, the process of cooling is much accelerated by the candles being kept in constant motion through the air. The number of revolutions which the wheel must make in order to complete one operation must obviously depend upon the state of the weather and the size of the candles; but it is said in moderately cool weather, not more than two hours are necessary for a single person to finish one wheel of candles of a common size. Upon the supposition, therefore, that six wheels are completed in one day, no less a number than 7776 candles will be manufactured in that space of time by one workman."

The almost universal preference given to dipped candles on account of their comparatively low price, especially by the poorer classes of the community, has made their manufacture rather the most important and extensive branch of the art.

The spirit of emulation engendered among the producers by a healthy competition, is indeed commendable, where it leads to results favorably affecting the interests of the consumer, and more particularly, where the consumers are of that class whose limited means require every protection against excessive charges for the absolute requisites of life. The influence of its aggregate consumption upon the price and quality of an article, is in no instance more evident than in the manufacture of dipped candles, for the very extended demand for this product enables the chandler to bestow more care and attention in facturing it, whilst the enterprise of his equally intelligent neighbors of the same craft, compels him, in order to retain his share of patronage, to lower the price to the minimum profit. Instead of the greasy uncouth looking dips of former years, our more ambitious chandlers now send forth quite a respectable looking candle, and the attention occupied in still perfecting this branch of manufacture will furnish further desirable improvements. Among those of late which have already come to be known, are several very important suggestions, with a view to bettering their burning properties. First of these is that of Mr. Palmer, recommending in the manufacture of mould as well as dipped candles, the application of two similar plaited wicks in the same candle, and in such a manner that they will both turn in the same direction out of the flame. This he effects in making the wicks, by forming one or more strands into a linked or looped plait, as represented at Fig. 64, the strand or strands being composed of several yarns, either loose, or formed into a cord, or otherwise combined.

Figs. 64, 65.

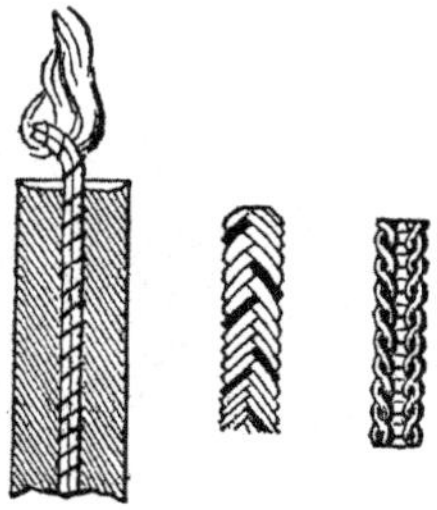

"The machinery used in making the linked plait, is of the kind called 'warp lace machinery;' it should be of a coarse gage, and have only one row of guides; that is, one guide to each pair of needles on the needle bar; each guide is to lap over two needles, in order to form one loop thereon—which, being carried under the beards of the needles, the work that was previously on their stems is forced off, and thus each pair of needles will make a linked plait or cord."

Machines for plaiting and braiding wicks are made by the machinists of this city and New Bedford.

The wicks are strengthened or gimped by binding a thread or yarn around them in a spiral direction, as seen at Fig. 65. The face or bending surface of one wick comes against the back of the other, and they will thus incline out of the flame in the same direction.

The Messrs. Doudney's invention, of a later date, consists also in the use of plaited wicks, but made of coarser cotton yarns and fewer in number, each wick being limited to forty threads. I will here remark that a singeing of the thread, before its conversion into wicks, deprives it of those little adherent fibres, which are a great detriment to the capillarity of the wick, and an obstruction to the ascent of the melted tallow.

It is contended, by these patentees, that the plaited wick used in "vaxeme" and stearic candles, so advantageously, on account of requiring no snuffing, and presenting other desiderata, have failed in applicability to tallow candles, owing to their mode of manufacture; they having in the first place been too small, and when increased by an additional quantity of the same description of fine yarn, failed to yield a better result. Such wicks, plaited with 100 to 120 fine yarns to each, become overloaded, fall, and consequently cause the candles to gutter or run away on one side. Now, the Messrs. Doudney propose making their plaited wick of a coarser cotton yarn, and fewer in each wick, making up the necessary substance for burning tallow by the increased coarseness of the yarns used having only enough twist to retain them in the form of yarn whilst being worked into a plait, it being desirable to have the cotton soft in the plaited wicks. The cotton yarn must be bleached, and of such size that one pound avoirdupois will measure from 5000 to 5050 yards. If the thread or yarn be finer, then its applicability is proportionally lessened. For both mould and dipped candles the sizes most preferable are the following. "For 'short fours,' the wicks are formed of three strands, each composed of thirteen yarns; for 'long fours,' and 'short sixes,' three strands of eleven yarns each; for 'long sixes,' and 'short eights,' three strands of eight yarns each; and for 'long eights' and 'short tens,' three strands of six yarns each. If a greater degree of light is required from any of these sizes of candles, the number of yarns is increased, or coarser yarns are used; but coarser yarns are used in all those cases where the increased size of the wick would require more than forty yarns, of the degree of coarseness above mentioned.

"In making the wicks of such yarns, the strands are to be plaited together tightly, not loosely, and thus made, when used for dipped candles, are each supplied with a thread loop."

Passing on to the year 1845, we meet with a description of Waller's newly contrived apparatus for making wicks of two plaits of three strands each, and which, when joined together, are rendered applicable to dipped candles. The wicks are plaited together "with two sets of three strands, in such manner that two plaits of three strands each are produced and joined together at their selvages, whereby the central portion of the combined wick is made thicker than the two outer edges or selvages. Such plaited wicks are made by a braiding or plaiting machine."

We will not enter into a description of a braiding machine, but only explain how the tappets work to cause the pipes to interplait their strands so as to produce two three strand plaits joined together.

Fig. 66 shows an outline sketch of three braiding

Fig. 66.

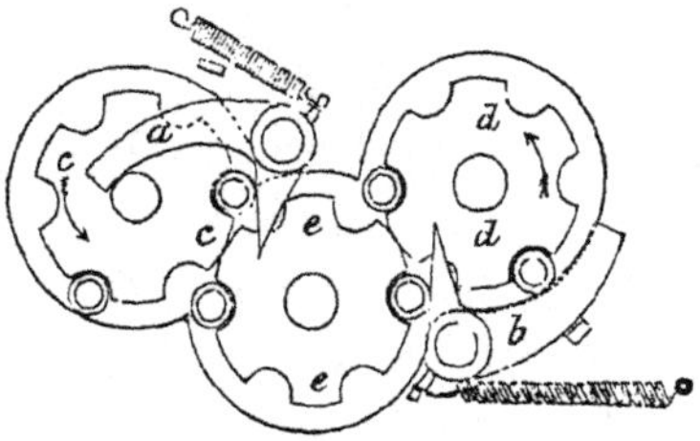

heads or tables, each having six notches there being six pipes; the only peculiarity of the arrangement is in the use of the tappets *a*, *b*, constantly drawn on by their springs, as shown. In the revolution of the heads, *c* and *d*, in the direction of the arrows, they will each produce a three strand plait, and these plaits will be joined together by means of the head *e*, into which the pipes for the time being, that are producing the selvages or edges, pass, so as to link their strands, and by so doing join the two plaits.

The other working of the machinery will be well understood by the workman experienced in the management of such apparatus, or will be explained by the machinist who builds it.

One feature in this invention is, that in applying the plaited wicks to dipped candles, the cost of fastening loops at the upper end of each is dispensed with, and iron wires or rods substituted. These wires, which should be pointed, are run through the upper end of each plaited wick, thus suitably suspending the wicks for the process of dipping.

It is proper to state, in regard to these foregoing improvements, that though they are the invention of respectable and ingenious English artisans, we, personally, know nothing positive either of their merits or demerits, not having as yet had the opportunity of seeing or testing a candle wick'd in any of the ways above described. However, as it is only by catching at and testing all improvements, that a manufacturer, unwilling or unable to be himself the architect of better plans, can hope to attain to any success in his art, we recommend the trial of each of these methods, as there is apparently much reason for their success. The essays can be made in a small way and at little expense, by forming the plaits with the hand. Thus can an approximate result be obtained sufficiently accurate to enable a judgment as to whether the improvement is an advantage at all, and if so, whether of sufficient value to justify the outlay for machinery to carry it practically into effect.

CHAPTER XLV.

PROCESSES FOR IMPROVING THE MATERIAL OF DIPPED AND MOULD CANDLES.

WE proceed to make known several methods much used for *improving* the quality of tallow for dipped and mould candles. The accuracy or efficiency of any of them is not positively guaranteed, though we know that in this city, the candles of a certain chandler always have the preference for exportation. An examination of the tallow of which they are made conclusively shows that its solidity does not depend upon a partial extraction of the oily constituent by pressure, as is done for "summer" candles, but by an admixture of some material with the tallow in the process of refining, which, at the same time that it imparts a solidity, beautifies its appearance, and makes it particularly applicable for such candles as are to be shipped to southern climates. I have myself made essays not dissimilar to the after mentioned, which resulted in partial success, and would possibly have turned out still more favorably, if they had been conducted with greater care and accuracy. There is nothing to be lost by trying, provided the experiments are made in a small way, for even though you may not realize your expectation, some valuable knowledge may be acquired, which will repay the cost of the essays. It will be as well to keep a laboratory-book, and record therein each experiment that you make, consecutively, noting the circumstances under which it was conducted, and the results obtained. In this way may gradually be formed a book which will be very useful for future reference, and thus, in new experiments, a repetition of preceding errors will be avoided; moreover it will serve as a guide-book to furnish new suggestions, by which, while old errors are avoided, new improvements may be devised. A store of information and experience can thus be hoarded up to give character and reputation to the chandler, and to furnish its invaluable assistance in improving and facilitating his operations.

The results of the aforementioned experiments upon a small scale, that is, each upon a pound of suet, are as follows:—Three and a half ounces of alum dissolved in a pint of hot water, is the most proper portion to give the *filet* during the melting. The effect is to promptly stiffen and clarify the tallow. The better mode to be pursued is to finely powder and sift the alum, and dust it in four equal portions, at intervals of fifteen minutes, over the tallow just as it is about to melt. This alum, in its passage through the melted tallow, carries down all its impurities to the water which was added as the *filet*.

Cream of tatar (bi-tartrate of potassa) employed in the same manner as alum, produced even a better result. It should be very finely powdered and intimately mixed with one-fourth its weight of finely pulverized calcined borax. The borax in union with the cream of tartar renders this latter salt soluble in the water of the filet when it reaches it, this being a point of importance. The powder is dusted on similar to the alum. The suet drawn off when cool, is white and dry.

There have been many other modes suggested for improving the quality of candles, the most available of which is that followed in Germany. Take twenty-four pounds of mixed suets, coarsely minced, and place it in a caldron of boiling water, and proportionally as the water evaporates, supply the deficiency. Strain the contents of the kettle through a clean cloth, after which heat the suet for half an hour with two pints of pure water in which one ounce and a half of alum, two ounces of potash, and eight ounces of common salt have been dissolved. The whole being again strained, the same quantity of water is added; and, moreover, to every eight pounds of tallow add half an ounce of saltpetre, half an ounce of sal ammoniac, and one ounce of alum, all finely powdered and thoroughly mixed together. The mixture being heated, or rather gently boiled, until bubbles cease to form, the surface becomes smooth, and a small transparent place is perceptible in the centre; the fire is then withdrawn, and the whole left to cool. As taken from the caldron, the lower surface of the blocks is covered with the precipitated impurities, to separate which, and give a final cleansing to the tallow, it must be melted anew with two drachms of fine saltpetre, and after skimming off the scum that arises to the surface, the clear tallow is run into proper vessels. It is said that the candles made of suet thus rectified endure two hours longer than ordinary candles, and when furnished with wicks, half and half cotton and hemp, soaked in alcoholic solution of camphor, last still longer, and without running or guttering.

Another process for improving tallow, very similar indeed to one of the preceding, is that which provides for the melting of the suet together with one-fourth its weight of water. It must be well stirred, and skimmed of all dirt that may arise during the heating operation. The water is added to prevent the suet burning at the bottom of the kettle, but it must not be poured in whilst the suet is melted, else a commotion will ensue, and the hot material be ejected to all points. The melted tallow being strained through a cloth or fine hair sieve, is replaced in the kettle with the same quantity of water in which have been dissolved, to every twelve pounds of tallow, three-quarters of an ounce of saltpetre, three-quarters of an ounce of sal ammoniac, and one ounce calcined alum. The mixture, being slightly boiled until the surface becomes smooth and free of bubbles, is then drawn off and cooled. The solid blocks, after the dirt is scraped from their bottoms, are again melted carefully and settled, so that all impurities may deposit. Drawn off clear and cooled, it is then fit to make superior candles.

We have introduced the preceding remarks preparatory to writing of moulded candles, because the tallow refined by the foregoing processes, is equally applicable to those as to dipped candles.

CHAPTER XLVI.

MOULDED CANDLES.

THESE candles differ in their mode of manufacture from the dips, in being cast at once into a mould suitably arranged, with the wick in its centre, and shaped to receive the melted tallow; while the former acquire their full size by successive layers imparted by several consecutive dippings. They are symmetrical, and the manner of procedure will be readily intelligible after we have described the instruments requisite in the operations of moulding. This kind of candle carefully made of good white tallow burns and looks well; but when factured negligently and of impure common material, they are a nuisance to the most indifferent housekeeper. The melting kettle used is similar to that described at page 329. It is mounted in a brick furnace, by the side of which is placed the tinette *b*, Fig. 67, similar to an iron bound

Fig. 67.

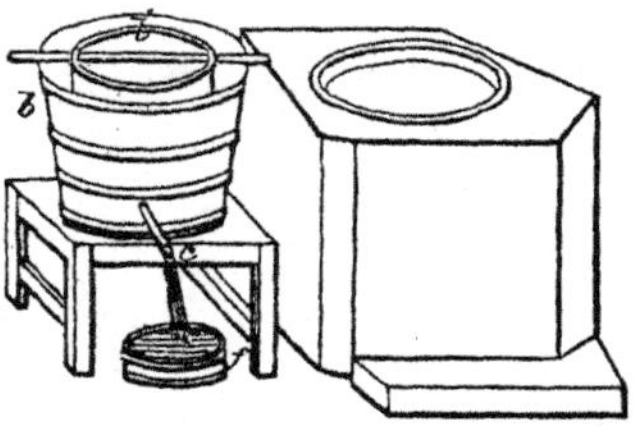

wooden cask. This tinette is placed upon a stool sufficiently high to allow the free passage of the burette (feed-can) *d*, Fig. 68, under the cock *c*. This burette is filled

Fig. 68.

with melted tallow by placing it in the tub *f*, which is there to catch any stray drops of suet that may leak from the exit hole, and would otherwise be wasted. The workman is elevated upon a step in order that he may the more easily handle the pail H, Fig. 29, in dipping the tallow from the kettle and passing it over upon the sieve I. This sieve is in the form of a kettle, pierced with innumerable holes throughout its bottom. Directly opposite to each other are two handles, through which passes a stick of wood, which, itself resting on the edge of the tinette, in turn holds up or suspends the sieve.

The mould stand, or frame, Fig. 69, is formed of an

Fig. 69.

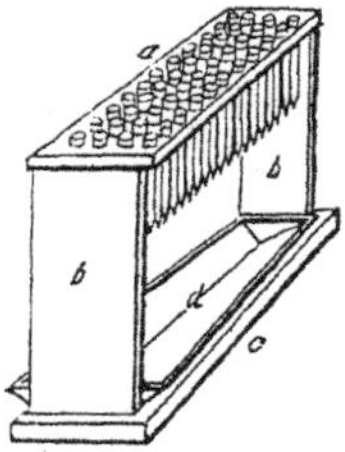

oaken top two feet long by nine inches wide. It is pierced throughout its length with three ranges of holes into which the moulds, slightly conical, enter easily up to the *culot*. The top of the stand is supported at each end by two upright pieces of plank, *b*, fitted by tenon and mortice, both in the top and the base piece or pedestal *c*. As the top of the stand has to support the whole weight of the moulds and their contents, it should be made of good solid walnut or oak wood. Between the legs or side pieces are placed the egouttoirs, *d*, for the reception of any portions of suet that may accidentally fall from the top of the stand.

Instead of wooden tops, the mould stands sometimes have metallic bed plates, in which the culots of the moulds are soldered. This is a much neater arrangement, and, coupled with the present plan of having an additional cross piece B, Fig. 77, (bored correspondingly and proportionably as the upper bed plate,) to receive the tips of the moulds, and thus render their position more steady, makes the apparatus both complete and convenient.

There should be a large number of these stands, furnished with moulds of assorted sizes; one portion being filled with sixes, another with fours, another with eights, another with twelves, and so on to complete the series requisite to make all the numbers of candles called for in the market. The number in one stand of the sizes above named should not be above thirty-six, otherwise there will be much loss of time in cooling them.

The burette, Fig. 68, is the can from which the moulds are filled with liquid tallow. It is supplied from the tinette, Fig. 67, through the cock *c*.

The moulds, Figs. 70, 71, 72, are slightly conical tubes made of iron, bronze, pewter, or a mixture of lead and tin, and occasionally of tin alone. They are even made of glass in some countries, and but for their fragility, these, uniting all the desirable qualities, are far the most preferable. As before said, there should be a liberal number of moulds on hand, first because the variety of sizes requires it, a different dimensioned mould being necessary for each number of candle; and secondly, a large stock of the same size is requisite, especially in extensive establishments, for as the candles cannot be drawn until they are entirely cooled, much delay and loss of time would be occasioned in working with a too limited number of stands.

Fig. 70.

Fig. 71.

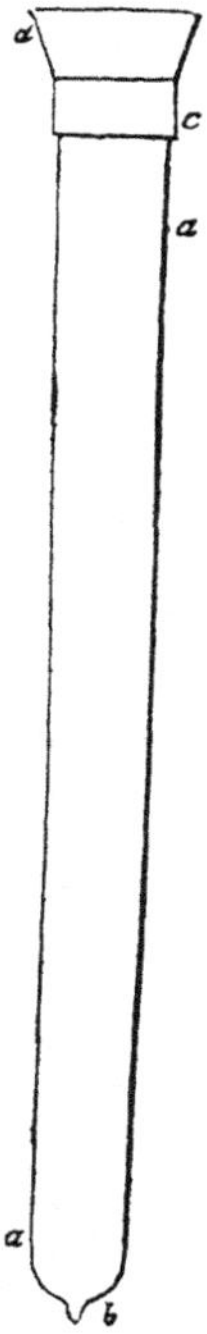

Fig. 72.

Fig. 73.

Each mould consists of two parts, the shaft or body, and the top piece or *culot*.* Fig. 70 shows these two parts separate. The shaft *a a* is conical at its lower end, *b*, to give form to the tip of the candle.

The mould is pierced, at *b*, with a hole sufficiently large to allow the entrance through of the wick. The shaft *a a*, is widened at its upper end for the reception of the barrel *m*, of the culot *d*. The culot is a kind of small funnel through which the liquid tallow runs into the mould. Fig. 71 shows these two pieces united together, and Fig. 72 represents the position of the wick

* This description refers to the *French* moulds, of which the figures above are representations. Those used in this country are spoken of on the next page.

in the mould previous to the pouring in of the tallow. The greater diameter of the mould at its upper extremity is indispensable, in order that the candle, when it has cooled, may be easily drawn from it. At the top of the culot will be observed a small iron hook *n*, soldered upon the side, and sometimes supported by a small iron or copper triangle *o*, soldered interiorly to the side of the culot, and enabling the hook *n*, the better to resist the tension of the wick. This hook should be exactly in the axis or centre of the mould, so that the wick may be uniformly enveloped with tallow throughout its whole length and circumference. The moulds of metal are bored out as accurately as possible, by machinery, so that their interior shall be perfectly true and finely polished.

Figs. 74, 75, 76, represent the different numbers of candles most generally cast, viz., 4, 5, 6, &c. It will be observed they are of a graceful shape, the tip especially being very neatly formed. These were run in Dyre's moulds, the superiority of which consists in their being burnished by a vertical instead of rotary motion, and thus freed of those rings and imperfections which in common moulds offer a resistance to the free egress of the candle. The composition too is of a better kind, and more intimately combined, whilst the workmanship is more elaborate than that upon moulds generally. They consist of only two pieces, the shaft and the tip, the latter being sometimes made of brass, or bushed with brass, so as to resist the abrasive effect of the threading wire and wick pegs, but they are so easily corroded, that experience has rendered them obsolete. The metal for tallow candle moulds Mr. Dyre composes of tin and lead —the latter in proportion less than that used by other pewterers, because, as he says, though costing a little higher, the moulds are then improved in a greater ratio.

For mould candles he makes five sizes, viz., 16s, 12s, 10s, 8s, and 6s.

For sperm candles, the moulds are made of the same metals as for the aforementioned, with the addition of a little hardening. The usual size for these are 4s, 5s, and 6s, as represented by the figures.

Figs. 76, 75, 74.

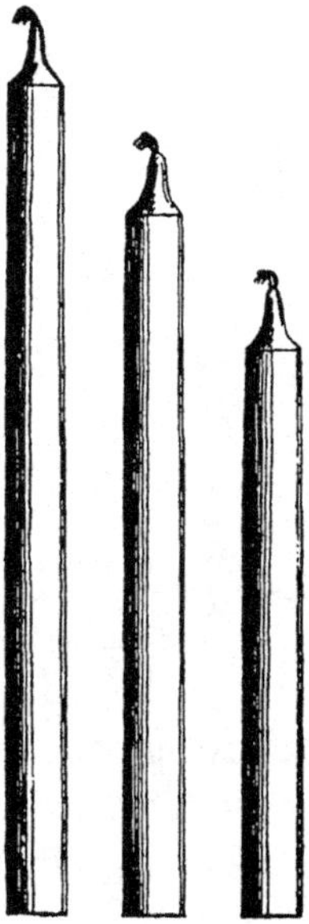

Fig. 74 is a 6 candle, and its dimensions are $8\frac{5}{8}$ inches in the length, and $\frac{7}{8}$ inch in the diameter of the shaft; the tip in this, as in the rest, being 1 inch high.

Fig. 75 is a 5 candle of $10\frac{3}{4}$ inches length, and $\frac{7}{8}$ inch diameter in the shaft.

Fig. 76 is a 4 candle of $12\frac{3}{4}$ inches length, and $\frac{7}{8}$ inch diameter in the shaft.

The moulds used for the stearin and stearic candles are of the same size but made of tin and antimony; lead being omitted because it not only smuts these candles, but prevents their easy withdrawal. These moulds should be made with care, and with as little metal as possible, and some manufacturers, in order to remove all impediments to the rapid cooling of the liquid candle, have the moulds turned smooth on the outside. This seems expensive, but as there is no more metal bought than is actually required, the amount of their cost does not much exceed that of the same number of rough heavy moulds Occasionally, the dimensions of the moulds are altered to suit the directions and taste of the purchaser, but whatever their form, those for mould candles should never contain more lead than is necessary to render tin free of grain and easy to be worked. Some factors add lead even to the stearic moulds, but Dyre effects the same result, which the addition of that metal would produce, by other and better means.

Fig. 77.

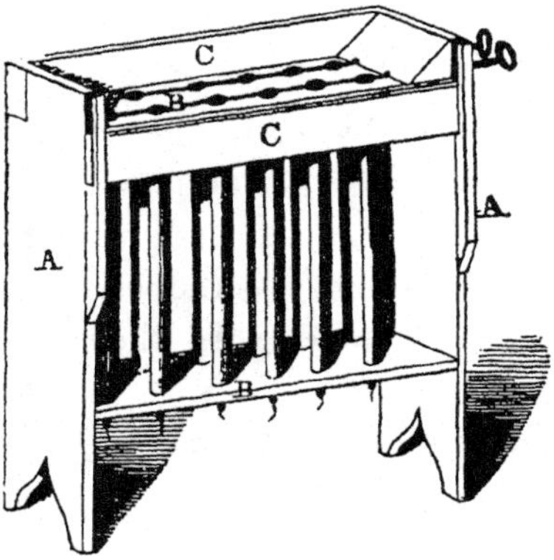

Fig. 77 exhibits a stand of twelve of these improved moulds. I am indebted to the kindness of the manufacturer, T. I. Dyre, Jr., Philadelphia, for the privilege of copying it. A A are the two upright ends of wood, which support the two cross beds B B, that serve to hold the moulds in an upright position. The upper bed is of metal, this material being better than wood, to which the redundant tallow sometimes adheres so tenaciously as to cause a delay in its removal. C C are the broad side pieces, which, with the upper ends of the uprights, form a receptacle for the melted tallow when the moulds are being filled. One of these side pieces slides in a groove so that it can be moved downwards to afford facility, first in threading the moulds with wicks, and again in removing with a knife the excess of tallow which congeals upon the metallic bed. The whole arrangement is very convenient, and is well worth the additional cost above that of the commoner stands of moulds.

At *d*, are seen the wires which traverse the upper bed of the stand, immediately over the mouths of the moulds, for the purpose of suspending the wicks. Each stand is of a convenient size, and by an expert workman can be rapidly and dextrously furnished with wicks, in manner as follows, supposing the wicks to be already cut and prepared, previously to commencing the operation.

For this purpose is used a strong iron wire needle *a a*, (Fig. 73,) sufficiently small to pass freely with its catch *b*, at the lower end through the small aperture in the tip of the mould, and crooked into a ring at the upper extremity so as to admit the indicator (finger), for the purpose of facilitating its movement. This needle is several inches longer than the mould. The workman lays the frame upon its side on a neighboring table, and holding in his left hand a quantity of wicks, with the right takes the wire needle and introduces it within the mould. As soon as the catch appears at the exterior mouth *b*, of the hole in the tip, the looped end of the wick is hooked on, and is carried through into the mould by immediately drawing back the wire needle. A small pointed wire running across the diameters of the moulds, in each row over the whole length of their upper bed, is passed through the loop of each wick as it is drawn up, and serves to suspend the wicks and prevent their falling back into the moulds, when the frame is placed in an upright position. This manipulation is generally done by a female, who, as soon as every mould in the stand has been strung, stretches and straightens each wick, by pulling at the lower end and fastening it in the hole of the tip with a little wooden peg made and kept for the purpose. The stand being in an upright position, and the loops being also adjusted over the upper wires by which they are suspended, so as to exactly centre the moulds, all is in readiness to receive the melted tallow.

To succeed well, the temperature of the tallow must neither be too high nor too low when it is poured in the moulds. If too hot, it heats the mould, and thus causing the candles to "stick" or adhere tightly, renders it difficult to draw them; and when drawn, the candles are, as the workman says, *tavelées*.

In such case the moulds must be doused with warm water after they have cooled, and the candles dextrously drawn out. When, on the contrary, the tallow is too much cooled, the candles are not uniform in appearance, but granular looking.

The proper moment for casting the tallow is, when a congelation of its surface commences near the sides of the kettle in which it is contained.

The moulds are filled from the burette, (Fig. 68,) and being stoppered at the mouth of the tip, there is no waste of tallow by leakage. So soon as the burette becomes empty it is refilled from the tinette. For mould candles most chandlers do not use the wooden pegs to keep the wicks distended in the moulds; in such cases, after having poured in the tallow, that portion of the wick which remains beyond the tip of the mould, must be laid hold of with the fingers, and straightened by pulling before the tallow congeals, else the wick will shrink up and thus injure the quality of the candle. As one of the general properties of caloric is to dilate bodies, so accordingly the tallow as it cools occupies less space than whilst hot and liquid. For this reason the mouth of the mould, which, after semi-congelation of the contents of the mould, is not so full as it was at first and ought to be that the candle may be even and of full length, must be replenished with more melted suet. The moulds should not be handled, but set aside in the frames to cool. When the tallow has entirely congealed, the wires laying upon the top bed of the mould stand, and serving to suspend the wicks, are carefully drawn out, the overplus of tallow scraped from this bed with a wooden spatula or paddle, and the wooden pegs, if they have been used, removed from the openings in the tip. This done, the stands are placed upright, and the workman goes along with a kind of bodkin in his hand, and after pressing against the bottom of each candle with his thumb to loosen it, an effect recognizable by a simultaneous snapping noise as the candle becomes detached, catches this bodkin in the loop of the wick, and draws out each candle successively from its mould. Or, from those moulds different from Dyre's, and such as have been described as having culots, the candles are readily drawn by raising this culot. Some chandlers pursue the better plan of cutting them at the junction of the culot with the shaft of the mould, and then pulling the candles attached to the culot. Fig. 78 represents a candle as drawn from the French mould. Those from Dyre's moulds are shown at Figs. 74 75, 76.

Fig. 78.

In all the manipulations of moulding candles, a little experience will impart surprising dexterity. The rapidity with which a stand of moulds can be threaded with wicks by one girl, dextrous in that particular branch of the work, is almost incredible. The amount of moulding to be accomplished, therefore, in a day, is in proportion to the activity and expertness of the manipulators.

Tallow candles, either dipped or moulded, as made from ordinary material, are always more or less yellowish when recently factured, but acquire a whiteness by age. But this whiteness is only skin deep, if we may use the expression, and is not of that desirable brilliancy which the ambitious chandler takes a pride in imparting to his product. The best means to provide against any imperfect appearance is to make the candles of tallow refined by some of the aforementioned processes; but when this is not done, then other means are resorted to for remedying the neglect.

Most factories are so built as to have an adjoining lot, which can be green sodded to serve as a plat on which to expose the candles in the open air to the influence of the dew and shade but free from the rays of the sun, and for this purpose, light frame sheds solidly constructed, should be built over them and furnished with a movable canvas covering, so that it can be put on and taken off at will. The bleaching thus acquired is only superficial, however, the white layer being very thin, and if yellow tallow has been used, is very soon overshaded by the inner strata. The best way then, as above said, is to make the candles of tallow previously bleached, by any of the methods heretofore given. When such suet has been used, the candles, on exposure to air, acquire a brilliant whiteness, and improve in quality. The rationale or theory of this bleaching by exposure is not definitely known, or rather has not been satisfactorily explained.

The atmospheric air assisted by the influence of light produces the change of color, perhaps, by reason of its oxygen, which being given up to the coloring matter of the material, unites either with it or its elements, carbon and hydrogen, and thus effects a bleaching of the candle.

It has been said that tallow candles whiten by age; this is true, but there is this inconvenience and disadvantage in keeping them, and that is, they become rancid and smell unpleasantly after a time. When there is no compulsion to make forced sales, it is well to keep them in boxes lined with brown paper, or, better still, to envelop them in packages of two, four, or six pounds each, and preserve these packets in clean closed cases. The whiteness which they thus acquire is more durable than that obtained by exposure to air. Some chandlers have carried into practice, for whitening tallow, that process followed for bleaching wax, viz., ribboning and spreading it upon cloths, but the tallow is thus deteriorated in quality, and the candles, made therefrom, run. This mode, therefore, is not applicable to tallow, and has consequently been abandoned.

In extensive establishments where the quantity of material to be worked up is large, every improvement which tends to facilitate the operations and lessen the amount and expense of labor, should receive consideration proportional to its merit. Acting under this rule, below are presented descriptions of several contrivances having for their purpose the more ready, rapid, and economical moulding of candles, but as to their respective advantages we urge nothing positive, preferring to leave to the judgment of the intelligent chandler, the practical applicability of either or all of them. The first machine for this purpose that we shall make note of is arranged so as to produce 396 candles at a time. The moulds are of corresponding number, and ranged in a stand, similar to the ordinary mould frame, on the top of which is a second stand, but ledged and of cast iron or composition metal, and pierced with holes corresponding with those in the wooden stand beneath. In each of these holes is lodged a *godet* or kind of small funnel, to which is attached the wick which descends into the moulds. The metallic stand is so arranged, that by the aid of a winch which moves a cogged rack, it can be lowered or raised at will. When the candles are about to be cast, the metal stand is lowered until the godets touch in the top of the moulds, and the fluid tallow being then poured upon this stand, runs through the godets into the 396 moulds. After perfect congelation, the metallic plate is raised up by the winch, and with a long knife, the candles are all cut at a single dash. In warm weather, the moulds are cooled simultaneously in a vessel of cold water, elevated so high that the moulds filled with suet may dip their whole length in the water. Figs. 79 and 80 show a front

Fig. 79.

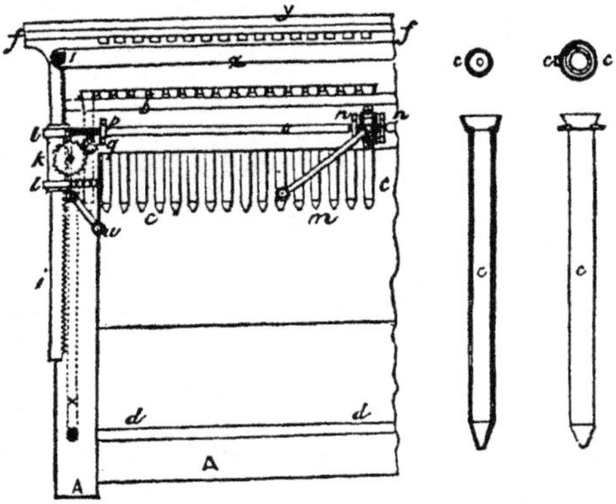

elevation and plan of the apparatus. We have suppressed in these figures about one-half of the breadth of the machine; they are elongated on the right by being drawn out. All the suppressed portion is similar to that part shown by the figures. The winch *m*, is in the centre of the total breadth of the machine. A is the frame work of oak wood. The table or stand *b* also of oak wood, mounted upon the feet and upper cross pieces of the frame, is pierced with 396 holes in 12 rows of thirty-three each, and intended for the reception of the metallic moulds *c*, one of which is represented also by *c*, in elevation, vertical cut, and in detail by the figures.

The platform *d*, is for the support of the copper vessel *e*, containing water for the cooling of the moulds filled with liquid suet.

The metallic table *f*, is pierced throughout its surface

with 396 holes corresponding in size and position with those in the wooden table *b*, as is seen at one of the corners of the stand or frame *f* (Fig. 80). The holes of the metallic stand are intended to receive the godets or culots

Fig. 80.

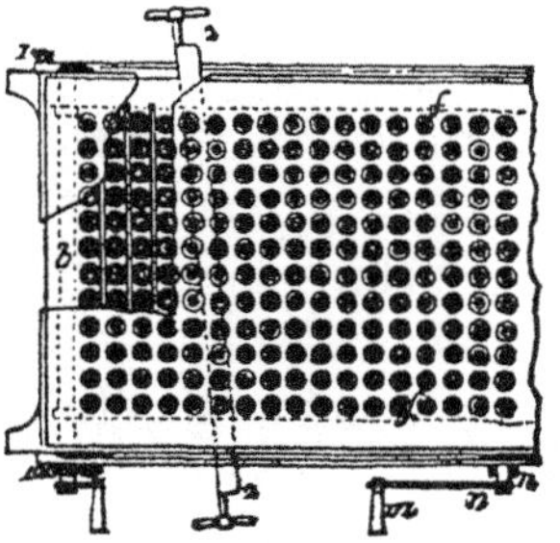

g, each of which is fitted as shown at Fig. 82, in the middle, so that it will be immovably retained in its place at the time of the drawing of its candle.

The metallic stand or frame *f* is mounted upon the upper ends of four racks *i*, gliding in copper boxes, *k*, fastened by the iron stays *l*. These racks are made to assume an ascending or descending motion by means of a winch, *m*, the axis of which carries a square tip that catches in or works the two cogs *n*. Each of these cogs

Fig. 81.

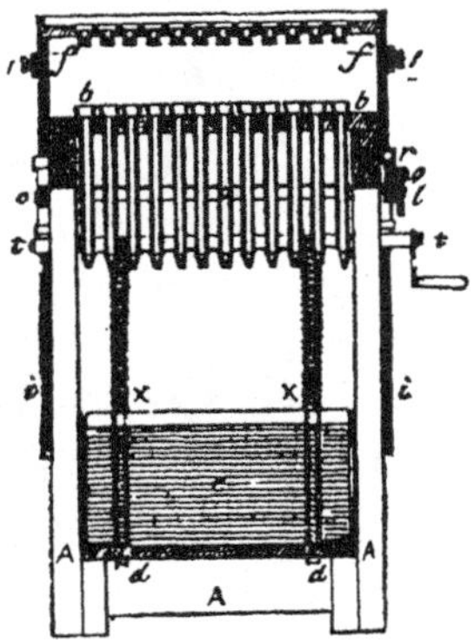

is fixed at the end of the two support beams *o*, which at the other extremities carry screws *p*, catching in or working an oblique toothed cog wheel *q*, at the centre of which is a pinion that governs the wheel *r*. The axis *s*, of this wheel (Fig. 81) has two pinions *t*, which catch or work in the teeth of the racks *i*, and thus create an ascending or descending movement of the frame *f*, accordingly as the winch is turned. When the moulds *c* are to be cooled, the vessel *e*, filled with water, is raised by the aid of two winches *u*, the axis *v* of which, carry two pinions that catch in the chains *x*, the ends of which are fixed to the water bath.

The iron frame *y* forming the ledge of the metallic table *f*, confines the melted tallow when poured in, and prevents any waste by overrunning. The two iron plates *k*, retained horizontally by two screws fastened by eared nuts *i*, serve as guides to the knife 2, which, when the candles are drawn out of the moulds, cuts them all beneath the godets *g*.

The manner of working this machine is as follows. Commence by placing the moulds in the holes of the wooden stand *b*, and the godets in those of the metallic stand *f*. This done, then lower the upper stand, by properly turning the winch *m*, until the godets are entirely enclosed by the upper and widened end of the moulds *c*. Arrange

Fig. 82.

the wicks *g* (Fig. 82) in the usual manner, and retain them in a proper position by the hook in the centre of each of the culots. All being ready, the melted suet is then poured in upon the table *f*, whence it runs into each of the moulds.

When the material has cooled and hardened, either in

cold weather by the atmosphere, or in warm weather by the water in the bath, the metallic table to which the candles are adherent, is lifted up and the candles detached by a knife which is kept for this purpose, but not attached to the machine. The operation finished, the godets are removed and cleansed in water, before repeating the filling of the moulds. That the result may be perfectly successful, it is important that the metallic stand *f*, should always be sufficiently warm to prevent a too rapid congelation of the liquid suet. To this end, the circumference of the upper table exceeds that of the lower stand some three inches, and upon this excess or edge is placed a tin box two inches in height. The admission of a constant current of steam into this box communicates thereto, and from thence to the metal stand, heat sufficient to maintain the proper fluidity of the tallow. At one corner of this steam box there should be fitted a stop-cock, through which to discharge the condensed steam as it accumulates.

Another apparatus, which, with the aid of a single workman, is said to produce 600 candles in 30 minutes, is represented by Figs. 83 and 84. It has the form of a six footed table, seven feet in height. Its upper surface or top consists of a hard wooden board *a*, of fifteen lines thickness, and pierced with 600 round holes, in which are placed the candle moulds, fastened by two screws through the small catches which they carry about five lines above their grand opening.

At eight inches below this first board, there is a second, *b*, pierced as is the upper similarly and numerically with holes, in which, sets a part of each mould, merely that it may be kept true and in an upright position. Finally there is a third board *c*, which contains 600 funnels, corresponding with the moulds in the stand beneath. The top surface of this last board is divided off into four equal

Fig. 83.

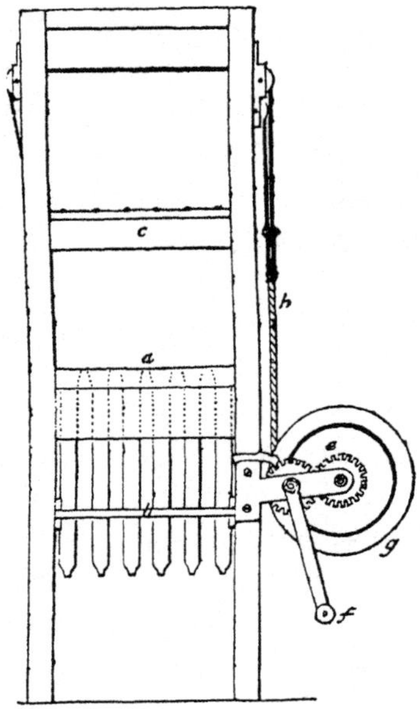

Fig. 84.

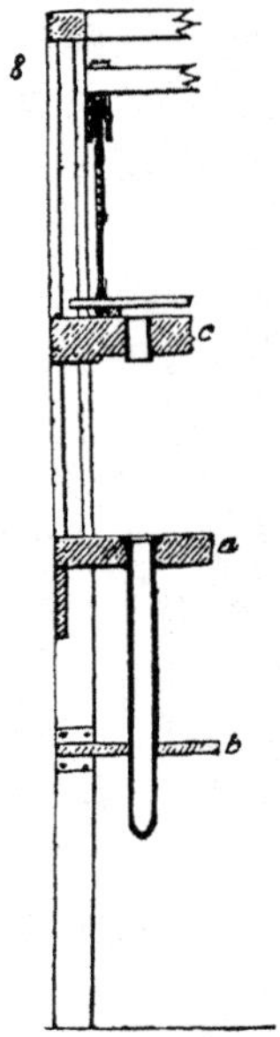

parts by the iron frames upon which are placed the rods suspending the wicks that are to pass into the moulds. Into these frames, by means of a funnel for that purpose, is poured the quantity of tallow necessary to fill the number of moulds placed thereunder.

To raise and lower this board or plateau, and by these means simultaneously draw all the candles from their moulds, there are fixed upon the upper extremities six iron jack screws *e*, eighteen inches in height, and fastened against the six feet of the table. These screws are propelled by turning the winch *f*, and in this movement or motion the wheel *g*, revolving upon the axis of one of the wheels of the jack screw, winds the cord *h*, and lifts the board or plateau;—the candles are then drawn from the moulds, and hang in the air from the plateau to which they are suspended, and from which they are now to be cut with a knife.

At the moment that the plateau or board *c*, is being stripped of the candles which it holds, and of superfluous suet, which has been poured thereon in casting the candles, cover the plateau *a*, so as to protect the empty moulds from falling particles of tallow and other foreign matters.

The above apparatus much improved and added to, is shown in front view at Fig. 85, and in plan at Fig. 86.

Fig. 85.

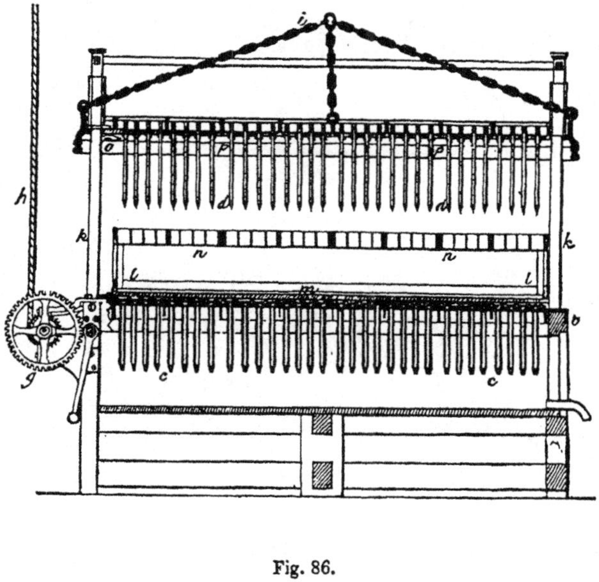

Fig. 86.

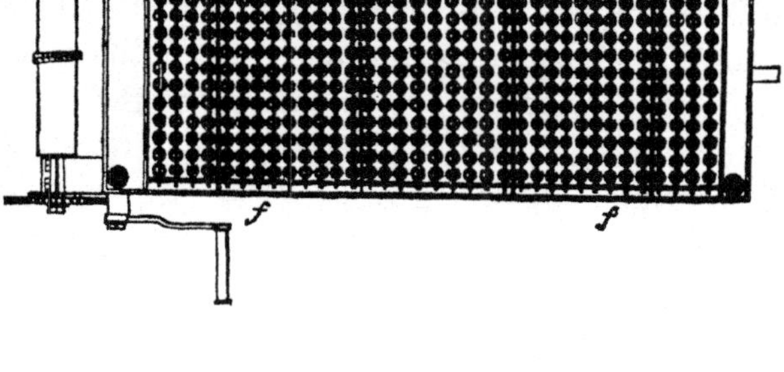

The top or upper table *a*, is seen in plan at Fig. 87, and horizontally and longitudinally at Fig. 88. The lower table *b*, is shown in plan by Fig. 89, and horizontally by Fig. 90. These two tables are pierced with

Fig. 87.

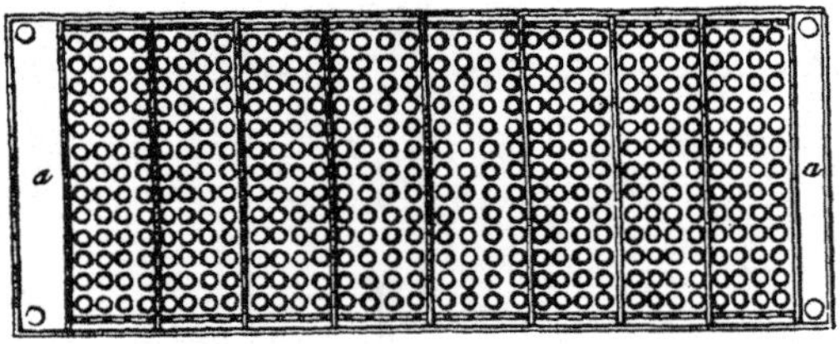

Fig. 88.

Fig. 89.

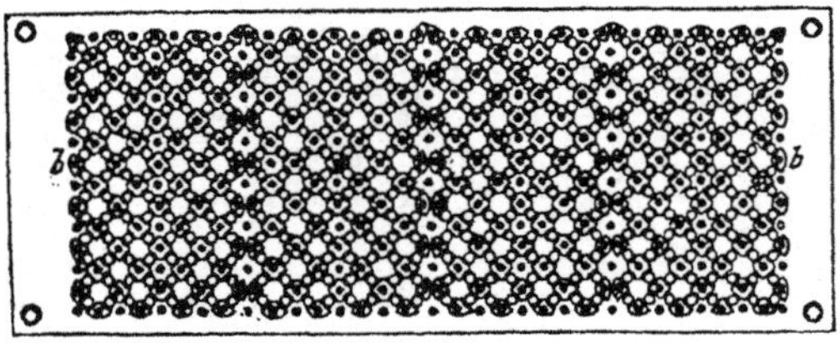

Fig. 90.

holes; those of the lower receive the moulds *c*, and those of the upper table the culots *d*, which, penetrating the orifice of the moulds, prevent, during the casting of the candles, the overflowing or spilling of the suet upon the under table. The moulds are caught and lodged in the lower table by iron hooks *e* (Fig. 90), fastened by screws, the cleft heads of which are seen in Fig. 89. There is a space left between the table and the hooks in order that when the candle is being drawn, the mould, as soon as detached therefrom, may fall by its own weight, and allow the candle to escape without resistance. The wicks are strung upon iron wires or rods *f*, (Fig. 86) laid level upon the upper table and notched at proper intervals to indicate the position for the wick. Each wick is furnished with an iron needle, and in this way is obtained the facility of threading as many wicks at a time, as there are moulds in the width of the table. The drawing of the candles is effected by means of a windlass or winch *g*, fastened to the foot of the machine, and upon which runs a cord *h*, that passes upon pulleys, and is attached to a chain *i*, fixed to the centre and each of the four corners of the upper table. To avoid the oscillating motion of the lower table, there are adjusted four iron columns *k*, which pass into or through four copper clamps fixed upon the upper table. The cutting of the candles is accomplished by raising the upper table so that there can be placed under, a wooden frame *l*, laying upon a wooden plateau *m*, in order that no particles of tallow or dirt may fall back into the moulds. When this frame is properly set, the upper table is lowered and each candle drops into a square *n*, formed of tape, so that at the moment the candles are detached with the knife, they shall not, in falling, become bruised or broken.

The knife *o* (Fig. 85), which is used for this operation, slides upon two iron strips *p*, placed at a proper height, and is removed after having effected its purpose.

In order that the candles may with facility be drawn at all seasons, the moulds are assembled together in a closed chest, framed between the feet of the machine, through one side of which enters a cullendered tube *q*, to lay its whole length, in the interior of the chest. A current of hot air or steam, expands the moulds and enables the candle to glide easily out; this application being most requisite in hot and damp weather. With the above apparatus one man can run 416 candles within the space of ten minutes, and draw them all simultaneously in a few seconds.

Mr. Morgan's machine for moulding candles is another contrivance for hastening the manipulations of casting and increasing the product of candles in a given time. It is so constructed that with a sufficient number of frames of moulds to allow the requisite cooling, the action of the apparatus may be continued for any required period, and at a saving of both labor and time;—the cotton wicks being threaded through the moulds at the same time, and by the same action as that which expels the candle. Fig. 91 represents an end elevation of this improved machine; Fig. 92 a front view, and Fig. 93 a plan

Fig. 91.

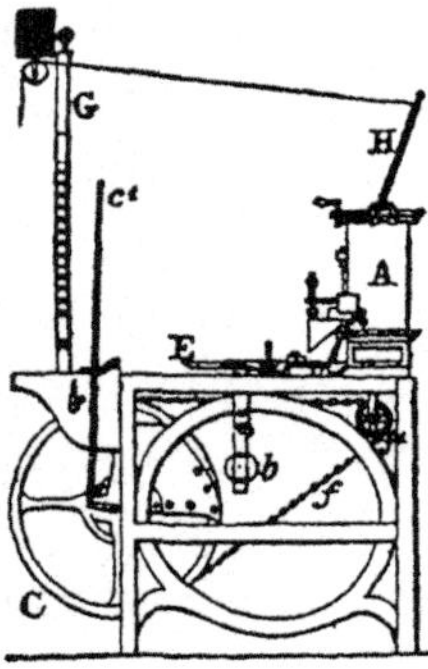

Fig. 92.

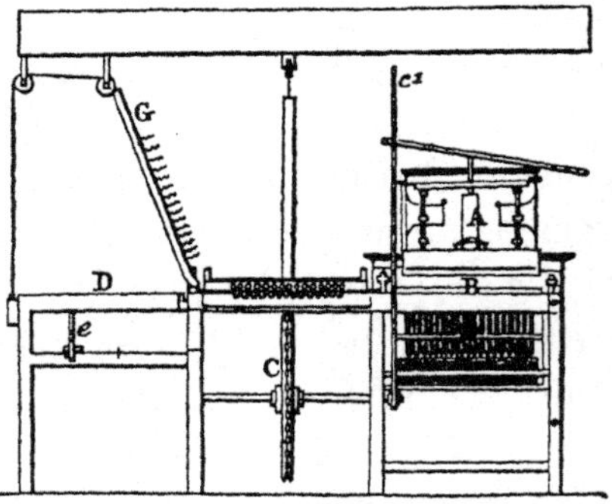

Fig. 93.

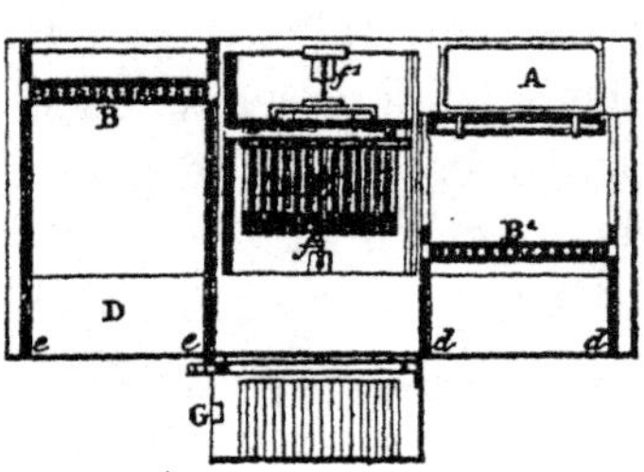

Fig. 94.

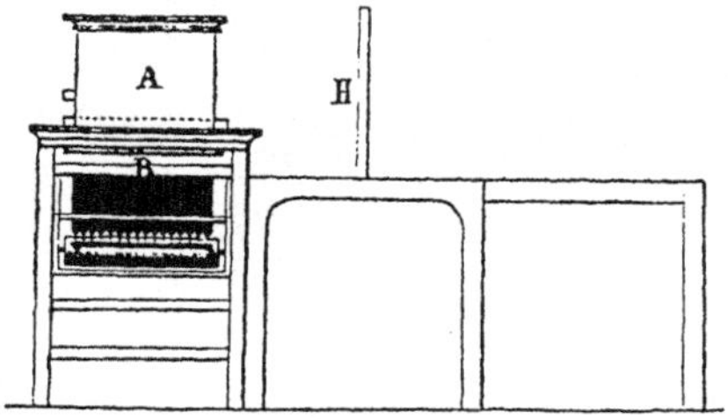

of the same; Fig. 94 is an elevation seen from the back of the apparatus in the opposite direction to that represented at Fig. 92. In these figures A represents the vessel or reservoir in which the melted tallow or other material for making the candle is contained, and B exhibits a series of moulds, for shaping the candles in the usual manner; but as the arrangement and construction of these moulds in which the candles are formed, vary from those of ordinary construction, this part is shown on an enlarged scale at Fig. 95, and some of the detached

Fig. 95.

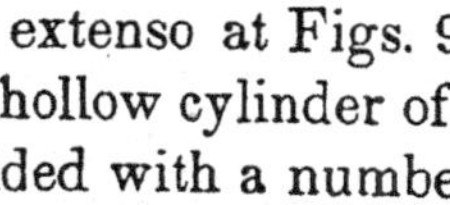

parts still more in extenso at Figs. 96 and 97. In Fig. 95, *b b* represent a hollow cylinder of tin, opening longitudinally and provided with a number of bobbins of cotton or other wick corresponding with the number of moulds to be supplied. These bobbins revolve freely on a shaft which passes the whole length of the cylindrical case *b b*, and gives off the cotton wick as it is required for the formation of the candles, as hereafter explained.

Fig. 96 shows an elevation of the upper end of one of the cylindrical moulds, as given at B, in the other figures; and Fig. 97, a plan view of the same in an inverted position, so that the interior of the mould is there represented.

Figs. 96, 97.

In these two figures, 96 and 97, it will be remarked that the top is not constructed of one piece, as in ordinary moulds, but consists of the part *b* 1, which is a portion of the cylindrical side of the mould, and the part *b* 2, the sliding or movable portion. This latter, *b* 2, is hollow, for the passage of the wick, and fits closely to *b* 1, when the tallow is poured into the mould to form the candle. But as soon as the candle is cold, and in a condition to be removed, instead of being drawn out in the usual way, it is, by the arrangement of this apparatus, forced out by pressure applied to the extremity of the part *b* 2, through which the wick passes from the cylindrical case *b b*; the wick that goes out by a hole in the side of *b* 2, following the course of the candle as it is forced from the mould, re-cottoning the mould for a successive candle.

Fig. 98 exhibits a series of nippers or pincers which open and shut together by the action of the lever C, and

Fig. 98.

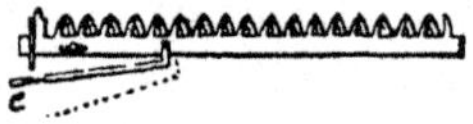

Fig. 99.

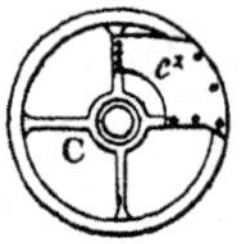

hold the wick in the mould (at the end opposite to that of its entrance at the hole in *b* 2), in a perpendicular position when the tallow is introduced.

The separate and detached parts of the apparatus being described and represented at Figs. 96, 97 and 98, we proceed to make known the mode of working the whole apparatus. "Supposing a set or frame of moulds, as represented at B, to be regularly wicked, and each wick held accurately in the centre of the mould by means of the series of nippers shown at Fig. 98, the moulds are first taken to the position shown at B 1, (Figs. 92, 93, 94,) where they are supported perpendicularly on the small straight edges or railway *d d*, as seen at Fig. 93. In this position they are run forward until they come perpendicularly under the reservoir A, where the tallow is applied in the usual manner. The candles being in this state, the moulds are run along on the railway *d d*, to any convenient position which the railway may be conducted to, where they remain until they are sufficiently hardened to be taken from the moulds, at which period the nippers at Fig. 98 are removed. When the candles have perfectly congealed, the moulds are brought to the position shown at B, (Fig. 93,) where they are placed on a railway similar to that shown at *d d*, on the other side of the machine. Here they are pushed forward in the direction indicated by the black arrow, until they arrive at the hanging table D, which vibrates on the joint *e e*, and is then let down, but immediately returned to the longitudinal position by the operator, carrying along with it the frame of moulds B, in which position it is held by a catch, shown at *e* 1, in Fig. 92; from the horizontal position given at D, in Fig. 93, the moulds B, are removed in the direction of the dotted arrow, until they arrive immediately opposite the series of rammers E, as separately represented at Fig. 100, where the cylindrical case *b b*, is removed by the binding the jointed frame, as seen at Fig. 101, to be out of the way of the rammer E.

Fig. 100.

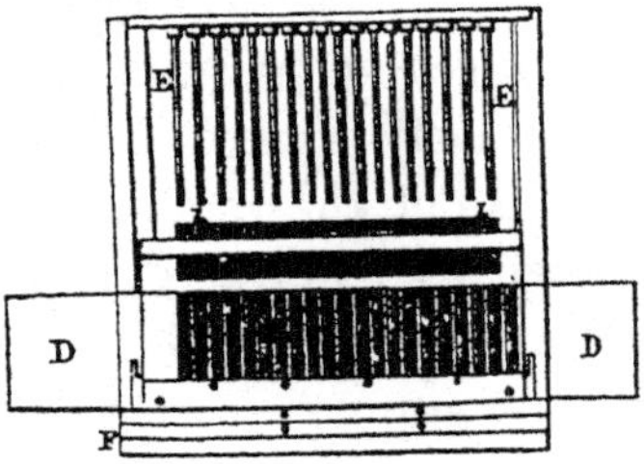

Fig. 101.

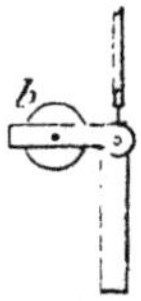

This series of rammers E, moves freely in a horizontal direction, supported on straight edges at each extremity, and is moved by the partial rotation of the wheel C, as represented at Fig. 91, where *f* shows a band or chain passing over its periphery, and round the guide pulley *f* 1. This chain or band *f*, is attached to the series of rammers E, so that any vibration of the lever *c* 1, which is fixed on the axis of the wheel C, is imparted to the rammers E, in a horizontal direction. The moulds being in the position shown at Fig. 101, the next duty of the operative is to vibrate the lever *c* 1, in the direction of the arrow at its extremity, (see Fig. 91,) and thereby force the series of rammers E, into contact with the sliding part *b* 1, of each of the moulds, and thereby displacing or pushing out the candles which are received into the grooved table F, raised up in the exact position to receive them, by the action of the scroll piece *c* 2, attached to the wheel C, and on which the grooved table F is supported. The candles being thus forced from the moulds by the rammers, are immediately secured and held stationary by depressing the lever G, which is provided with a series of like number of small convex pieces of pewter, formed of a section of the candle mould, which are attached to slight springs, as seen at Fig. 92. This lever is held down by a small catch. From what has been said of the frame of moulds B, it is obvious that the same action of the rammers E, which displaces the candles, will carry down to the moulds a fresh supply of wicks for the succeeding candles; and at this period, while the finished candles are secured on the table F, the nippers *f*, shown at Fig. 98, must be reapplied, after which the finished candles are cut off and disposed of."

"The next duty of the operative is to replace the lever *c* 1, in the position shown at Fig. 91, which retires or carries back the rammer E, along with the sliding top of the moulds 2 *b*, to their former position, and the moulds are cottoned or wicked ready for a fresh supply of tallow. This series of rammers, shown at E, are formed of separate hollow tubes, supported in the cross piece *g g*, each of which tubes is provided with a small spring having a slight projection on its inside, by means of which, when the rammers are pressed against the sliding part of the moulds marked 2 *b*, the spring gives way and catches firm hold of the notched part, as shown at Fig. 96, and is thereby enabled to bring it back to its former position, when the candles are forced from the moulds as soon as the rammers are retired, and have brought back the sliding tops 2 *b*, of the respective moulds; the springs at their extremity which had held the part 2 *b*, are relieved or lifted up by a second series of rammers or rods, which pass up the interior of the hollow rammers, as already described. These second series of rods are fixed in a similar cross piece marked *h h*, in Fig. 93, which, as soon as the rammers are retired from the moulds, is forced forwards by means of the lever H, and thereby the caps 2 *b*, and the whole of the moulds marked B, freed from any connection with the rammers E. At this period the moulds are passed forward to the railroad *d d*, and replaced in the position shown at B 1; the tallow from the reservoir A, again supplied to them and the process already described repeated to any number of times."

"Summer" candles of tallow called "Vaxeme" lights.— It is the custom with some chandlers whose trade is mostly with the West Indies or southern parts, to perfect and harden the tallow before making it into candles, by extracting a portion of its olein. The candles thus made are called "summer candles," by reason of their being better able to withstand a summer heat than those made of unpressed tallow. The mode of procedure is to throw the rendered suet into a clean wooden cistern and treat it with a current of steam for some hours, and after having allowed a sufficient repose, to draw it off free from the subsident water, into cooling tubs, and leave it to stand for several days in a temperature of 70° to 80° F. By this means it undergoes a chemical change, predisposed by the steaming, and that is the crystalization or granulation of the solid constituent of the tallow separately from the more fluid part. This granular mass, distributed in linen bags and slightly pressed, gives up much of its olein component, whilst the residual harder portion is improved by the loss. It is of these cakes of tallow left in the bags, that candles are made in the usual manner of moulding.

As the process for preparing tallow for the "Summer Candle" is exactly similar to the preparatory steps followed in the facture of lard candles, by the plan of the author of this work, we refer the reader, for particulars in detail, to Chap. LVIII; only stating, that this article of candle is much superior to the common tallow lights, both as regards burning qualities, and appearance. The liquid olein which runs out in moderate temperatures, takes a mushy consistence, and is sold to the leather-dressers, for whose purpose it is even more appropriate than the tallow itself, which is the article they generally use. As the price it brings is equal to its original cost, the expense of preparing the Vaxeme candle becomes very trifling, and is more than compensated for by the excess per pound which they bring in the market, over the ordinary mould candles.

CHAPTER XLVII.

STEARIC ACID CANDLES.

UNDER this denomination only such candles can be properly classed as are made from pure stearic acid resulting from the decomposition of soaps prepared with fats containing stearin; but in common parlance it is made to apply to all candles produced by saponification, regardless whether the constituent of the grease whence they are obtained, be stearin or any other principle. Thus, for instance, the block or saponic acids of coco butter, palm oil, and suet, anterior to the present writing, have each been indiscriminately styled by the chandlers stearic acid. We have endeavored to furnish a remedy for these misnomers by giving the true composition of each grease respectively, so that there may be no excuse hereafter for the ignorance of applying a term, universally, to the solid component of every fat, regardless of the difference in properties that may exist between one or more of them. Chevreul was the first who gave the idea of separating the oily from the solid constituents of fats; and his suggestions, afterwards verified by experiment, and enforced by Braconnot, led to the practical application of these hints by Cambaceres, who, in 1825, obtained a patent for making what he then called "*oxygenated bougies.*" The fact demonstrated by Chevreul of there being two solid matters (stearin and margarin), and one fluid (olein), in tallow, induced, at first, the effort of separating the latter by mechanical means, or pressure founded upon the physical property possessed by oil, of a perfect imbibition by brown paper. This means was not, however, favorable to the manufacture of the stearin upon a large scale; and when the further researches of Chevreul proved that the unequally fusible constituents of the fats were convertible, by saponification, into other distinct but equally combustible bodies, with a fusing point much higher than when in their original state, and on this account much more applicable as a material for illumination, the above method of Cambaceres was adopted. Whilst the art of manufacturing these candles was yet in its infancy, the product was not sufficiently perfect to obtain an equal competition with other kinds of illuminating material; subsequently, however, when the French society for the encouragement of the arts offered a premium of four thousand francs for the most economical candle, and the best mode of making it, the spirit of emulation excited among the chemists resulted in the development of such improvements in the process that a reduction in their cost was effected, whilst the quality, instead of being correspondingly depreciated, was advanced to such a degree that the hardness of the candle is equal to wax, and its illuminating power but little inferior. The application, too, of the plaited wick, which counteracts the too rapid suction of the melted fat, so perfects the combustion as to render the use of the snuffers unnecessary.

The objects which it was intended that these candles should fulfil have been more than realized by those manufactured at the present day in the United States, where this branch of trade has attained greater perfection than elsewhere. The commerce of the article has increased so rapidly within the last few years, that in in every section of the country there are a greater or lesser number of factories.

Since the publication of Gay Lussac's process for manufacturing the stearic candles, there have been many modifications without any sensible alteration of the plan; but such is the dexterity with which our workmen perform the various manipulations, that, in the economy and superiority of this kind of candle, we excel the world.

In the factories of Messrs. Verden and Werk, and of Gross, Dietrich and Lang, Cincinnati, Ohio, there are manufactured daily from fifteen hundred to two thousand pounds. With respect to the latter establishment, I obtained information directly from Mr. Lang, through whose kindness I was permitted to view their works. In return I am pleased that I can say in truth that a more economically arranged establishment, proportional to the amount of its product, I have never seen, although it has fallen to my chance to have passed through many. The brisk demand for these candles, however, prevents that care in their manufacture which is necessary to produce a handsome article; and, consequently, the "star" brand of Cincinnati is not equal to Hancock & Mann's "adamantine candles," which, it must be said, rival in every respect those of any other make of stearic candle which have as yet appeared in our markets. Their laboratory at Baltimore is constructed upon the most extended scale, and embodies every improvement up to the present time, and as their prices are as low as those of other factors it proves that capital expended in multiplying the facilities for manufacturing, is a judicious outlay. Messrs. Coffin & Landell, of this city, by pursuing the same policy, in the fabrication of soaps, have been enabled to possess one of the most elegant and conveniently arranged soaperies in the country; and by their systematized arrangement of manufacture, they compete successfully with the oldest manufacturers in the trade. A more limited capital than either of these aforementioned firms will of course restrict the number of conveniences, but under such circumstances, the better way is to contract the size of the establishment, for more work can be accomplished in a given time with a small place properly fitted than in a larger one partially wanting in the necessaries of the process.

There is one drawback to the profits of this mode of making candles, and that is the loss on the large amount of oleic acid obtained. To be sure this is a most excellent article for the soap-makers, but then there is so great a quantity thrown in the market, that its price is depreciated far below its original cost in the state of tallow. The amount of olein in one hundred parts of tallow is about twenty-four parts, but in many establishments where the stearic lights are made, through some mismanagement, the quantity reaches seventy parts, leaving but thirty parts of solid constituent. Now as this is contrary to what should be, of course there must be some means of rectifying the disadvantage. It is not a general result, for Mr. Lang tells me that their stearin product reaches sixty per cent. of every hundred pounds of tallow worked in their establishment. No wonder, then, that some of our eastern factors complain of the slight profit on these candles. They may well complain, when, for the raw material, say one hundred pounds of suet, which costs them seven dollars and fifty cents, they only obtain, in its manufactured state, an excess of one dollar and thirty cents: viz., thirty pounds stearic acid at twenty cents, equal six dollars, and seventy pounds of oleic acid at four cents, equal two dollars and eighty cents; without making any deductions for loss, wastage, or expense of manufacture. Reverse the result, however, in a measure, so that you obtain Mr. Lang's product: thus, sixty pounds solid constituent at twenty cents, equal twelve dollars, and forty pounds oleic liquid at four cents, equal one dollar and sixty cents, why then you receive thirteen dollars and sixty cents for your first outlay of seven dollars and fifty cents; thus receiving the liberal margin of six dollars and ten cents from which to deduct the loss and expenses. Now as Mr. Lang's intelligence and skill enable him to have so great an advantage over his fellow laborers in the same art, the question is, how does he acquire this advantage? The author, though he has visited that factory, is unable to record the experience of the proprietor, even were he willing to commit a breach of gratitude in return for the kind invitation to pass through his laboratory; but as there is no hindrance to his own views in the matter, he recommends a strict compliance with the preceding method, which has been carefully arranged after much practical experience and observation. In the first place the amount of liquid fat is lessened by crystalizing or granulating the saponic acids in shallow pans and at proper temepratura; and in the second place by subjecting the pressed oleic acid to such a temperature as will cause the perfect separation and complete deposition of its margaric and stearic constituents. This plan will increase the product to above fifty per cent., and its success only depends upon the careful manipulation and management of the operator.

Where nitric acid is to be had at low rates, the residual oleic acid may be converted into elaidic and other acids; being thus solidified, it will consequently bring a higher price, as an ingredient for soap, to the manufacture of which it is partially applicable.

To commence the process for stearic candles:—

Take say 1000 pounds of common tallow or pressed lard, called in commerce "stearin," and throw it into No. 1 of a series of large iron bound slightly conical cedar cisterns, each of which is heated by a current of steam conveyed therein through branch pipes leading from a main conduit of larger size, attached to the boiler. When melted, pump it off into No. 2, and add 150 pounds of good lime previously slaked, and made into a whitewash with 200 gallons of water, and steam away steadily for five or six hours, under a tight cover, which should fit closely over the cistern. This process is called the saponification of the tallow, but it is, more properly speaking, a decomposition, by which it is made through the influence of the stronger base lime, to put away that which its constituent acids are naturally combined with, viz., oxide of glyceryl or glycerin. In this manner we get oleate, stearate and margarate of the oxide of calcium, (lime,) which, in union, make an insoluble soap, the formation of these salts being requisite as a preparatory step towards predisposing the more solid acid constituents of the tallow to separate from the oleic acid or fluid portion.* To

* It is especially necessary that the lime should be perfectly caustic, and for this purpose that which is fresh and well burned must be selected; for unless every particle of it will slake to powder, it is more or less impure and should be rejected. The stirring shaft recommended is quite advantageous in effecting the saponification of the fat, for by its use not only is the process accelerated, but the incorporation of the ingredients is by this means so intimate a decrease in the proportion of lime and consequently a saving of acid, will result. For instance, tallow contains about 88 parts in 100 of fatty acids, hence it is easily calculated that 100 kilogrammes of suet require 9 kilogrammes 35 of lime for their saponification, and further that this quantity of lime will need for its saturation 125 kilogrammes of acid at 66° Baumè. These are Dumas' proportions, and though they differ widely from those given by the author, yet he inclines to their correctness, for they are founded upon the regular and definite combining equivalents of the tallow and lime and acid and lime. The excess of lime ordinarily used is doubtless rendered necessary by the mismanagement of the operative, and can be done away with by the use of the revolving agitator, which is a great and economical auxiliary to the proper completion of the process. It may be as well to add 10 per cent. to this last given

know that this saponification is complete, a test portion taken out with a ladle or spatula, should, on cooling, appear smooth when scratched with the nail, and snap with a cracking noise, when the sample is broken in two, and be pulverulent under the pestle. This operation sometimes requires only four hours, but time is no criterion to go by; the test portion must guide the judgment as to when the saponification is complete. When this saponification is conducted in an ordinary cistern, the soap formed is left undisturbed until the next day, and then after the water has been drawn off, is shoveled out, mashed to powder, and thrown into the cistern or vat No. 3. But as all this requires both time and labor, which add to the expense of the process, you do away with vat No. 3, for this purpose, and so construct cistern No. 2, that the operation can be continued therein without transferring its contents to another vessel. Vat No. 2, therefore, must be constructed with an internal arrangement, by which, at the same time that the saponification is facilitated, an effect may be produced doing away with the necessity of powdering the resultant soap, and consequently its transfer from this to another vessel. Fig. 102 gives both an exterior and interior view of a proper saponifying vat. You see at *a*, the upright shaft of iron, turned by the level wheel above, in gear with another level wheel on the moving shaft, not shown in the figure. The power is communicated by the steam-engine, which, as the boiler is a necessary appendage to a soap and candle factory, creates no additional expense for fuel. This upright shaft bears several arms *d*, fur-

quantity of acid, which, when applied, should always be diluted with 20 times its volume of water. Where the plan is pursued of shoveling out the lime soap and pulverizing it by machines, the best apparatus for the purpose is a pair of revolving channeled iron rollers.

Fig. 102.

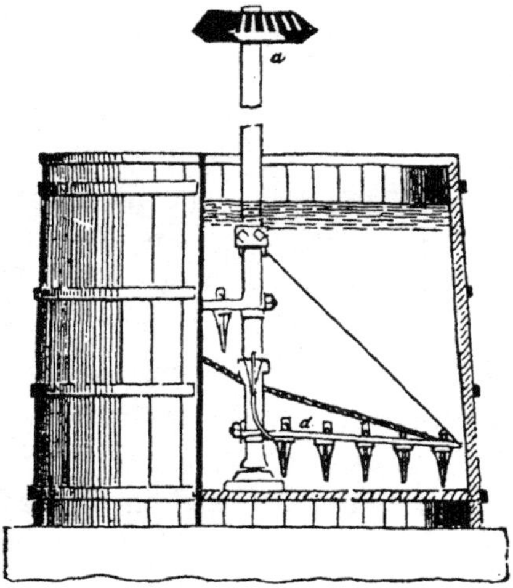

nished with large teeth. The tun or vat is of cedar wood bound with strong hoops of iron, and its contents are heated by means of a spiral tube laid on the bottom, perforated with numerous holes, and connected by a pipe with the high pressure steam boiler of the establishment. The upright shaft, and the teethed shaft, also must be of wood, for it is absolutely necessary that the absence of metal from the interior of the vat should be complete as far as possible, all fastenings being made of copper, where metal is positively requisite. After the addition of the milk of lime to the melted oil as before told, this shaft is made to revolve during the whole term of ebullition, or rather until the saponification has been effected;—that is known, when the above agitator has been used, by the same signs as those mentioned previously. The operation finished, the steam is stopped off and a quantity of cold water added, gradually at first, whilst the stirring shafts are in constant motion. The whole is soon converted into a granular mass, and after being well washed, is ready in the same vat for the next step and without being transferred to another vessel. To effect this washing there should be a vent hole at the bottom of the saponifying tun, and plugged up with a movable wooden stopper. There should also be placed behind the hole a piece of brass or copper wire cloth, so that when the water is being drawn off, there may be no waste of the soap by an escape of small particles. By withdrawing this plug, the water first added can be drawn off; when it is out, the opening should be closed up and the mass washed with fresh water; after another evacuation of the water repeat the washing until the soap has been three times cleansed. In this way you get rid of all the glycerin, and put the soap in a state to be acted on without hinderance, at the next stage of the process. This next stage consists in the decomposition of the insoluble soap composed of the oleate, margarate, and stearate of lime, and it is effected by the action of a strong acid, which, according to a rule of affinity, seizes on to the lime and eliminates the fatty matters as uncombined acids. Therefore, to proceed practically, you introduce into the vat a large quantity of water, and immediately afterwards 250 pounds of sulphuric acid, 66° Baumé, diluted with 2000 pounds of water. The application of steam must be in a slow and uniform current, and through copper pipes. The mixture must be constantly stirred by the motion of the shafts, and the warming, for the heat should not be sufficient to boil the mixture, continued until the decomposition of the soap is complete, that which is known by the entire disappearance of all the granules and lumps, and the rise of fatty matter to the surface. During the warming, the steam should occasionally be stopped off for a while, else the stuff is liable to become tinged with color. This tub is uncovered during the process, but after completion the top is placed on this wise / \ so as to leave a partial opening. The whole apparatus is then allowed to repose until the fatty acids have entirely separated from the water and other ingredients of the lower stratum, and occupy the upper portion of the vat. The plug being then taken out from the evacuating hole, the substratum of water carrying with it the saline matters (sulphate of lime), and excess of lime not acted on during the operation, is drawn off and the hole again plugged up. As the fatty acids are more or less contaminated with salts, you must now throw over them a portion of warm water (heated, by the waste steam from the boiler, in a cistern kept exclusively for that purpose), and agitate the whole for some time by the revolutions of the teethed shaft. That done, after having remained a sufficient time to settle, the saline water is drawn off as above directed, and this washing with hot water again repeated several times, or, until the fatty acids have become entirely free of alkaline or saline matters. The washed mixture of oleic, margaric, and stearic acids is now to be drawn off into tin pans of sixteen inches length, ten inches breadth, and two inches depth. These pans, when filled, should be set upon shelved racks, arranged in and around a room specially appropriated for the purpose, and the temperature of which should not exceed 90° nor descend below 70° F. In this room these filled pans remain for several days, or until their contents become granulated, that is, until the margaric and stearic acids have undergone a kind of crystallization, which does not befall the oleic acid except at a much lower temperature. When the granulation is complete, or rather when the cakes in the pans have solidified, and the surfaces appear sweated and dotted over with drops of oily matter, and yield slightly to the pressure of the finger, an oiliness being imparted thereto at the same time, then they are ready to be emptied out for pressure. To this end it is requisite to have suitable press cloths. These are generally square pieces of canvas duck linen, such as is used for sacking bottoms, but of a better quality, and to be selected with judgment. The size is such that one of the cakes in the pan can be entirely enveloped in each cloth. Additional to these cloths there must be a fellow to each, corresponding in size and to act as a jacket or outer wrapper, and made of twine cloth, similar to that used in the sperm candle factories. The cloths being ready at hand, an inner and outer one are taken together, and placed upon a table, the twine cloth nearest to the top of the table, and one of the pans emptied thereon. The workman now closes the canvas cloth carefully around the cake, and again in turn envelops the cake already covered with the cloth, in the other twine wrapper. And so he goes on until all are wrapped up. But this job is not done in a heap, for each cake, as it becomes covered, must be placed in the press which should be prepared to receive them. This press is generally an ordinary cold hydraulic press, such as is made in New Bedford for the sperm candle factors, but for greater efficiency the pattern most used is that of Maudslay, an English engineer. It is shown in plan at Fig. 103; in side view of pump at

Fig. 103.

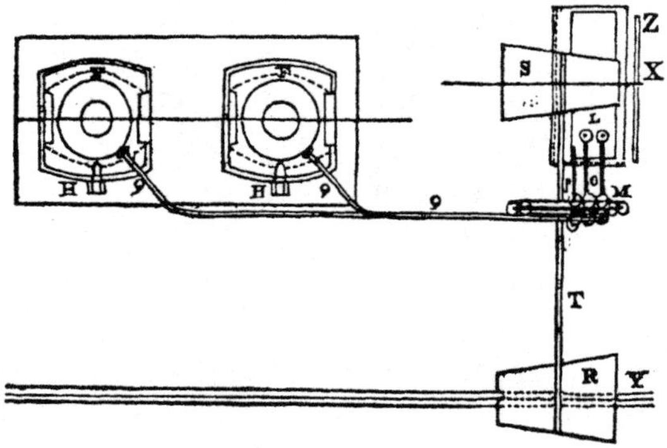

Fig. 104, and in elevation Fig. 105, where the same letters refer to like objects. We are indebted for our description to Ure's Dictionary of Arts and Manufactures.

Fig. 104.

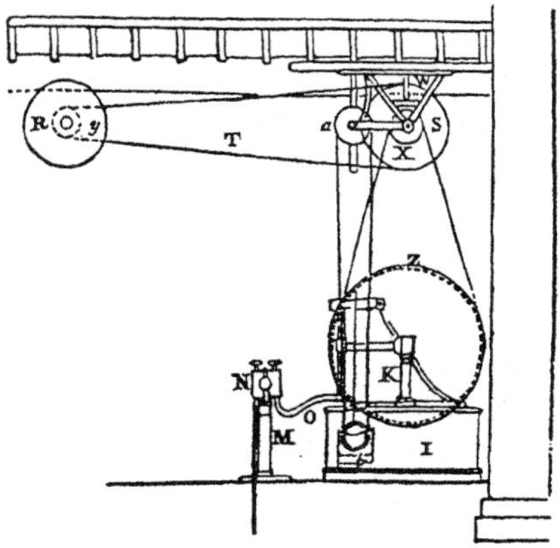

Fig. 105.

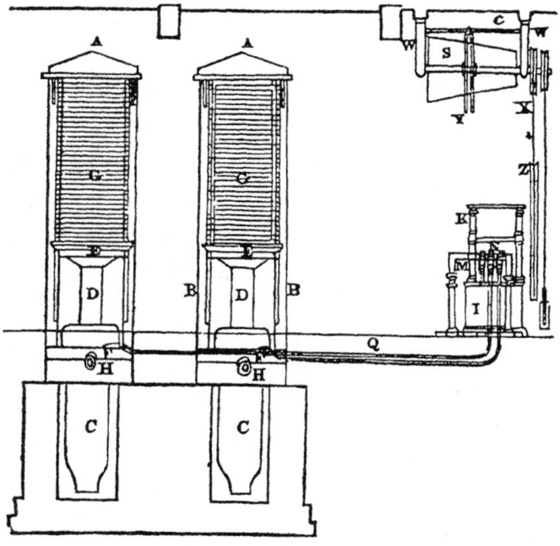

"A A are two hydraulic presses; B, the frame; C, the cylinder: D, the piston or ram; E, the follower; F, the recess in the bottom to receive the oil; G, the cloths with the material to be pressed, having a thin plate of wrought iron between each; H, apertures for the discharge of the oil; I, cistern in which the pumps are fixed; K, framing for machinery to work in; L, two pumps, large and small, to inject the water into the cylinders; M, a frame containing three double branches; N, three branches, each having two stops or plugs, by which the action of one of the pumps may be intercepted from, or communicated to, one or both of the presses; the large pump is worked at the beginning of the operation and the small one towards the end; by these branches, one or both presses may be discharged when the operation is finished; O, two pipes from the pumps to the branches; P, pipe to return the water from the cylinders to the cisterns; Q, pipes leading from the pumps, through the branches, to the cylinders; R, conical drum, fixed upon the main shaft Y, driven by the steam engine of the factory; S, a like conical drum to work the pumps; T, a narrow leather strap to communicate the motion from R to S; U, a long screw, bearing a nut, which works along the whole length of the drum; V, the fork or guide for moving the strap T; W W, two hanging bearings to carry the drum S; X, a pulley on the spindle of the drum S; Y, the main shaft; Z, fly-wheel with groove on the edge, driven by the pulley X; on the axis of S, is a double crank which works the two pumps L; *a* is a pulley on the end of the long screw U; an endless cord passes twice around this pulley and under a pulley fixed in the weight *b*; by laying hold of both sides of this cord, and raising or loosening it, the forked guide V, and the leather strap T, are moved backwards or forwards, by means of the nut fixed in the guide, so as to accelerate or retard at pleasure the speed of the working of the pumps; *c* is a piece of iron with a long slit, in which a pin attached to the fork V, travels, to keep it in the vertical position." The principle upon which the hydraulic press is constructed does not vary with the form of the machine, and the above described apparatus only differs from the usual shape in being more conveniently arranged and improved in working facilities. To Joseph Bramah, of Piccadilly, has been yielded the credit of applying the hydrostatic pressure as a powerful agent in many kinds of machinery; but it properly belongs to Pascal, who invented it as far back as 1664, and announced it as a new mechanical power. The term hydrostatic is derived from the two Greek words υδωρ, water, and σταω, I stand; but these apparatuses are not restricted to the use of water, as the application of other dense fluids, for instance oil, creates a pressure of immense power. In the hydraulic or hydrostatic press, the water is forced by a strong pump into an accurately bored, internally smooth iron cylinder, of great strength and thickness, where it acts upon a much larger piston, and pushes it against the matter to be compressed; consequently, the force with which this large piston is propelled is in power exceeding that acting on the forcing pump rod proportional as its surface is greater. For instance "if the small pump have only one-thousandth of the area of the large barrel, and if a man, by means of its lever handle, presses down its piston with a force of five hundred pounds, the piston of the larger barrel, in virtue of the hydrostatic principle of equal pressure in all directions, will rise with a force of a thousand times five hundred pounds, or more than two hundred tons." Two pumps generally accompany each press; one rather larger than the other, to expedite the motion of the ram at the beginning of the pressing, when the resistance is slight; and the other, or smaller one, to give a slower but more powerful impulse, when the resistance is augmented. A pressure of five hundred tons may be obtained from a well made hydraulic press, with a ten inch ram, and a two and one inch set of pumps." The fundamental law upon which the action of this press is based is that which regulates the doctrine of the equilibrium and pressure of fluids, and is as follows: "*When a liquid mass is in equilibrium under the action of forces of any kind, every molecule of the mass sustains an equal pressure in all directions.*"

The press being filled up with the bags aforesaid, a wrought iron plate intervening between every two, pressure is applied, *gradually* at first, and continued until liquid ceases to run and ooze from the enclosed cakes. In this way, nearly all the oleic acid and a small portion of the margaric which had not crystallized, are expelled. This granulated material presses as well as spermaceti, and the operation is conducted in much the same manner. When the pressure is down so tight that it can force no more liquid through, then the press must be unlocked and the cloths taken out and opened. The contents will be found in the form of solid cakes much thinner than when first put in, and so firm as scarcely to yield to a heavy pressure of the finger-nail, but they can still be rid of more oil. As they are unwrapped the workman throws them into a clean iron-bound wooden cistern similar to No. 1, wherein they are melted by the introduction of steam in a continual current for six hours, when, after sufficient repose for the water to settle, the superstratum of stearic and margaric acids is drawn off into tin pans of about eight to ten gallons capacity, and of form as ordinarily made for domestic purposes. These pans, when filled, are carried either into a separately appropriated room or into the same apartment, a portion of which can be allotted for the purpose, occupied by the smaller pans of the first granulation. But this time the temperature must range between 80 and 90° F. After having stood a day or more, the contents of these pans will be found solidified and granular when dug into; this is an evidence of the completion of the congelation, and proves the fitness of the blocks for the next operation. Proceed then by emptying the contents of these pans into a large clean wooden, solidly built chest, where they are divided into lumps with a hatchet. These lumps are then shaved by large knives, fixed on the face of a fly-wheel, called a rotary rasping machine, such as is used in the sperm candle factories, and made by any of the machinists at Nantucket or New Bedford. They are thus reduced to a mealy powder which falls down in a large wooden box to which the rasp is attached. The workman fills this meal into twilled woolen bags. The whole bulk, when the ends of the bag are turned over to prevent the escape of its contents at the mouth, should equal that of the first pressed cakes, in every dimension but thickness, in this instance it being requisite that the bags, when filled, should not measure more than half an inch that way. Instead of these woolen bags, there are many manufacturers who prefer those made of hair; either will answer the purpose, and perhaps the better way would be to use both, the hair cloth as an outer covering for the woolen bags. The hair cloth is not of that kind used for covering sofas, but a thick heavy web, formed like, and similar in appearance to the twine bags before mentioned. They are made in New Bedford, I believe, though most of those used here were imported from France or England. But to proceed with the process. Each bag as it is filled must be, as in the first pressing, arranged in the press with an iron plate intervening. But the press for this purpose differs from those used for the previous pressings. The remnant of oily matter contained in this meal is much more difficult to be got rid off; hence, to facilitate its removal, the re-granulation and rasping of the first pressed cakes. It should be so arranged that the atmosphere in the immediate neighborhood of that part of the press containing the bags, shall have a temperature of 75° to 85° F., whilst the iron press plates, before being put in, should also be heated up to that point, either in a steam or hot air chest. It is of different construction, too, being horizontal, and capable of exerting great power. Fig. 106

Fig. 106.

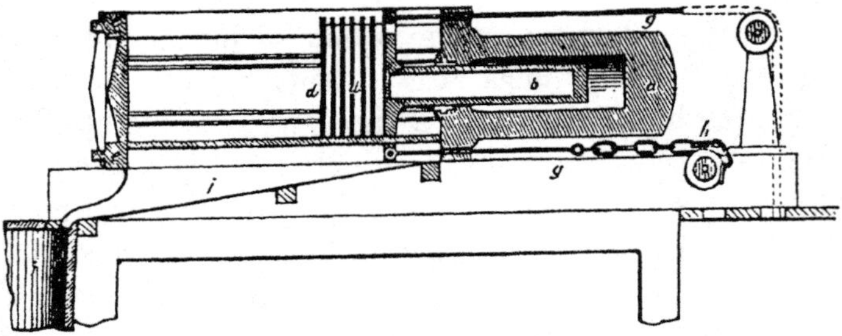

gives a longitudinal section of the hydraulic press used for the second pressure. The drawing is from Dumas; *a* is the cylinder of the press; *b* the ram or piston; *j j j j* the hair and flannel bags enclosing the mealy stearic and margaric acids, to be exposed to pressure; *d d d d* wrought iron plates previously heated and placed between every two cakes to facilitate the discharge of their residual oily matter; *e e* solid iron end of the press, made to resist great pressure; it is strongly bolted to the cylinder *a*, so as to resist the force of the ram; *g g* iron rods, for bringing back

NOTE.—These horizontal presses are made at the present time with a heating apparatus attached; the arrangement being such as to present no inconvenience to the operators, and to occupy as little space as possible. The heat is generally obtained from steam, which is furnished through pipes, to a metallic casing appropriately fitted to the press, and otherwise properly furnished, so as to be completely effective.

the ram *b*, into its place after the pressure is over, by means of counter weights suspended to a chain, which passes over the pulleys *h h*; *i i* a spout and sheet iron pan for receiving the oily fluid which flows from the bags. This press should be worked through with a full of bags, about ten or twelve times daily. The pressure must at first be very slow, but gradually increased until the greatest power is applied. The margaric acid, having a lower melting point than the stearic acid, is by this hydraulic power assisted with the warmth conveyed by the surrounding steam chest, almost entirely expelled, together with such residual oleic acid as the stuff was not freed of by the first pressing. The press being driven to its full power is allowed to stand tight set for some ten or fifteen minutes upon the bags, after which it must be unlocked and emptied out. To this end the workman takes out a bag at each time (laying the intervening plates in a trough adjoining the press), and throws it upon an appropriate table, where it is unfolded, relieved of its contents, now in the state of a hard flinty cake of wax-like substance. This cake must be carefully taken out, dexterously rid with a knife of its comparatively moist edges, which, owing to the supersaturation of the portion of the bag adjacent thereto, could not get rid of its more fluid portion. These parings must be allowed to fall in the rasping box to undergo another pressure with the next batch. The stearic acid, as it comes from this last press, is styled "hot pressed;" and the cakes, previous to being finally cleansed, must be entirely rid of their outside skins or coats, with a knife, and these scrapings saved for the next operation. Each cake thus cleared of dirt, is thrown into a very clean cedar tub, smaller than the rest, and the whole batch melted by steam, previous to being subjected to the final cleansing, three or four inches of water having been added to the vat at the commencement of the operation. This final cleansing is effected by adding to the melted stearic acid about five pounds of sulphuric acid, diluted with ten pounds of water, which abstracts any remaining alkaline or other impurity, and carries it down to the bottom. After the mixture has stood long enough for the perfect subsidence of the acidulated water, this latter is drawn off by means of a cock in the lower part of the vessel, and the stearic acid washed three several times with hot water, each addition of water, upon settling, being drawn off as was the dilute acid first added. The cleansed stearic acid, now transferred to tin or copper pans, on cooling, takes the form of blocks, corresponding in size with the vessels containing them, and which, when broken, exhibit a highly crystaline texture. These blocks, before being fit for conversion into candles, must be externally well rid of all dirt, with a knife, and then melted down with a very small proportion of white wax, in order to counteract the crystalization to which this fat acid is so prone, and which renders it unfit, in such state, to be formed into candles. There are some chandlers who destroy the crystaline texture by fusing a thousandth part of arsenious acid with the stearic acid, but in this way the health of the inmates of the apartment where candles so treated are burned, is jeoparded by the volatilization of the admixture, which is a deleterious poison. Wax is much the better material, and even a minute proportion of very fine white marble dust will effect the same result, which can be attained to, however, in another way, as will be told, without the agency of any addition. This final cleansing of the stearic acid implies the additional bleaching of the candles, after they have been moulded, but the plan now adopted by those cognizant of the method, is to whiten the material previously to casting it into candles. To do this, the pressed cakes, freed of their outer coats, instead of being treated with sulphuric acid, are melted in pure boiling water, and after fusion, the steam being then shut off, to every hundred pounds of stearic acid are added four pounds pure aqua fortis, and the mixture, by means of a twirl, kept in constant agitation for an hour or more, and then allowed to cool. This cistern should be of clean cedar wood, and iron bound. To its interior should be fitted a movable twirling shaft with arms, similar to that used for flotant soaps, and represented by Figs. 19 and 20. This twirl revolves upon a pivot fastened to the bottom of the vat, and is put on or off just as the operation requires it. The acidulated water having settled, is drawn off through a cock in the lower part of the tub, and the fatty acid subjected to three several washings with warm water, heated by the introduction of a current of steam. The twirl shaft should be used upon each addition of water, except the last, but not until the mixture has been heated up and the steam shut off. Intervals are allowed for the settling of the water after each washing, and it is drawn off as before, through the cock. The last washing is accomplished merely by the addition of water and steaming, and in this instance, the upper stratum of liquid fat is racked into clean pans, the contents of which, when cold, are blocks of pure stearic acid, brilliantly white and clean. These blocks, afterwards scraped on the outside with a knife, are then ready to be melted and moulded into candles. It is much better to have a vat expressly appropriated for the bleaching of the stearic acid; then the twirl shaft can be so arranged as to derive its propulsive force from the steam engine of the establishment. The figure given represents one that is moved by hand power; to go by steam it must be fitted up with cogs in a manner similar to the agitating shaft of the saponifying tun, p. 433. The brass cocks should be placed in the sides of the vat, in such positions, that in drawing off the liquid fat or the subsident water, neither may be commingled with any particles of the other. These cleansed blocks to be converted into candles, are now melted up in a perfectly clean enameled, or bright copper kettle, at a temperature of 212° F., and to prevent any tendency to crystalization, stirred with a stick or wooden paddle, until the liquid fat assumes a milky appearance, when it is run into the moulds previously heated in a hot room up to 125° F. In this way the candles on cooling separate easily from the moulds. I recommend Dyre's moulds for this article; they are so bored and polished that there is less resistance offered to the exit of the candle, and being otherwise made with greater care and of better metal than those had elsewhere, are generally preferable; especially as they are not more, if even as costly, as those made at New Bedford. The candles, upon being drawn from the moulds, are rubbed quickly with a silk or woolen cloth moistened with alcohol or ammonia, and thus highly polished, previously to being packed in boxes with hard pressed *white* paper.

When the candles are to be bleached, after having been moulded, they must be spread out horizontally upon frames, and exposed to the light, occasionally running over them with the hand so as to turn them and present a new surface for action. But the whiteness thus obtained is only superficial, and not of that purity and brilliancy desirable; so resort must be had to chemical means to remedy the deficiency. The most successful method is by the action of chlorine, and the mode of procedure is as follows. In a very tightly jointed, strong wooden chest, are placed the candles suspended to strips of wood properly ranged in the interior, and hanging by their wicks upon wooden pegs, distant from each other far enough to prevent the contact of any two candles. The rows are ranged one over the other until the chest is filled. The cover fits on hermetically, and besides being fastened down by clasps, has its joints and cracks closed over with glued cloths. At the lower end of one of the sides of the

NOTE.—The moulds used in France for the stearic candles, says Dumas, are made of an alloy of one-third tin, and two-thirds of lead. When they have been threaded with wicks, previously to being filled with melted fat, they are heated in a warming chest of sheet iron, divided off into a number of compartments, each of which holds eight moulds ranged side by side. This chest is then placed in a water bath kept hot by a jet of steam. Below is a drawing of the arrangement. Fig. A shows it in detail, and Fig. B in plan. The tin moulds *m*, placed in the receptacles *n*, are kept at the proper temperature by means of a water bath *o*, which thoroughly envelops the whole apparatus, and which is heated by the steam tube *p*. The heating of the moulds facilitates the drawing of the candles, and prevents the disagreeable crystaline appearance they might otherwise have.

Fig. A.

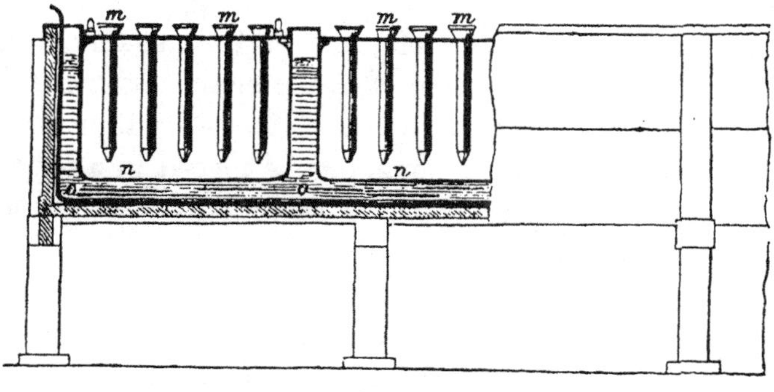

Fig. B.

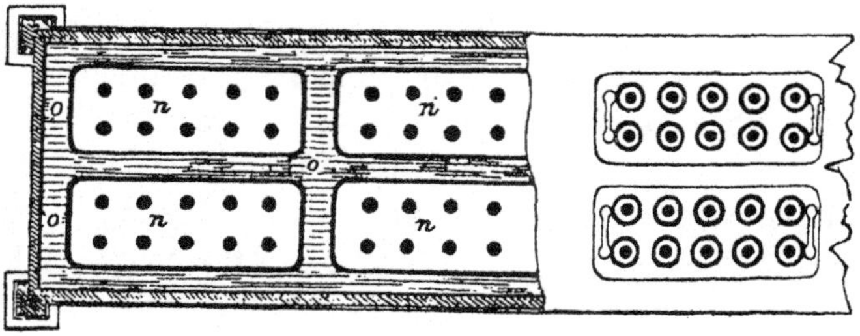

boxes enters to the interior a porcelain or leaden tube connected with a leaden still or retort in which is generated, and from which proceeds the chlorine gas. On the opposite side of the box must be two windows eight or ten inches square, (the glasses of which are puttied in upon the outside,) through which the progress and completion of the operation can be seen and determined, without the inconvenience of opening the chest. In twenty-four hours, the candles most generally acquire a brilliant whiteness, though at times the operation endures from two to three days. After having been removed from the box, they should be transferred, (hanging as they are,) to an airy room, that the adherent smell of chlorine may escape as much as possible. A square chest of forty inches clear, in each of its dimensions, will easily contain 7000 candles, which can be bleached at a single operation.

The chlorine gas is obtained by placing in the leaden or glass retort a mixture of three parts of common salt with two parts of peroxide of manganese, and pouring upon it two parts of sulphuric acid, diluted with its own weight of water. The retort must be placed in a gently heated sand bath, and the gas as it escapes passes through the porcelain tube into the bleaching chest.

We cannot positively give the rationale of the bleaching operation, for it has not been determined experimentally in what manner the color is decomposed, though in this instance it is highly probable that the chlorine combines directly with the hydrogen of the coloring matter to form hydrochloric acid, and thus destroys the color.

The wick formerly used for these candles was of cotton yarn tightly twisted around wires of lengths and diameter suitable to the size of the candle to be made. These wires thus enveloped, being inserted in the centre of the moulds, as wicks are generally, upon the cooling of the materials, when withdrawn, left a hollow cylindrical aperture entirely through the middle of the candle. I have neither seen nor tested a candle thus wicked, but would infer that, if this mode of applying the wicks was not so tedious and troublesome, it presents other advantages which would lead to its universal adoption. As the size of the wick must be proportional to the fusibility of the candle material, therefore, in order that the suction and consumption of the melted fat in the cup may be complete, whilst the requisite number of threads or capillary tubes are supplied for the purpose, the too rapid ascent of the liquid grease is prevented by their being plaited.

As decidedly the best stearic candle made in this country, I have selected one of Hancock & Mann's "Adamantine lights," to furnish the wick of which the description is about to be given. My reasons for so doing are, that the superior illuminating power of their candle, proves conclusively, that their essays to that end have resulted successfully, both as to the arrangement and size of the wick, and dimensions of the thread requisite for the proper burning of stearic acid. Moreover, the wick ensures a perfect combustion of the material, and for these reasons we present it as a model, though the same wick may be, for aught we know, not dissimilar to that adopted by other manufacturers. It is a plaited wick consisting of three strands of twenty-eight threads each. The threads are fine and cleanly spun, being free from imperfections or tags, so that there shall be no hinderance to the perfect combustion of the burning material from the formation of "*thieves*," those inconveniences which bend the wick over, and cause the candle to run and gutter.

It is very requisite that the cotton should be uniformly and finely spun.

The improvement in the brilliancy of the flame, and

combustibility of the wick to be obtained by such means, has induced a most prevalent practice with manufacturers, of giving the wicks for these candles a previous saturation with one or more appropriate chemical salts or admixtures.

Some chandlers use for this purpose an alcoholic or hot aqueous solution of boracic acid;* this, however, imparts to the flame of the candle more or less of its characteristic green flame. The most preferable mixture, perhaps, is the following solution, into which the wicks, after having been dipped, must be dried in the air.

Pure hydrochloric acid - -	1 pound
Muriate of ammonia - - -	4 ounces
Water - - - - - - -	12 pounds

The liquid oleic acid which runs from the first pressing, and still contains much of the solid acids, must be barreled and put in a cool cellar where it is allowed to remain until perfectly granulated. The barrels are now unheaded, and their contents pressed as before, to extract all the residual stearic acid, the liquid oleic acid running through, being, together with the thick liquid of the second pressing, reserved for conversion into soap.

Cambaceres has proposed to render the manufacture of stearic acid less expensive, by obtaining, instead of the useless sulphate of lime when the lime soap is decomposed by sulphuric acid, a more valuable salt of alumina, as an accessory product; but as it is impossible to saponify with alumina, the tallow must be converted into soap by potash, the resulting soaps decomposed by a solution of alumina in caustic potash, and from the gelatinous aluminated soap the fatty acids separated by acetic or sulphuric acid; thus is formed the acetate or sulphate

* The presence of free acid serves to neutralize any alkaline matter which may be adherent to the threads.

of alumina, either of which is a more or less useful salt. The ley remaining after the separation of the aluminous soap is solution of caustic potash, and can be employed for fresh operations; scarcely a tenth of the potash being lost in each repetition. This plan has not been tried by the author, but is submitted to the chandler who can soon test its efficacy and economy by an essay on a small scale.

There are many manufacturers who granulate their tallow by steaming, and then press it previously to saponifying it. In such case the liquid oil running from the press is mostly olein, and being admirably fit for their use, commands a good price from the curriers and leather-dressers, a good fortune not happening to the oleic acid, obtained in subsequent operations. In the granulation of the cakes after first pressure, so soon as the liquid fat cools down to 150° F., there is also frequently stirred in eight pounds of common, eighty per cent. alcohol to every one hundred pounds of fat; but it is doubtful whether the benefits resulting from this addition compensate for its cost. A thorough experience and dexterity in the manipulations of the process preceding, will enable the production of as good and handsome a candle as need be desired. Sometimes common sperm taught presses are substituted for the hydraulic in the first pressing, but this is bad policy; they cost much less, to be sure, but the candle obtained is correspondingly inferior. No, the better way is to have the very best apparatus; it is more economical, greater facilities being thus afforded for conducting the work, and at less expense. There should be to each factory two upright and one horizontal hydraulic press, each furnished with a convenient quota of press cloths, bags and iron plates. There is a kind of knuckle or knee press made in New Bedford or Nantucket, said to be very effective, and may answer for first pressing, but their use cannot positively be recommended, as we speak only from a description furnished by an acquaintance, not having personally seen a drawing of one of them.

The advantages which the stearic candles have over those other species ranging in price near their cost is, that they give more light with a less consumption of material, and burn without running or smoking, or requiring to be snuffed. They moreover have no disagreeable smell, are perfectly dry, hard, and of fusibility never lower than 120° Fahrenheit.

A well made stearic candle should be of a beautiful marble-like smoothness exteriorly, glossy and somewhat translucent, not of a positive whiteness, rather tinted with a scarcely perceptible shade of blue, so that it may appear similar to spermaceti. Besides being perfectly dry and inodorous, their wick should be so appropriate and properly arranged, as to ensure a perfect combustion.

All good specimens possess the above named qualifications, and the best article manufactured in the United States, proves that they are not merely incidental properties, but are natural to this species of candle, when properly made. Hancock & Mann's Adamantine lights burn with a flame that sheds a brilliant light.

CHAPTER XLVIII.

STEARIN OR LARD CANDLE, BY C. MORFIT'S PROCESS.

In the years 1842 and 1843, when the public attention was being called to the applicability of lard for the manufacture of lard oil, and the conversion of the residual stearin into stearic blocks for candle use, I was requested by Mr. Ellsworth, the able and energetic commissioner of the Patent Office, at that time, to furnish a paper upon the subject for his annual report to Congress. Being then professionally occupied at my laboratory, I had no time to comply with his wish further than to forward him a copy of Gay Lussac's process, which he published. A few months following, however, when more leisure ensued, my attention was again called to the matter by a gentleman who desired my instruction as to a practical method of separating the lard oil from the lard, with a view to commencing the manufacture of that article. The information was furnished and exemplified by an actual performance of the whole process; after which, and some matured reflections, it occurred to me that as lard consisted of thirty-eight parts of solid constituent in every one hundred, some means might be adopted for its perfect separation without resort to saponification; and in this way procured a perfectly free stearin, which is, when pure, unalterable in the air, and contracts no bad smell, as tallow does. The object in doing away with the saponification was to obviate its expense, whilst, at the same time, a candle might be made, if not equaling sperm, much better, handsomer, and more durable than the disagreeable greasy tallow candle, and consequently, particularly fitted for the use of workshops, printing-offices, and the humbler classes of the community. The design was to furnish an article of handsome appearance and superior quality at a low price :—so low that without competing with sperm, they would entirely drive the tallow candles from the market. The sequel will show how far the author succeeded in his intentions and efforts.

My experiments were based upon the entire solubility of olein in spirits of turpentine, whilst the solid constituents of the lard, though also possessed of that property, are nevertheless greatly impaired in the exercise of it at *sufficiently low temperatures.* The great disadvantage to be overcome, was in the sale of the large quantity of lard oil, yielded as an incidental product; this, however, no longer exists, as that article has become a standard one in commerce; moreover, for the purpose of making candles, the solid residual cake can now be purchased from those who manufacture the oil exclusively. The results exceeded my anticipation, though not until after much outlay of both time and money; success was nevertheless obtained; that was the grand object, and the following report of the committee on chemicals of the Franklin Institute,—to whom was presented for exhibition, in 1843, a sample of the stearin in every progressive stage of the process, up to the finished candle—afforded me additional pleasure.

"No. 351, a lot of articles from Mr. Morfit, exhibiting the various processes in the manufacture of stearin from ard; the specimens are highly interesting, and show in a very clear light the method of the manufacture. Mr. Morfit claims the discovery of a new method of purification without saponifying, about which claim the committee know nothing, but consider Mr. Morfit (a very young man) as deserving all credit for his unremitting industry in perfecting and extending this important branch of manufacture, and, therefore, recommend for his interesting display, a certificate of honorable mention."

It may be as well to state that my candles were in competition with the superior adamantine lights of Hancock & Mann, before spoken of, which, though nearly twice as costly, were not considered worthy of, and did not receive, a higher premium.

The following is a rough estimate of the weekly yield of the manufacture by my process.

Wages 2 men, each $10	$20 00		
" 4 " " 6	24 00		
" 6 boys " 2	12 00	per week	$56 00
Wear and tear of apparatus, 20 per ct. per week on $3500			14 00
Rent $10. Clerk, &c., $10			20 00
Interest on $1000, 1-8 per cent. per week			12 00
			$102 50
3000 pounds pressed lard will make 1200 pounds candle block per day; say 18,000 pounds per week. Cost 7¼			1305 00
Material 60 cents per 100 pounds, on 18,000 pounds			108 00
			$1515 50
The 18,000 lbs. will produce a loss = to 360 lbs. at 7¼ =		$26 10	
7056 lbs. solid fat or stearin at 12 cts. =		$846 72	
10584 lbs. semi-fluid fat at 5 cts. =		$529 20 =	$1402 02
Nett weekly gain			$113 48

In these calculations it is considered that either tallow or pressed lard has been used, though the product yielded by the former, it may be as well to say, is far more deficient in beauty than that from the rough stearin. The ultimatum to which I limited their cost in commencing the experiments, is not exceeded, for allowing one cent per pound additional for moulding the block into candles, they then sell for but one cent per pound (thirteen and a half cents), over the retail price of mould candles, without at all taking into consideration their superior excellence. With these introductory remarks we proceed to make known our plan.

There are frequently in the market lots of lard, quite dark and dirty in appearance, offered for sale at tempting prices; and relative to such articles a word or two of advice is necessary, when about to be purchased for conversion into stearin candles. If upon examination the darkness of color is occasioned by the lard having been scorched or burned in the course of its preparation, touch it not; it will cost you more than it is worth to bleach it even should you succeed in the attempt—a doubtful matter in three instances out of five. But if the tinge is only owing to the presence of dirt, then buy the lard, if it can be had at a price proportionably low, because it regains its whiteness in the process of being worked into oil and candles. There is a great deal of what is called "stearin" also offered for sale by our merchants. It is the solid residue of the lard after the expression of its oil. This answers every purpose for making candles, except that in working this article you lose the profit that might accrue upon the oil, which would be obtained by commencing the process with lard itself. This stearin is of about the consistence of tallow in summer time, and is more or less contaminated with coloring matter, which must be got rid of by one of the usual bleaching processes, or by the application of the chameleon mineral in the manner before directed. If the color does not yield to the blanching influence of this agent, then such stearin is unfit to be converted into candles, and must be rejected. The best way is to select good solid corn-fed lard, and by preference, that prepared in Wilson's tanks, for reasons which will hereafter be assigned.

The first step after the selection of the lard is to transfer it from the kegs or barrels into a clean wooden cistern, of adequate capacity, and therein melt it by the application of steam through a pipe attached to the steam boiler. This cistern must be iron bound, and in order that there may be no splashing over of its contents, should only be filled three-fourths of its depth. The current of steam must be continued for six or eight hours, at the expiration of which time it is stopped off and the vat of boiled lard allowed to stand for three or four hours; by this repose the water, condensed from the steam thrown into it, by its greater specific gravity subsides, and being unnecessary to the granulation of the lard, is to be drawn off through a spigot, the inner end of which, that its aid may be effectual, it is best to have resting immediately on the bottom of the boiling cistern. The water rushing out carries with it all the subsident foreign matter of the lard, which is by this operation purified, whitened, and disposed to granulation. As there is more or less of filth still adjacent to the bottom of the cistern, it is better to draw the steamed lard through a cock placed some two or more inches from the bottom, so that what runs through may be clean, white and uncontaminated. The vessels into which it is racked are called cooling tubs; they should be of capacity not less than five hundred nor more than two thousand pounds. In these tubs the boiled lard must remain undisturbed for twenty-four hours, when, through a little gimlet hole at the bottom, fitted with a movable plug, such water as was retained by the warm fat and has then settled, is drawn off. The hole being closed up, the tubs are left undisturbed for several days, or until the lard has congealed, and exhibits a granular appearance. This granulating room should have a temperature between 40° and 60° F.; if it is higher, the cooling is too slow; if lower, the congelation occurs so rapidly as to endanger the perfect granulation of the mass, that which is necessary for the success of the operation.

From this thoroughly granulated mass, the oil, portions of which are even now seen floating around the granules, is separated by ladling it into bags to a thickness of almost two inches, the size otherwise not being important, and folding them so as to prevent the expulsion of any of the solid matter. Some factors, instead of bags, use large wrought iron cylinders, cullendered with small holes throughout their whole circumference, and lined interiorly with canvas. This, however, is a wholesale way of operating, very proper where followed for obtaining the oil alone, but not cleanly enough when the residual stearin is intended for candles, and after all does not possess a great advantage over the bags on the score of economy. Each bag, as it is filled, is put into the press with an iron plate intervening between every two; when full, it is then locked by gradual pressure at first, and increased as the oil becomes more reluctant of expulsion. The temperature of the pressing room should never be higher than that at which the lard was granulated. The oil expelled by the pressure is caught in suitable vessels, and transferred thence into large covered tin stands.

This oil is the "LARD OIL" of commerce, and when good lard has been used, is very pure and white, but of a brown color when it has been pressed from burned or scorched lard. It is considered much superior to sperm or olive oil for greasing wool and machinery, and being more economical in price, is very largely used. But it should not be bleached for these purposes by chemical agents, as it will be apt to retain enough of the material to make its application disadvantageous. When the lard from which it is expelled is good and not burned, the steaming produces all the necessary bleaching, even when it is to be used for lamps.

Winter lard oil is made by subjecting this oil to the influence of a cold temperature, and when it has granulated, re-pressing it in bags as before. The oil running through is able to bear a much lower temperature without congealing, than if it had not undergone this second granulation and pressure. The solid matter in the bags can be mixed with the stearin which is to be made into candles.

When lard prepared by Wilson's plan of steaming is used, then the reboiling and granulation are unnecessary, for all lard rendered in tanks and by steam is sufficiently granular as it comes to the market, without the necessity of any reboiling, and is ready to be transferred from the kegs to the bags, and in them to the press. This is a great advantage, because, independent of the saving in labor, you gain all that would be otherwise wasted by a re-steaming, as that brand of lard is always white and clean. Sometimes ordinary screw presses are used for pressing out the oil, but the most economical and perhaps best, is the lever press, easily and cheaply made by properly adjusting a long oaken beam over a frame arranged for the reception of the bags. By these means the pressure goes on gradually even during the night, and whilst the attendant is absent.

When the pressed cakes cease to give any more oil the press is unlocked, and the bags emptied of their contents. This pressed matter is thrown into a clean cistern and steamed again in the same manner as at first, and after six hours ebullition, allowed to stand for twenty-four hours, when, after the water is drawn off beneath, it is run into tubs of five hundred pounds capacity each, which tubs are to remain stationary in a continued temperature ranging between 50° and 70° F., until the whole mass is thoroughly granulated. The centre of the mass being sometimes rather mushy, must be stirred up well with the rest, but before this is done, the residual water which has subsided must first be drawn off as before, through the gimlet hole in the bottom. The granulation is known to be perfect when a lump from near the sides of the tub can be handled without softening, and is strictly granular in its appearance. With this grained matter fill up canvas bags, precisely in the same manner as described for the first pressing of the stearic acid, and submit them to the action of similar presses. The bags must be filled to a thickness of only one inch, and the temperature of the room in which they are pressed should be between 60° and 70° F. When all the oil that can be driven out in this way is expelled, then the press is unlocked and emptied. In all instances, where the press is used, apply the pressure gradually at first, else the bags will be apt to burst, and the material worm through the meshes of the threads. The oil which runs out at this pressure, contains a goodly quantity of stearin and margarin, and consequently will not remain liquid even at a moderate temperature, and so upon being congealed at a gradual temperature is ready to give up another portion of its solid constituents by pressure as before. This oil, however, brings nearly its original price as an ingredient for fine soaps, and perhaps the economy would be as perfect in directing it to that course. In case it is re-granulated and re-pressed, the solid constituent remaining in the bags is saved for the next operation, and the oil sold for greasing wool, or burning in lard lamps, its fluidity not being sufficiently thin to fit it for sperm lamps, or for machinery.

To proceed with our process, we will state that by this

first hydraulic pressure of the stearin, it is rid of so much of its more liquid portion, that when taken from the bags, it is in hard cakes, which break with a cracking noise. These cakes, as they are emptied from the bags, are first weighed and then thrown into a very clean jacketed copper kettle, similar to Fig. 13, where it is to be melted anew by the heat of the steam let in between the sides of the inner and outer kettle, and kept at the temperature it is thus raised to, for about four hours, taking care, however, that the liquid fat at no time gets beyond 200° to 212° F. After the four hours heating, it is allowed to stand until its temperature has fallen to between 110° and 140° F., when to every one hundred pounds of the weighed cake must be thoroughly incorporated by stirring, eight pints of spirits of turpentine of *recent* distillation. The liquid fat thus prepared, must now be drawn off into tubs of about five hundred pounds capacity each, and placed in a temperature ranging between 45° and 55° F., and there allowed to remain for several days, when it will have hardened into a granular mass. This granulated mass must now be bagged and pressed in the horizontal hydraulic press, precisely in the same manner as is effected the last pressing of the stearic acid, using similar materials for the bags and envelops. This thick oil that is expressed must be thrown in an iron kettle with some water, and boiled, until the evaporation of all the turpentine, and the absence of its smell; thus it is made fit for sale to the soap boilers. But in those localities where fuel is plenty, and the price of spirits of turpentine is costly, it is better to have a copper still purposely erected for its reception, mixed with a portion of water, so that when the head is luted on and heat applied, the turpentine may be driven over into a receiver, and thus recovered and saved. The greater volatility of the turpentine enables its separation from the fat in this way without injuring that body, which, upon being dipped out, cooled, and separated from the water, is as saleable for conversion into soap, as it would be if rid of its spirits in an open caldron.

When the whole extent of pressure has been applied to the bags, so that no more liquid fat will ooze out, then the press must be unlocked, the bags taken out, and their contents immediately emptied into an iron bound vat where it is to be steamed until entirely rid of the presence and odor of turpentine; afterwards it is allowed to settle so that the small quantity of subsident water can be drawn from the bottom through a brass cock. I have said that it should be immediately steamed after pressure, because if the cakes stand too long, the turpentine they contain is prone to change its state by the absorption of oxygen from the atmosphere, and render the future operations of cleansing and bleaching difficult. The liquid fat, whilst still warm, is drawn off into a perfectly clean iron pan, and then bleached by digestion with a little water and a very small quantity of alkali.* The scum arising to the surface brings up all impurities, and as soon as the liquid becomes clear and free of milkiness, the operation is finished. The stearin is now strained through linen cloths into tin pans, and when cool forms blocks of wax-like consistence and appearance.

The bags for the turpentined fat must be kept exclusively for that purpose, and steamed out at least once every fortnight.

In winter the candles into which these blocks are converted are readily drawn from the moulds, when cast in the usual way of moulding candles; but in summer the tenacity of their material is such that they adhere to the

* This mode of bleaching is practised as directed in the chapter upon Sperm candles.

metal and are very difficult of separation. To obviate this the stearin must be run in at a temperature between 150° and 180° F., and the moulds immediately and rapidly cooled in water; the moulds being heated to the same temperature as the fat.

Another mode is to cast the candles in separate moulds without regard to its temperature, or that of the moulds, and, when they have cooled, by a little dexterity, plunge the mould into hot water and draw the candle at the moment of the expansion of the metal. In this way the candle loses but little of its polish, and by practice a girl will acquire sufficient expertness to run and pull a large number daily.

The wick which I have used is represented by Fig. 107. It consists of seventy fine spun cotton threads, gimped or tied around with thread by a machine similar to that for wrapping bonnet wire. I am not satisfied with this wick; it truly answers to afford as good a light as is shed by a sperm candle, but there is little doubt that the burning property of stearin, when properly aided, is nearly equal to that of stearic acid. The essays in the adaptation of a suitable wick were numerous but not conclusive, and when my attention was called from the subject to other matters, success had become complete only so far as to decide upon this wick in preference to all others that I had tried. It is still, however, imperfect, and can be materially improved

Fig. 107.

The conveniences with which the author conducted his experiments were very limited, yet the candles produced were of great beauty, perfectly dry, brittle and inodorous. The light which they emitted was very nearly equal to that of a sperm candle, and so perfect was the combustion, that the cup formed at the tip, near the burning wick, was not once so much surcharged with melted fat as to cause a running or guttering of the candle. The appearance of the candles and even the taste, too, of the material so nearly resembled white wax that it was with difficulty the difference between the two could be distinguished. This wax-like appearance is only peculiar to the candles from lard; when tallow is used they are equally hard and good, but their appearance is more like that of a very white compact polished marble. This plan has never been carried to any large extent into practical operation, because the limited means of the author have prevented that step, whilst none of the several propositions from capitalists have been liberal enough to induce an acceptance on his part.

Having now exposed my plan, it must in justice be said, though the fact is to be regretted by me, that the application of the spirits of turpentine, conscientiously thought to be, until a late research, original with the author of this work, is due to Braconnot, who, as far back as 1821, presented specimens by his plan, under the name of *céromimème*. I had not, in the course of my reading, seen this fact stated, and it was only lately that I met with it. Had I been familiar with it years ago, a great deal of time and money would have been saved, while I would have been relieved from a disappointment which is somewhat unpleasant. Still Braconnot's plan of operation is neither so explicit nor perfect as mine, and, therefore, while strict equity is accorded to him, may I not claim some merit for so modifying, perfecting and simplifying the process, as to render it applicable to the production of pure stearin candles to any desired extent?

CHAPTER XLIX.

SPERMACETI CANDLES.

In cleanliness, brilliancy of light and beauty of appearance, the sperm candles are only second to the wax bougies. Their beautiful transparent whiteness renders them particularly applicable for parlor use, and though of higher cost than the ordinary kinds of candles, they are, considering their advantages, scarcely more expensive. As met with in the market, however, they are not universally of good quality; some brands are more reliable than others, for the excellence of the candle depends upon the care and skill bestowed in its manufacture. The requisites for a perfect candle are, entire freeness from oil, a proper apportionment of the wick, and a skillful manipulation in moulding, so as to produce a diaphaneity of the stock. Nantucket and New Bedford are the great manufacturing marts, but there are also large quantities made in this city, and for many reasons these latter are preferable. Whether the greater intelligence of the manufacturers here more than counterbalances the experience of the Eastern operatives, is a matter left for decision: this is certain, however, the Philadelphia brands are, as a general thing, surpassingly excellent. Their mode of manufacture is as follows.

The crude spermaceti of the whale, as it reaches us from the whale ships, is either directly bagged and subjected to a taught press, or it is first steamed and granulated. Without previous granulation the sperm, even after having gone through all the subsequent operations, is greasy, less transparent, and of inferior quality. The process of steaming, therefore, is an advantageous step. The *modus operandi* is the same as directed for the stearin in Chap. XLVIII. The grained mass having been transferred to bags, and folded so as to form thin cakes, is placed between wrought iron plates in a taught (screw) press, and subjected to pressure until liquid ceases to flow from the pressed cake. This liquid, which passes through after having gone through the proper rectification, is the winter sperm oil of commerce.

When the screw press has been carried to its fullest power, and no more oil will ooze from the filled bags, they are to be taken out, and their contents emptied into a clean cast iron kettle and melted. The kettle, with its charge, should be kept at 125° to 150° F., for some hours, and at the end of that time the liquid sperm must be drawn off into hogsheads and placed aside in a mild temperature to granulate a second time. It is then bagged and pressed as before. The oil obtained in this stage of the process is "Spring oil."

After this pressure, the bags are taken out and their contents again melted as above directed. The liquid fat having remained at a gentle heat for one or more hours is then to be ladled into tin conical pans, each of five or six gallons capacity. When the blocks thus formed have cooled, they are to be placed upon a rack and left for a day or more to "sweat" and "ripen." This done, their surfaces are coated with oleaginous drops, and they are then to be transferred, one at a time, to the sloping hopper D, of the rasping machine represented by the figure annexed.

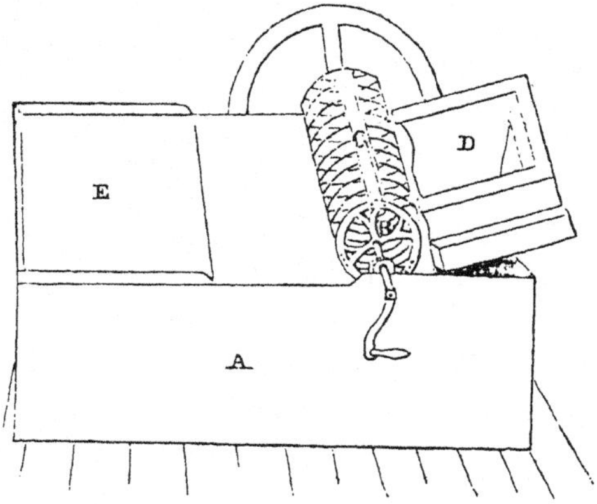

There they come in contact with the knives C, traversing the circumference of the wheel B, kept in constant rotation. In this way the blocks are reduced to a mealy powder, which falls in the chest A beneath. It is now ready for a re-pressing, in order to expel all residual oil, and for this purpose, is filled into cloths with twine wrappers, and folded to a less thickness than in the first pressing. The shelf E is placed on the end of the chest, both for convenience and for facilitating the operation of bagging. The cloths thus filled, are now arranged as before, between iron plates, but in a hydraulic press of greater power. Some factors use an upright, and others an horizontal apparatus. The former is shown at Figs. 103, 104, 105; the latter, which is preferable, is seen at Fig. 106. In cold weather, or when the expulsion of the oil is to be complete, the presses are surrounded, as in the pressing of stearic block, by a steam chest;—the product obtained in this way is called *hot pressed sperm* in contradistinction to that which is *cold pressed*, and procured without artificial heat.

When the compressed cakes cease to yield oil they are to be taken out of the press, uncovered, and placed aside to be purified. The oil from this last pressing is worked over again to obtain No. 2, or tight pressed spermaceti. The head matter sperm is never entirely freed from oil by pressure; therefore, when it reaches the refining pan, the small portions which it retains are to be neutralized, as it were, by the addition of about 66 per cent. of pressed "body matter or foots," which is usually pulverulently dry. The cakes from the last pressing are hard, dry and sonorous. Their purification is founded upon the feeble action of dilute alkalies upon sperm itself, which unite with its oleaginous and other impurities, and throw them to the surface in a thick froth or scum. The cleansing and whitening of the sperm are effected by melting it in a perfectly clean iron kettle, at 160° to 180° F., and dosing it with potassa solution of 45° B., as its temperature decreases. The proportion of ley is about one and a half gallons to forty of liquid sperm. The heat must be steady and moderate when the ley is added, because a *rising temperature will soap the whole batch.* This is called "physicking" the kettle. After the thorough incorporation of the ley by continued stirring, and sufficient repose to allow a complete deposition, it is dextrously bailed out with a dipper and thrown aside as material for "refining kettle soap." The scum and impurities which rise to the surface, are to be removed as fast as they appear. The physic having been all bailed out, the kettle is heated anew with a *steady fire of oak wood,* to about 230°, or 270° at furthest, and then washed over with water, added pintwise and carefully, throughout the surface. The time required for this operation is about four hours, and the proportion of water, say four gallons; but a much better criterion by which to judge the completion of the process is the perfect limpidity and transparency of the melted sperm. The washing must be continued until no more scum arises and the spermaceti is clean and white. In the meantime the fire must be allowed to gradually decrease, and when the liquid fat has stood for two or more hours after the completion of the washing, it must be gradually dosed with a bucket of cold water, thrown in and over the fat, a cupful at a time. The suspended particles of dirt and the residual impurities are thus carried to the bottom with the water, which also serves the double purpose of protecting the sperm from the heated metal. The supernatant stratum of melted sperm is then to be transvased into clean tin pans (free from rusty spots), where it congeals into perfectly white and pure crystaline blocks.

These blocks are the purified candle stock, but its crystaline structure must first be destroyed, else the candles moulded from it would be brittle. Some chandlers accomplish this by adding three per cent. of virgin white wax; others by casting the candles at a fixed and uniform temperature, for instance 140° F. The cavities which occur as the candles cool and contract, must be filled by fresh additions of melted tallow, or else the excess upon the upper bed of the mould-stand should be sufficient to supply these fissures as they gradually form. The shape of the moulds and the proper construction of the stand are exhibited by figure 77 and described at page 403.

The temperature at which the candles are moulded has an important influence upon their beauty of appearance; for example, in the winter when there is less difficulty to detach them from the moulds, the spermaceti can be run whilst very hot, and yet the candle, instead of being brittle and imperfect, will be beautifully diaphanous and wax like in appearance.

Wicks.—I have taken a candle, of the brand of "A. H. Campbell, Philadelphia," as my guide. Its superior burning qualities has entitled it to this preference. It is wicked with a doubled strand of eighteen very fine threads.

For what are called the "patent sperm" candles (which are compositions of sperm and wax, or stearic acid) the braided wicks are used.

Coloring.—For ornamental purposes the sperm candles are sometimes colored. This is readily done by triturating a very minute quantity of the pigment with oil, and incorporating it thoroughly with the melted sperm. Carmine is used to impart a red or pink shade; chrome yellow gives a lemon tint, and Prussian blue furnishes an azure color.

CHAPTER L.

COMPOSITION CANDLES.

THIS indefinite term is given most generally to candles made from block consisting of spermaceti and wax, or these two ingredients with a little tallow. The two following recipes are for the legitimate composite candles; but of late days the term has acquired a wider signification, being made to apply to all those kinds of candles consisting of one or more material without regard of their name or nature. The term first originated with an ingenious manufacturer, who, having a stock of dirty colored inferior candles of some particular kind, hit upon the expedient of getting rid of the unsaleable lot by dignifying them with the title of composition candles, a ruse which proved successful, for his credulous customers eagerly relieved him of them under the notion of having bought an article somewhat better than that usually purchased at the same price.

The exact constituency of the different kinds of composite candles we are ignorant of, as they are mostly the subjects of patents, and secret to all but the inventors.

The individual title of each varies with the whim of the patentee, so that we have very nearly the same composition, to be sure, a little modified in the proportions of each ingredient, under the names of "German wax," "Pearl wax," "Venetian" and "Patent" candles. It is improper to burden our book with directions as to inferior and unimportant mixtures, so we confine ourselves to those only which are likely to afford the results intended by the combination, viz., economy and improvement in burning properties.

Parlor bougies, similar to Judd's "Patent candles."—The handsome appearance of these candles, and their similarity in that respect to those of wax, has induced the title for them of bougies, which, properly speaking, is only applicable to candles exclusively of wax. Their mode of manufacture is as follows:—

Melt slowly over a moderate fire in a well tinned copper kettle, seventy pounds of pure spermaceti, and to it add piecemeal, and during constant stirring, thirty pounds of best white wax. By increasing the proportion of wax to fifty pounds, the resulting compound is much more diaphanous. The bougies moulded of this mixture are not of as lengthy durance as candles made from wax alone. These bougies are often tinted of various colors, rose, yellow, light blue, green, &c. For the red shade, carmine or Brazil wood, and alum are used. The yellow is made with gamboge, the blue with indigo, and the green with a mixture of yellow and blue. The bougies are sometimes perfumed with essences, so that in being burned, they may dispense an agreeable aroma.

Experience has shown that a more transparent and elegant bougie is made by adding only six and a half pounds of wax to one hundred pounds of pure dry sperm, and the candles made of the block formed in these proportions closely resemble the "Patent candles" of Judd.

Transparent bastard bougie, by Debitte of Paris.—To compose one hundred pounds of stock, take ninety pounds of spermaceti, five pounds purified mutton suet, and five pounds wax; melt each separately over a water bath, and to the whole then mixed together, add two ounces cream of tartar, and two ounces of alum in very fine powder, and whilst stirring it constantly raise the heat up to 80° Centig., (176° F.,) then withdraw the fire and allow the mixture to rest until it has fallen in temperature to 60° Centigrade, (140° F.) When the impurities subside the clear liquid composition must be drawn off into clean pans. Of this cooled block, candles are made which not only look well, but burn well. The suet is in just such proportion as will be a benefit rather than an injury.

The product of these admixed ingredients is not equal in beauty to that by the preceding process, but its quality and good appearance is more than proportional to its cost, which is much less than the aforementioned composition.

CHAPTER LI.

PALMITINE LIGHTS, OR CANDLES FROM PALM OIL.

THE process given below is that followed by Newton, the inventor of these kinds of lights. It is founded upon the same principles as that by which the stearin candles are made, the palmitine of the palm oil being the solid constituent thereof corresponding with the stearin of tallow. By saponification, palmitic candles can be made: the procedure being exactly similar to that for procuring the adamantine light. What their qualifications are, the author is unable to say, not hàving seen a sample, but from the properties of the material there is no reason to doubt its efficacy as an illuminating agent. The principle upon which the inventor bases his discovery is, as he says, the hitherto unobserved fact that the liquid and solid constituents of palm oil exist naturally in a state of chemical separation, although their particles are in mechanical contiguity, and that therefore nothing else is requisite to effect their complete separation but a purely mechanical process, unassisted by any ingredients which would effect a chemical change in the oily or fatty matters, as is the case when saponification takes place. The process consists in causing the solid fats to crystalize by first heating and then gradually cooling the raw material; afterwards submitting it to pressure for the removal of a certain portion of the olein, and then granulating it as before, a second time, so that another portion of olein may be ejected by a second pressure. By this means the separation of the solid and liquid fat is effected without resort to the process of saponification. The mode of procedure in detail is as follows:

Take the raw palm oil, or that which has been previously bleached, and heat it by the introduction of steam to about 212° F., and maintain it at this temperature for about an hour, and after that time run it off into wooden vessels, where it is allowed to cool very slowly. The congealed or granulated mass is then to be packed in cold woolen cloths, in quantities varying from sixteen to thirty pounds each bundle, according to the capacity or size of the press. The liquid running out from these bags under the influence of hydraulic pressure, is nearly pure olein, but as palm oil is generally considered as containing about sixty-nine per cent. of olein, and thirty-one per cent. of palmitine or solid matter, it follows, that as but one-half of the liquid portion is expelled by the first pressure, the other part must be still contained in the solid cakes enclosed in the bags. In order to facilitate the further separation of the remaining portion of olein from the palmitine, the cakes of fatty matters must be emptied from the bags and again heated in a tub by steam to about 212° F., for three or four hours, after which it is again run off into wooden vessels, where it is left to cool slowly, and granulate a second time. The grained mass is again bagged in woolen cloths, with horse hair envelops, and these bags subjected to the action of the horizontal press in manner as aforetold, for the last pressure of stearin, p. 442. During this pressure the olein running off carries with it also small quantities of the solid fat, and hence is particularly applicable for conversion into palm soap. The cakes that remain in the bags will be found firm and solid, and are now to be bleached, if the oil from which extracted was not previously whitened. This is done by placing them in a tub or vessel, and melting them at a temperature not exceeding 212° F. When the whole is liquid, it is to be run out into a trough which is kept supplied with a running stream of clear cold water, in the proportion of about two parts of water to one of palmitine, so that the latter always meets a quantity of water about double its own bulk, and by this means it becomes almost instantaneously solidified, and converted into small solid flakes. The water carrying with it the palmitine in this state is run into a vessel whence the flakes are removed, and exposed to the action of the air, and light of the sun, whereby they become perfectly white in a very short time. The bleached palmitine is then to be re-melted by steam, and to the heated fat is added sulphuric acid in the proportion of two pounds to every one hundred pounds of palmitine, for the purpose of cleaning it of any extraneous matters. After the fat has been cleansed in this manner, it is removed to another vessel, where it is again melted in water with the addition of the white of five eggs to every one hundred pounds of palmitine. The whole must boil for about an hour, and be kept well stirred, after which it should be left to itself, that all the dirt and impurities may subside. The palmitine is then run into pans, and when solid, is a fine white clear block.

The wicks for these candles, Mr. Newton says, should be soaked for twelve hours in a solution of one part of boracic acid in twenty-four parts of dilute alcohol, then pressed gently and dried.

The "palm wax," a beautiful wax-like fat, the properties of which were described when writing of palm oil and palmitic acid, with the latter of which it is identical, is procured and worked in a manner very similar to that for stearic acid.

Hempel & Blundell have published a long process for making the palmitic candles, but there is much of it superfluous, we prefer recommending our plan followed for making the adamantine lights. Runge, of Berlin, claims to have been the first to procure the palmitic acid on a large scale, and at an economical rate, and the samples exhibited by him, it is said, were of a brilliant pearly whiteness.

Coco nut oil candles.—The coco stearin is separated, by pressure, from the butter in the manner and after the process that is prescribed for the stearin candles, p. 453, omitting the spirits of turpentine. The candles made of coco tallow are superior in appearance to moulded tallow candles, and, besides, yield a greater proportional light. Their smell, too, is not unpleasant, and they burn with a clear white flame, whilst the wick requires but little snuffing.

The resultant soap from the saponification of this butter, when decomposed by strong acids, yields its fatty acids in a free state. These, when treated as is directed for the stearic lights, will make a most excellent body for candles. The composition of this butter and its properties have been given in a previous chapter.

We now leave this part of our subject to treat of wax candles, the manufacture of which forms an entirely separate branch of the trade; still as it properly belongs to the profession of the general chandler, we will present a full and explicit direction as to the requisite manipulations.

CHAPTER LII.

WAX CANDLES.

It was of wax and tallow that candles were first made. Those of the former material have been superseded, almost entirely, by the spermaceti and the much improved stearic candles; so that now, they are scarcely used except in churches and extravagant families. They are chiefly made in two ways, by the hand and by the ladle, and in manner as will be fully described hereafter.

Wax being less fusible than most of the materials used for the body of candles, requires a small slender wick. All the manipulations requisite for a successful practice of the art will be found in the succeeding paragraphs.

The high price of wax candles, being about one hundred per cent. above the adamantine lights, which are nearly as good, if not as handsome, limits the extent of their consumption. The small supply needed in this country is imported from France, and the introduction of their process of manufacture in this book is in accordance with the author's desire that there shall be no link wanting in the chain of instruction which it is intended to complete. We proceed to speak of the consecutive operations, but, previously, acknowledge our indebtedness to the French of Normand for most of the information, and all of the drawings.

Selection and purification of the wax.—The composition and properties of wax have been given, so the purpose of this chapter is only to make known the manipulations requisite for the conversion of the natural yellow wax of commerce, into an article sufficiently pure and white to be formed into bougies or candles. Before proceeding to the details of the processes, it is proper that some knowledge should be imparted as to the means of distinguishing the admixtures with which the wax is very frequently more or less adulterated, and thus to afford an ability of judging its quality. Much of the wax comes from the western country,—some from the West Indies. It is yellow, and in pan formed cakes, dry, brittle, of a granular structure, and when chewed, does not adhere to the teeth, and, when pure, is entirely free of all resinous or suety odor. The wax should have an agreeable smell, and this is important, for the spirit of fraud has even taken possession of the farmers, some of whom send their wax to market sophisticated with a high per centage of resin or suet. To detect these frauds we here make known a method. When the wax contains resin, it sticks to the teeth when chewed, and the admixture is detected by the taste it imparts. The suet is similarly and as easily discovered. Besides, the addition of these bodies to the wax destroys its granular structure, and lessens its consistency. A little experience will enable an approximately correct opinion even by ocular examination.

Sometimes wax is also contaminated with the fecula of potatoes, which, when added after the wax is already adulterated, with resin and suet, restores its consistency and granulated structure; hence, here is an ingenious mode of defrauding which renders somewhat difficult the exposure of the deception. Delpich, an experienced pharmaceutist near Paris, has happily furnished a mode for detecting this fraud. Heat a certain quantity of the wax to be examined, with pure spirits of turpentine; the wax is dissolved by the spirits as are also the other greasy matters admixed with it, but the fecula is left untouched and subsides to the bottom of the capsule. Independent of the principal signs that have been above mentioned, to recognize the sophistication of wax, here are some other means which may be available in doubtful cases. 1st. Throw a small piece of the wax upon burning coals; a disagreeable odor and very thick fumes, betoken a fraud. 2d. Drop a small quantity, in a melted state, upon a piece of woolen stuff, and endeavor to remove it with alcohol. If the wax is pure the alcohol and a little rubbing effect its removal immediately, in the form of powder; whereas, on the contrary, when adulterated, it is very adherent, and leaves a spot. The wax is rarely employed in its virgin state, it being necessary for most purposes to purify and bleach it.—For this operation there is required, always, several apartments. Most candle manufacturers purchase their wax already whitened, from those who make the business of bleaching their exclusive occupation, but as our work is intended to be full and complete, we proceed with the proper processes for washing and bleaching the wax, in order that those living distant, who are compelled to bleach for themselves, may not be ignorant of the mode of doing it properly.

Washing and bleaching.—The bleaching-yard always forms a main portion of the wax foundry and refinery, and for the sake of being as explicit and intelligible as possible, we will describe the different apartments of the atelier or laboratory, and the tools and apparatus required for the operations. The laboratory consists of two distinct and separate apartments: the foundry and bleaching yard. The former is a large room, or hall, in which are arranged the furnaces, caldrons, and all such utensils as are indispensable to the process of melting and casting the wax. Fig. 108 represents in perspective an interior view of a foundry. The larger pieces of apparatus are apparent in

Fig. 108.

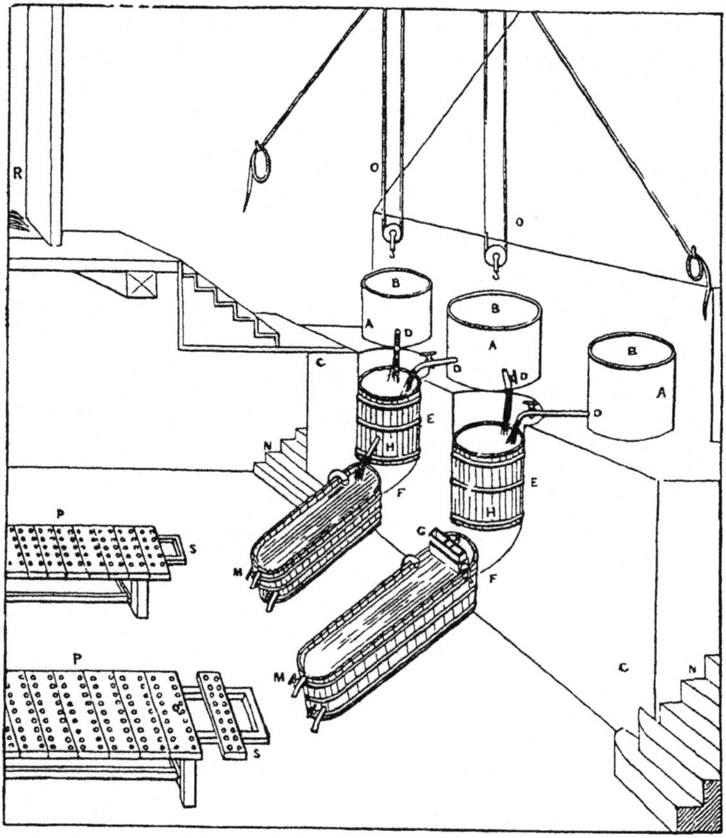

this figure; the smaller ones will be described in detail. Against one of the walls of the building, back of which there is a free passage for the workmen, are erected in the brick work, C C, the furnaces A A A, over which rest the caldrons B, B, B. These caldrons are formed of an half egg shape, and tinned throughout. The middle one has double the capacity of either of the other two, as it serves to feed both of them with melted wax. At four inches from the bottom of each caldron is soldered a copper tube carrying at its end a strong brass

41

cock, D; the centre kettle has two. The top edges of these kettles incline inwardly, that the wax which spatters against the sides may be driven back again. Between the caldrons, and under the furnaces, are placed the iron bound tubs, E E, supported by the brick platform F, of such a height that the melted wax can readily flow into the *greloir* G, which is beneath. These tubs have each two gully-holes, with movable wooden stops; one placed almost even with the bottom, the other a little above the third of its height. It is through this latter, H, that the wax runs to be grained. The lower hole serves as an evacuating outlet for the cleansing of the tub. The greloir G, Fig. 109, which will be described

Fig. 109.

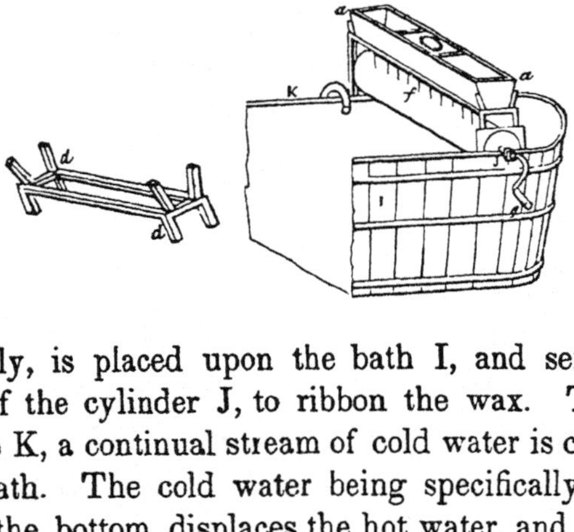

separately, is placed upon the bath I, and serves, by means of the cylinder J, to ribbon the wax. Through the tube K, a continual stream of cold water is conveyed to the bath. The cold water being specifically heavier goes to the bottom, displaces the hot water, and drives it to the top, where it escapes through the tube M. A similar greloir is placed upon the other bath. Two ladders or stair-cases, N N, are affixed to the two extremities of the furnaces, and lead to the magazine or store room R, in which is placed the ribboned wax. The tackle of pulleys O O, serves, by the assistance of three cords properly adjusted, to lift off the tubs E E, which frequently require to be cleansed, which, without these cords and pulleys, would be difficult to remove. In other parts of the atelier are arranged the tables P P, upon which are ranged the planks Q, indented throughout their surfaces with cavities, intended as moulds for the wax, as is seen at Fig. 110. Fig. 109 represents the

Fig. 110.

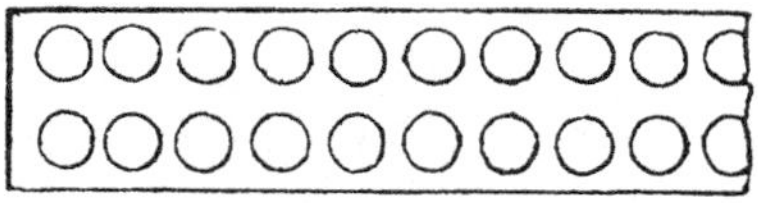

greloir, which is a tinned copper vessel *a a*, as long as the width of the bath, at the place where it rests. Its bottom has the form of a reversed gutter, and extends the whole length of the apparatus. In the bottom of this reversed gutter or trough there are numerous small holes of about two millimetres diameter, and at a distance apart of about thirteen millimetres. The wax runs through these holes in thin strings, and falls upon the surface of the cylinder, of which we will speak directly. It is easily conceivable why the bottom of the greloir is in the form of a reversed gutter, and that the holes are made in the more elevated part; for if any dirt should fall in the wax whilst running from the upper gulley-hole of the tub E, to spread upon the cylinder, these dirty particles remain below, and do not ascend to stop the holes and thus prevent the running out of the wax. For a like reason there is placed above the greloir a square cullender, Fig. 111, upon which the wax first

Fig. 111.

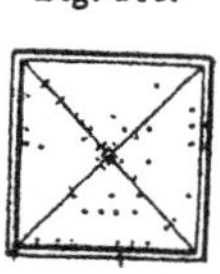

falls, and thence through the meshes into the greloir. This cullender or strainer is adjusted in a groove in the centre of the length of the greloir, which is entirely covered by two lids opening on hinges. At each end of this greloir is a cavity, of a similar form, into which are placed the hot coals for maintaining the fusion of the wax at the ends of the greloir, where it cools much sooner than in the middle. The greloir is supported by an iron frame-work or treadle *d d*. In some factories they do not use the square cullender, but adopt a more preferable method, discarding also the hinged covers of which we have just written. The entire opening is covered by a tinned copper plate about fifteen inches long and of breadth equal with that of the greloir. Three of its sides are ledged with a rim of twenty-seven millimetres height, and the fourth is jagged like a saw. This plate rests upon a sloping rabbet and upon it is placed an oval copper sieve, encompassed with a high rim, and tinned exteriorly and interiorly.

The advantage of this construction is that the liquid wax falls in the cullender, from thence spreads upon the plate of copper, and is conducted by the inclination or declivity of the plate, into the greloir, where it falls in thin layers; otherwise it would drop in flakes, and, by closing the holes of the greloir, stop its passage way. At

Fig. 112.

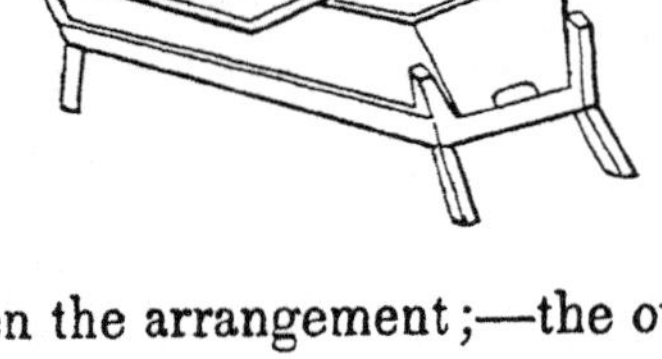

Fig. 112 is seen the arrangement;—the oval cullender *b*, the plate *c*.

Supposing the greloir to be placed upon the edges of

the tub, beneath must be arranged a wooden cylinder *f*, the iron axis of which has, at each end, a pivot resting in two notches upon the edges of the bath. The winch *g*, which fits to one of these pivots, serves to put the cylinder in motion. The larger the cylinders are, the better; they can be even a foot in diameter. The wax in falling through the holes of the greloir upon the cylinder, immersed about one-half in water, immediately congeals.

The cylinder is generally made of walnut wood well jointed and fastened together. Fig. 110 represents one of the boards in which are hollowed out the holes for moulding the wax. These holes are conical, five to six lines in depth, thirty lines in their upper or greater, and twenty-seven lines in their lower or lesser diameter. There are two rows in each plank, and so far distant from each other, that the two beaks of the feeder, Fig. 113, shall fall simultaneously in the centres of the two moulds when they are being filled. Fig. 114 exhibits

Fig. 113.

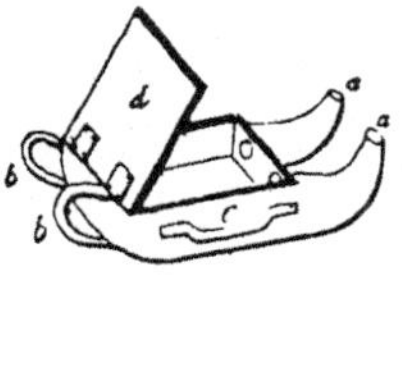

Fig. 114.

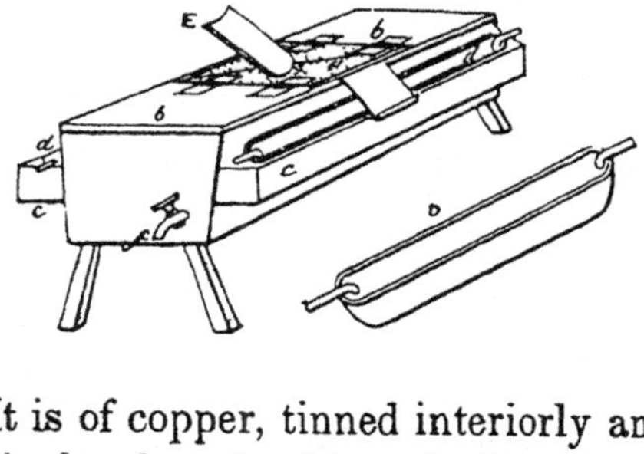

the *coffre à eculer*. It is of copper, tinned interiorly and exteriorly, and thirty inches long by fifteen inches broad, and formed something like a kneading trough. Its cover consists of three pieces; the centre *a* is a cullender exactly similar to that described on the other page; the end pieces being two hinged gates serving to hermetically close the chest, that the admission of dirt or dust may be prevented. At one of the corners of the trough and near the bottom is a cock *f*, and the two larger sides, throughout their whole outer length, have casings *c c* in which are placed the sheet iron warming pans *d d*, (the form of which is seen at D,) used, when filled with coals, for preventing the congelation of the wax. The coffre sets upon a trestle, and is movable in order that it can be substituted for the greloir when necessary.

When about to melt the wax, the coffre is substituted for the greloir under the cock of the tub, so that it may be filled with liquid wax, which runs at first upon a gutter E, thence into the strainer *a*, upon which is arrested all the particles of dirt contained in the wax. This coffre is a reservoir from which the workman draws a supply of wax for filling the moulds. It is for this reason that one or more of them are arranged near to the mould tables, so that it may be convenient for the workman to fill the feeder when it becomes empty.

Fig. 113 represents the feeder. When closed up it has the form of two clogs joined together. At *a a* are seen the two beaks or spouts; *b b* are the two handles, which can also be placed at *c*. The coverlet *d* is always kept closed whilst running the wax into the moulds, in order to prevent the falling in of dirt. The mould tables before spoken of, and represented at P P by Fig. 108, are strongly built, and from ten to twelve feet long. They have no top, but in its place a movable frame, five or six feet longer than the table itself, and overspreading equally on all sides. This frame is seen at S S. It is not to be entirely covered with mould boards, as there should be a foot at each end unoccupied.

Fig. 115 is a wheelbarrow carrying a wicker basket lined with linen duck. It is in this basket that the ribboned wax is placed as soon as it is formed in the bath; and when filled the workman wheels it into the bleaching yard.

Fig. 115.

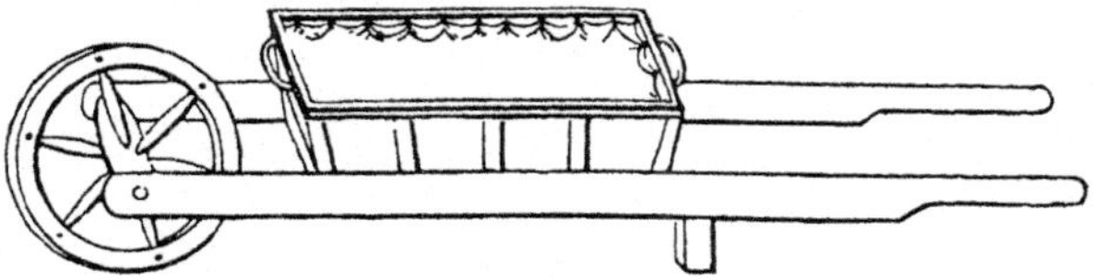

Fig. 116. A three forked shovel of white pine wood, and wrapped around with willow osiers, as is seen in the

Fig. 116.

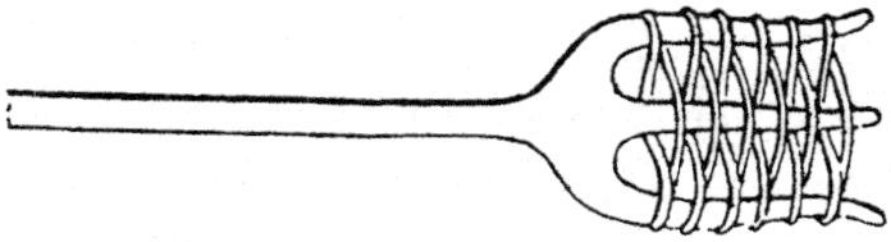

figure. This is used for shoveling or lifting out the ribbons of wax when they are too delicate and fragile to be otherwise handled. A lighter one, modified, is shown by Fig. 117.

Fig. 117.

Fig. 118 is a wooden spatula, or paddle, called by the workmen *palon*, and used for stirring the wax whilst it is in fusion.

Fig. 118.

Fig. 119 is a hair sieve for collecting the particles of wax, too small to be lifted out of the bath with the wickered forks.

Fig. 119.

The operations of the foundery and bleachery are connected so closely together that it would create confusion in the numerous details to attempt a separate description of the two ateliers. Different operations are being successively and some simultaneously performed in both apartments, so we continue our description, passing on to the bleachery or bleaching yard, and proceed to explain the different stages of the process consecutively, as conducted in either division of the factory.

For this part of the establishment an extensive field free of trees should be selected, for it is important that the sun's rays shall have full play upon the exposed wax, which, when ribboned, is spread upon cloths arranged as follows: The frame-work which sustains the cloths

Fig. 120.

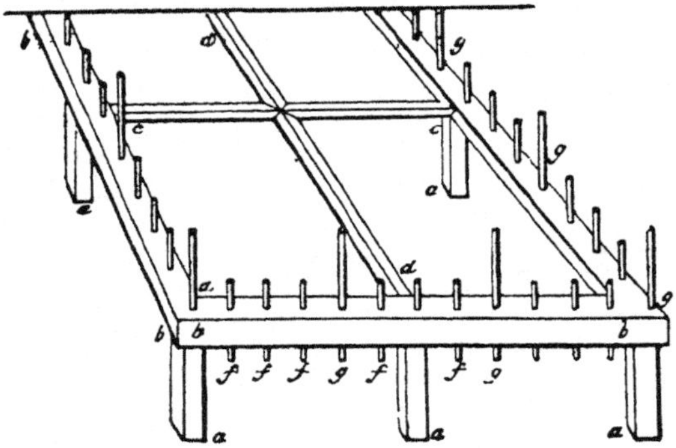

is of eighty feet length, by ten feet breadth. The supports *a a a* (Fig. 120), on which rest the squares two feet from the ground, are strong stakes, pointed at the end and well driven into the earth. See that the tops of all these pieces are uniformly and exactly two feet from the ground, and let there be three ten feet rows of nine stakes to each row, the distance between each being five feet, so that these twenty-seven stakes enclose a rectangular area of eighty by ten feet. Upon the ends of the three stakes is affixed a strong beam of three inches width and eighteen lines thickness, and parallel to the other stakes are fastened strong beams of similar dimensions.

The stakes or posts of the middle row are covered throughout their breadth by strips of the same height as those of the circumference, but only one inch wide. They are jointed in the centre, and supported by the posts and side pieces, and are beveled on the edges. Seven strips similar to these last are placed upon the three posts along the breadth. All being thus arranged, the strips which form the circumference of the frame are bored through with vertical holes of about an inch diameter. The holes are about nine inches apart: each of the side strips contain one hundred and three, and the end pieces thirteen, so that the whole number of holes is two hundred and thirty-two. In each hole is fitted a peg, tightly hammered in so as to be immovable. These pegs vary in length; some are of six inches projection above the strip, and the rest, being the lesser part of the whole, extend two feet beyond, and carry at their top ends a small iron hook which face the interior of the square. Here is the order in which the pegs are placed. At each of the four corners drive in a long one, then follow it with three short ones, and these again with a long peg, and so continue until the whole circuit is properly filled. Take care that the pegs are of good strong wood, and by preference of oak. Fig. 120 gives an end view of the frame, and will serve to elucidate our explanation. The strips or scantlings are seen at *b b*, the pieces *c c* and *d d* form the union stretch which connects the posts. The short pegs are shown at *f*, and the long ones with their hook nails attached, at *g*. The cloths that cover the frames are eighty by ten feet, and edged around with a hem or border of two feet height. Fig. 121 represents

Fig. 121.

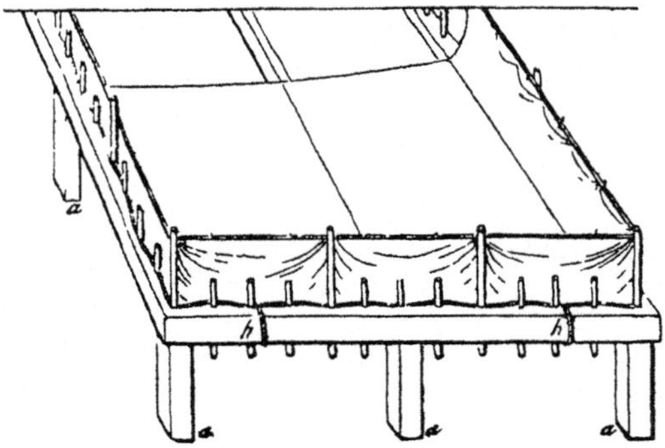

the manner in which it is stretched over by means of a cord and the pegs. Some strong iron rings are generally sewed to the edge of the cloth, and in position, so that they may be directly opposite the pegs. Fix this bag cloth over the frames in the same manner as a sacking bottom is stretched over a bedstead, being careful to fasten the ends of the cords to the posts in a manner preventive of loosening by the wind. In these two preceding figures we have only shown one half of the frames, because, independent of the want of room, the other portion being exactly similar to that part exhibited, it is unnecessary to give the whole.

The other apparatus used in the bleachery consists of

Fig. 122.

a small very light fork (Fig. 122) for spreading out the wax ribbons upon the cloths; a scoop and thin paddle for removing the wax when it is bleached, and duck linen sacks in which to collect it.

The first process is that of melting the wax; but, before we proceed with its recital, it must be told that the operations of bleaching can only be carried on during a portion of the year from April to October, as it is but slowly and indifferently effected in the remaining and colder months. Well, to proceed, the workman puts into one of the caldrons A (Fig. 108) a quantity of wax sufficient to cover one of the cloths destined to receive the ribbons proceeding from this melting. The wax should first be hashed up into small pieces, as was directed for the tallow at p. 328, and to every fifty kilogrammes there should be added in the kettle two or three quarts of water. The fire is then started under the caldron, and the wax slowly melted. The workman with a wooden spatula stirs the wax during its fusion, and continues the agitation until it has acquired sufficient fluidity, so that its dirty contents may better separate and deposit. The proper liquidity being attained, the cock of the caldron is opened, and the melted wax drawn off into the tub, partially filled with water. This tub is fitted with a plug, the management of which will be explained directly.

As soon as the wax has entirely run off, the tub is covered over, and the whole enveloped with woolen blankets, so that the liquidity of the wax may be maintained during the time it is settling, or, in other words, to ensure the retention of sufficient heat for the perfect separation and precipitation of dirt and other impurities, which, by their greater specific weight, fall to the bottom, together with the water added to facilitate the fusion. This is called refining the wax. The tub is left in repose for two or three hours or more, according to its amount of contents.

Having made this operation, just described, the first step of the process, consequently the tub should be placed ready for the reception of the purified liquid wax. In the subsequent stages the tubs are arranged as the wax commences to melt. Each time that the wax is emptied out to be ribboned, the tub should be cleaned, and in the meantime another substituted in its stead. It is well, whilst the wax is settling, to supply the bath which is below it with fresh water, (by opening the cock K, Figs. 108 and 109,) proportional as it becomes warm, and is emptied through the cock near its bottom.

Fig. 123 shows the two pieces used in this instrument.

Fig. 123.

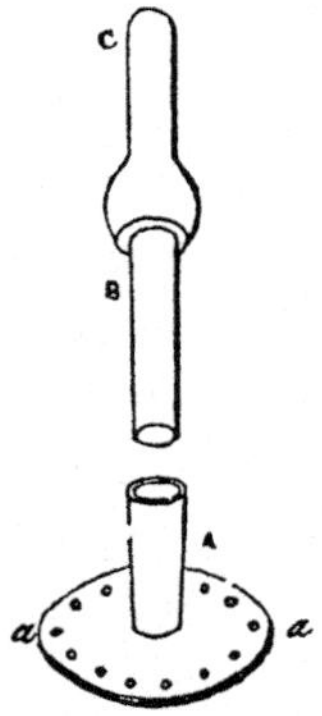

A, the gulley hole, properly speaking, is a tube of hard wood, threaded in and solidly fastened to an iron plate, *a a*. The tube fits and enters the gulley hole, and the iron plate is screwed to the tub. The channel is stoppered with a cock internally, for in this position it resists much better the pressure of the liquid wax; and for the purpose of drawing it off, there is used a cylinder B of hard, solid wood. It has a large strong handle C, by the aid of which it can be strongly driven or pushed in, accordingly as the wax is wanted to run out in a thick or thin stream, upon the greloir beneath.

The wax falls upon the cullender, from that upon the plate, and finally, into the greloir, where it occupies at first the two furrows or trenches at the bottom, and runs through the holes of the gutter in small threads, upon the wet cylinder beneath continually in motion. The thread of wax falling upon the cylinder becomes flattened, and on entering the water congeals, and forms a thin ribbon. The more rapid the motion of the cylinder, the thinner are the ribbons of wax. During this operation, the cock K, through which is furnished the cold water, is left open, and the heated water pushed out by the greater specific weight of the cold water through the escape pipe M. It is very necessary that the water of the bath shall be kept as cool as possible, for otherwise the wax, not being sufficiently cold, will ribbon badly, or perhaps adhere to the cylinder, and cake together.

The workman takes the fork and pushes the ribbons, as they accumulate near the cylinder, towards the end of the bath opposite to the tub, and then lifts them out into a basket. When this basket is full of ribbons, it is put upon the wheelbarrow, and conveyed to the cloth on which they are to be spread. They are now carefully taken out by a woman, and spread equally over the whole surface of the cloth, so that all the particles may have an uniform exposure.

This operation continues until the wax is all emptied out, and it requires about an hour and a half for every

1000 pounds of wax. The tub is then raised up behind by the aid of a lever, so that all its contents of wax may drain out through the exit hole.

That the steps may succeed each other rapidly, and without loss of time, a melting should be made every three hours, that is to say, you begin a melting three hours after the commencement of the previous fusion, and thus follow on uninterruptedly. The length of time for its exposure depends altogether upon how it is set out, and on the brightness of the weather; sometimes the ribbons require to be left on the cloths for ten, twelve, fifteen, twenty, and even thirty days, and more before they are turned over. This is done with a paddle which is pushed under, and made to turn them topsy turvy. A few days after the ribbons have been turned, they are stirred about upon the cloths with small forks so that they may present as much surface as possible, and this is called *regaler*. An experienced workman well knows the circumstances under which it is necessary to regaler the ribbons more than once or twice.

When by this operation the wax has acquired its first degree of whiteness, it is removed from the cloths with a scoop and thrown into linen sacks; these, when full, are carried into the magazine, piled up, and there left undisturbed for thirty or forty days. In this time it settles and becomes a matted rather compact mass, which by a second fusion and treatment acquires a greater whiteness than it would have otherwise.

The regrèlage or second fusion.—Light up the fire under the caldron into which are placed fifteen to twenty kilogrammes of the ribboned wax treated as above directed, and with it one quart of water. The wax should be thrown in gradually, and in small quantities at a time, by one workman, while another keeps the mixture of water and wax in constant agitation by continual stirring. The fire should be moderate and the addition of wax ceased so soon as the kettle is sufficiently full. This is a tedious operation and tiresome, because wax forms a kind of thick paste, which it is difficult to stir. The kettle being full, the fire is then increased, in order that the wax may acquire perfect liquidity, so that it will run easily through the cock. The stirring must be continued until the fusion is complete and all granules have disappeared; for, otherwise, there is a liability of the wax becoming scorched. It is now drawn off and put through the same process of ribboning and exposure as before, and as soon as it has acquired whiteness it must be transferred to linen bags and carried to the magazine.

Third and last fusion.—Some bleachers add to every 500 kilogrammes of fused wax three or four quarts of milk, but it is acknowledged that this addition occasions a loss by the considerable deposit that falls to the bottom of the tub; it is preferable to add cream of tartar or powdered alum, as has been directed for the suets. The fusion is made as described for the regrelage, in the preceding paragraph, but this time the wax is moulded and not ribboned. While the wax is reposing in the tub and depositing its impurities, the number of mould planks requisite to receive the whole of it, must be put to soak in the bath. The greloir and cylinder are lifted from their places; the coffre, Fig. 114, is substituted for them, and the casings are then filled with hot coal. When the wax has entirely cleaned or deposited its impurities, it is then drawn off into feeders through the exit cock of the coffre, and from thence served, by women, into the mould holes of the planks, on the tables P P, Fig. 108. They continue serving out the melted wax until all the cavities have been filled.

As fast as the wax becomes thoroughly congealed, the

small cakes are thrown into a tub of clean water, for as the holes were at first dampened, the wax does not adhere but detaches easily. These cakes are dipped out of the bath upon a pack thread sieve, which lets the water run through and retains the pieces of wax, that are then placed in the baskets and from thence transferred to the cloths, and there arranged side by side, for several days' exposure to the air and sun. Thus they become very white and dry, and when packed in clean boxes are ready for the market or conversion into candles.

The waste, which deposits at the bottom of the tub, is thrown into a cullendered vessel; after the water has drained off it is put by until enough has been collected to work upon for the separation of the residual wax. To every fifty kilogrammes of waste are added twelve pints of water, and the whole heated together until liquefied, and this state retained for six hours, to give sufficient time for the thorough deposition of the water and the impurities along with it. The fire being then extinguished, ladle off the wax which swims on the surface and transfer it to suitable pans. Thus are attained lumps of wax which are to be remelted and bleached in the manner aforesaid.

It must not be forgotten that the action of the wax upon copper prevents the use of vessels made from that material, unless they are thoroughly tinned inwardly and exteriorly, otherwise a greenish tinge will be acquired.

When it is windy, it is necessary always to have an eye upon the cloths, lest there be a loss of ribbons through negligence; as they are apt to be whisked off into the air far beyond recovery. To provide against this contingency, the cloth must be detached from that side of the frame facing the wind, and doubled over the other half. The wax is then protected between the folds. There often results another inconvenience—the intensity of the sun's rays is at times so great, that, concentrating upon certain portions of the wax, it becomes softened and mats together with the neighboring ribbons. When the wind has ceased blowing, the cloth folded over, as above directed, must then be restored to its original position, and the ribbons of wax spread over the surface.

In moulding the wax there is a certain experience required in order that it may have a pleasing appearance. It should be poured in whilst neither too cold nor too hot, otherwise the surface of the cakes will be wrinkled and full of holes, whereas it should be perfectly smooth.

OF THE MANUFACTURE OF BOUGIES—This is the French term for a wax candle in contradistinction to *chandelle*, applied to those from tallow. At the present time the word has a wider signification, it being extended to many composition candles of appearance resembling those of wax. It is in this chapter intended in its original sense, and as proper to candles exclusively of wax.

Before proceeding with the details of the manufacture of bougies, it is proper to give some insight into the mode of preparing the wicks requisite for this kind of candles. There are two sorts of bougies, or rather one species, the parlor light, subdivided into moulded and dipped bougies.

Of the wicks.—At Chap. XLII has been given a full treatise upon the material proper for wick and the method of preparing it, and whilst we know that it is the habit with some chandlers to mix with the threads of cotton those of hemp or flax, sometimes in the proportion of one-third, we condemn the practice, and in direct opposition to the opinion that induced the addition; for, instead of the hemp threads, as they aver, having more stiffness than the cotton, and being thus prevented from bending, the threads do not burn as lively as those of the latter, and the very curvation of the wick to be evaded invariably happens.

For the parlor bougies the very finest and most uniform cotton thread must be used in the manufacture of the wicks. Each wick should consist of two strands of sixteen threads each, making the total number of threads thirty-two. The comparative decrease in size is because the wax sheds a more brilliant and lively light than the fats. The *coupoir*, or wick cutter, is constructed in the same manner as that described at p. 359.

In describing the flambeaux, &c., we shall always take care to mention those whose wicks are of hemp or flax threads; the absence of any remark as to this, indicates the employment of pure cotton for the wicks. The chandlers, those that use hemp wicks, purchase the threads from the rope-makers, twist them slightly, and cut them six feet in length for the larger flambeaus. Each cut consists of eight threads, which, when doubled over and slightly twisted, forms the wick. A few white threads of cotton shaped into a loop and passed through the bend of each wick, forms the collet of the flambeau.

Some chandlers soak the wicks in alcohol before using them for the bougies or cierges, but this is useless, for the volatility of that body is such that it all escapes by evaporation, and leaves the wick as it was before being impregnated. Spirits of turpentine, too, is not unfrequently employed for the same purpose, yet with no better effect. The only benefit that might accrue from the use of it is the facility acquired in igniting the wicks from the ready inflammability of the turpentine; but this advantage is defeated, first, by the volatility of the essential oil, which causes its rapid evaporation; and secondly, by the disagreeable smoke and smell its application would occasion. Hence, also, the spirits of turpentine is a useless, and would be a deleterious admixture for the wicks.

Fine, uniformly spun, cotton thread, perfectly dried, answers every purpose without any previous preparation. The size of the wick must be proportional to that of the candle, so that during the burning of the bougie the melted wax at the top may be entirely absorbed by the wick and leave the cup which forms around nearly clean and empty. Well made bougies should not require to be snuffed, except where a "thief" occasionally forms upon the wick, and must be removed to prevent the candle from running. It is better to put out the candle, when necessary, by the breath than with an extinguisher, because the ashes or cinders are then blown away, whereas, in the latter case, they fall into the cup, and when the bougie is again relit, cause it to run.

The moulded parlor bougies are made very nearly in the same manner as are the tallow and other candles, with some difference, however, in the mode of preparing or soaking the wick, and in the material of which the moulds are cast; the manipulations requisite, therefore, will be described in full, when speaking hereafter of the manufacture of cierges. The moulds are made of the usual shape, but of glass. To draw the candle, they are plunged into hot water and immediately withdrawn; the dilation of the glass allows the easy exit of the candle, which must be dextrously pulled whilst the mould is being taken from the water.

Dipped bougies.—This kind of bougie is made in manner not dissimilar to that in which the dipped tallow candles are prepared, saving the difference that the wicks are placed *en ferret*, that is, fitted with a small tin tube so as to prevent the *lumignon* being covered with wax during the process of making the candles. The melted wax is then run on as will be explained directly. The placing of the wicks *en ferret* is as follows:—A workhand, generally a woman, sets before a table by the side of which is a box containing the ferrets or small tubes of tinned iron, which prevent the lumignon of the wick becoming covered with wax, and upon her knees places a packet of soaked wicks. Then with a brass wire catch in her right hand she threads each wick into a ferret just sufficiently large for its entrance. Be careful, however, in arranging the threads so that the ferret shall protect that part of the wick which it covers, during the process of soaking and dipping the wicks, which, when thus prepared, are handed over to the workman whose duty it is to form the bougies.

He then throws the wax, commencing about the centre of the wick, and thus covers the half of it and also the ferret. When by that means the end has acquired the proper size, the bougies are turned over, and by the other end, attached to the hoop, and in the same manner, he covers the other half of the bougie, until the whole candle has assumed a cylindrical form. When the bougies are presumed to be of the proper weight they are handed over to the next operative, who rolls them upon a table, as will be told in speaking of the cierges. There is this difference, however, that with the bougies the ferrets are removed as follows:—After having dampened the table so as to prevent the adherence of the wax, the workman rolls five or six of the bougies, and with the *couteau à ferrets*, a two edged knife of three decimetres length, and one decimetre breadth, cuts off about two centimetres from the height of the candle, in order to remove the ferret, and with another smaller knife the conical tip of the candle, leaving the wick, perfectly free of wax, a quarter of an inch above, so that it may form the collet of the bougie. After the tip is shaped then the candles are made of an uniform length by clipping off such excess as is designated, by the proper measures, for removal. All these manipulations must be prompt and systematic, there being a special workman for each step, in order that the work may proceed the more briskly, and whilst the bougie is still soft enough to be wrought with facility. Finally, the candles or bougies, being finished, are spread out upon white paper to be dried

Of the cierges à la cuillier.—The cierges differ from the bougies in being much larger and conical. They are mostly manufactured for the Catholic churches, who burn them at the performance of high mass, and on the occasion of their numerous religious festivals. These are made by furnishing the wax, from a cuillier or large ladle, to the wick, vertically suspended to an instrument called in technical language a *romaine*, a description of which will be given directly. The apparatus necessary for the operation is an iron furnace A, Fig. 124, enclosing the heating pan B, filled with coals. The furnace is

Figs. 124, 126, 127.

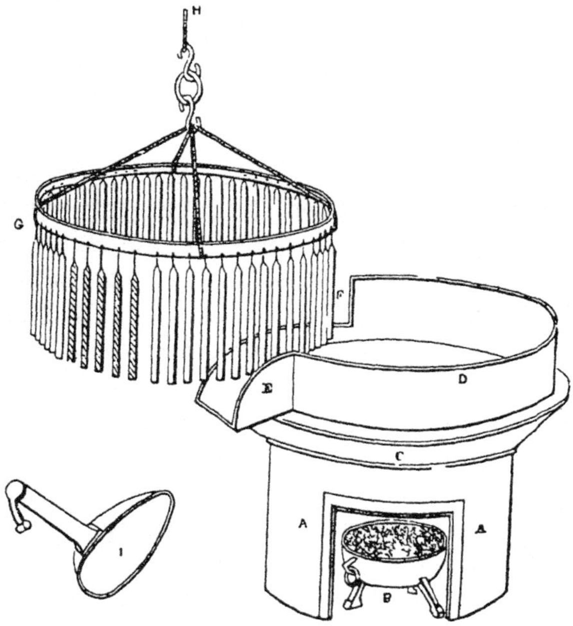

surmounted by a tinned copper basin C, strongly made of thick metal, and enclosed around on its top by a sheet tin ledge D, open at E, and also notched at F, for the ready entrance and exit of the cierges or bougies, accordingly as the hoop which suspends them is turned around. Fig.

Fig. 125.

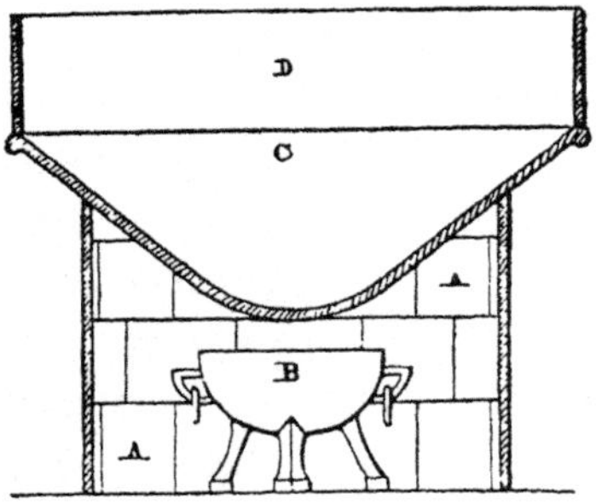

125 shows a sectional view of the four pieces united, as shown in perspective by Fig. 124.

There is required, secondly, an iron or wooden hoop, G, Fig. 126, sufficiently large for the suspension of fifty cierges upon the iron hooks placed throughout its circumference at a distance of one and a half to two inches apart. The circular hoop is hung by a rope H, an iron ring, and three cords, at such a height as will prevent the bougies touching the copper basin. It is raised or lowered to the proper distance by the rope H, which passes upon a pulley, and has its other end fastened to a fixed point.

The mode of constructing the *romaine* may be much simplified, for, instead of the hanging cords, there can be substituted six iron rods uniting in an iron socket fitted with a small wheel; it is borne by a cylindrical wooden beam, fixed upon a solid foot of the size of the hole of the socket, where it enters freely but not loosely. The beam has a similar socket pierced with two holes opposite each other and fastened by a movable iron bolt, so that it can be changed to a lower or upper hole accordingly as it is desired to raise or lower the beam. The socket of the hoop presses upon the socket of the beam, and in this way the hoop is kept always horizontal. The name of *romaine*, the French for steelyards, has been given to this hoop, because of its resemblance to a balance beam, there being suspended to each end of this beam a hoop smaller than that before described:

3d. A properly shaped ladle I, Fig. 127, with which to run the cierges or bougies: and—

4th. A cullendered iron plate K, Fig. 128, to be placed over the coals, in case of necessity, for the moderation of the heat, comprise the requisite apparatus.

Fig. 128.

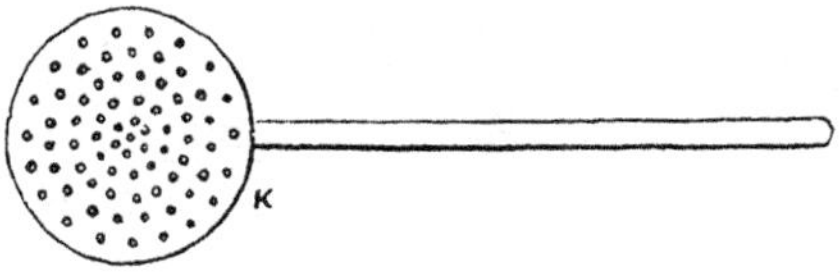

All being properly arranged, the workman, after having put a ferret on the end of each wick, hooks them to the ring, which, being held in the right hand so that the wick to be acted on shall be at the opening of the ledge, is dosed with melted wax from the iron ladle I, Fig. 127, replenished from the basin C, Figs. 124, 125, and held in and used with the left hand of the operative. When working upon the bougies or short cierges, the wax is lightly poured upon the wicks a little below their upper end, and thus are they consecutively treated; the excess of wax over that quantity requisite to entirely cover the wicks, falling and dripping back again into the basin, warmed beneath by the heated furnace afore-mentioned. It is necessary to make a remark here relative to the manner of running the wax. When the cierge being made is longer than can be conveniently handled by a man of ordinary height, the workman elevates himself upon a platform, so that his right shoulder is at the height of the hoop, and squeezing the wick, below the hook, between two fingers of the right hand, he imparts a twirling motion, whilst the melted wax is being poured on, so as to insure a uniform distribution of the wax over the circumference of the wick. He continues thus upon all the wicks which are placed around the hoop, so that when a revolution has been made, the wax that was poured or run upon the first candle has sufficiently cooled for the reception of a second layer; and in this way there must be ten or twelve repetitions before the cierge is finished. The three first jets should be given the whole length of the wicks; the fourth commenced a little below the top, and so on, decreasing the distance every jet, that the cierge may acquire a conical form. Except in case of a pressing demand, the cierge should not be completed in one continual operation, but, when nearly half finished, must be taken down from the ring and a cord passed through its collet and by that hung on hooks fastened to a strong board, and, thus arranged, left to cool thoroughly. This precaution is especially necessary in summer, for if in warm weather the running is finished at a single operation, the wax is liable to become separated from the wicks. Whilst one operative is removing the half finished cierges from the hoops, another replaces them with new wicks or with half made cierges that have entirely cooled, and are ready for the remaining quota of jets. A little experience enables the workman to run his cierge of any weight that may be desired, without unhooking it to test it by the scales. In order that the cierges may cool gradually, they should be put between cloths and covered with blankets, otherwise on rolling them they are apt to peel and become so hard as to be manageable with difficulty.

The rolling is done upon a walnut table with a wooden

planisher or small flat board, between which the cierges revolve, until they become even, smooth, and polished. Fig. 129 gives a perspective view of the planisher; Fig.

Fig. 129.

Fig. 130.

130 exhibits it in profile. The best wood for the planisher and top of the table, is that of the guaiac tree, which being very hard, imparts to the wax a brilliant smoothness and polish. The workman dampens the table with water occasionally, so that the wax, becoming warmed by the friction, shall not adhere to the wood. It requires a great deal of dexterity to roll the candles so as to make them smooth and round, by removing the inequalities which will inevitably happen during the running of them. Experience and care, however, gives great proficiency in these manipulations, which must be assisted occasionally, as before said, by wetting the table with water. Thus finished they are reduced to the proper and uniform length, with a knife, in manner as for the ordinary bougies. Bored at their base with a hole sufficiently large to fit upon the spike of the stand on which they are to be burned, the cierges are then ready to be packed for market These holes should be exactly in the centre, otherwise when the cierge is placed in the candelabra, it will not stand steady, and in order to counteract a proneness to cracking, the lower end should be bound around with a ribbon previously imbued with melted wax. These cierges are embellished and sometimes gilded; the ribbon binding, particularly, is generally colored red. Cierges are made, in France, bearing upon their circumference beautiful specimens of ornaments cast in relief.

Flambeaux differ from the cierges only in form, being made of wax and in the same manner. They are formed of four cylindric bougies, of equal size and equal length, united with melted wax, at first by pairs and then the two pairs together, and rounded at their base in the shape of an egg.

CHAPTER LIII.

PATENT CANDLES.

WITH a desire to present every improvement or attempted improvement, provided it possesses some merit, we have appropriated a chapter to "Patent Candles," intending to introduce under this head all such processes as have of late years been the result of invention. When we say all, it is intended to refer to such only as are either good in themselves, or present suggestions which may be valuable in aiding to remedy their own deficiencies, or to originate new and more perfect methods. We speak of each plan separately, and accompany them, respectively, with explanatory comments, without, however, deciding upon the practical advantages of any one of them, except where the author is thoroughly informed in the matter, and competent to give a judgment. The opening plan will be that for "azotized" candles, as the patentee calls them. Such a quack-like designation would be with scientific men rather prejudicial to a favorable opinion of the ability of Mr. Heard to concoct a valuable process. Whoever heard of tallow being azotized by nitric acid? this is not the effect produced; the change occasioned is that mentioned in the chapter treating of the action of acids upon the oils. The attempt to azotize fats with nitric acid, is like a late effort of a Philadelphia chemist, to oxidize nitric acid by the addition of peroxide of manganese. But to our subject:—here follow sketches of the patents; each tells its own tale.

"*Azotized candles.*"—Heard's process consists in the

use of nitric acid, very pure, and of specific gravity 1.500. The quantity he adds to each pound of fat, varies from a half drachm to three fourths of an ounce. When acting upon the best suet a half drachm of acid sufficed, whilst for the fats of soft consistency, the proportion had to be doubled. The suet is melted over a moderate fire, and the quantity of acid is then added, during continual stirring, until the fat has taken an orange tint, when the fire must be withdrawn, and the mass, when cooled, submitted to the action of an hydraulic press. An oily fluid is ejected by the pressure, and the solid suet remaining, though retentive of a yellow color, is easily bleached by the influence of air and light, and thus refined is formed into candles which are of good quality and do not run. The wicks are so prepared as to rarely require snuffing, and consist only of two strands, of size proportioned to the diameter of the candle.

When a wick burns, and its tip bends outside of the flame, it presents to the air the entire surface of all its volume, by the infinite number of its threads, and hence the oxygen component of the air acting upon the red heated carbon of the wick converts it into carbonic acid, and thus the wick is diminished sufficiently to continue a uniform combustion and illumination without the aid of snuffers. The use of the metallic wick is still more efficient, and its advantage is, that in burning, the carbonate of lead with which it is impregnated, becomes reduced to its metallic state by reason of the abstraction of all its oxygen, during the combustion, and a very small globule of lead is perceptible at the head of the wick, which, by its specific weight so bends the heated tip as to constantly expose it to the action of the air, and thus the light is kept brilliant and uniform by the continued removal of all redundant wick.

The same explanation applies to those other metallic

oxides with which wicks are sometimes impregnated. The preparation consists of one hundred parts of suet melted over a water bath, with which is thoroughly incorporated five to ten parts finely powdered subcarbonate of lead. A constant stirring with a spatula readily completes the mixture, and into this, whilst hot, the cotton wicks are dipped. Drawn out and cooled, they are then applicable to the foregoing, as also to the mould and dipped tallow candles.

The change produced by the aforementioned reaction is a crystalizable solid body called *elaïdine*, or a combination of elaïdic acid with oxide of glyceryl: this body being the result of the action of nitrous acid upon fat oils. A mixture of three parts of nitric acid of 38° B., with one part of nitrous acid, produces a similar solidification of the oil.

Only the fat oils experience a congelation from the action of nitrous acid; the siccative oils being only browned, without any sensible alteration by that reagent. As the oils which furnish the elaidine are generally constituted of glycerin and oleic and margaric acids, it is then evident that the elements of these two acids have contributed to the formation of the elaïdine, for it gives neither of them upon examination.

Candles with movable wicks.—Some years since, Mr. Bolts, an Englishman, invented a new mode of making candles and bougies, the products of which, it was said, equaled the best French candles in all the requisite properties. Mr. Bolt's candles are of particular forms, and are as represented by figures on next page. They can be varied at will, but the shapes most generally given, are those which we exhibit. The first is the most simple, Fig. 131, a solid cylinder or truncated cone, closely approximating the cylindric form. The body of the candle should consist entirely of combustible material, the wick

Figs. 131, 132, 133, 134, 135.

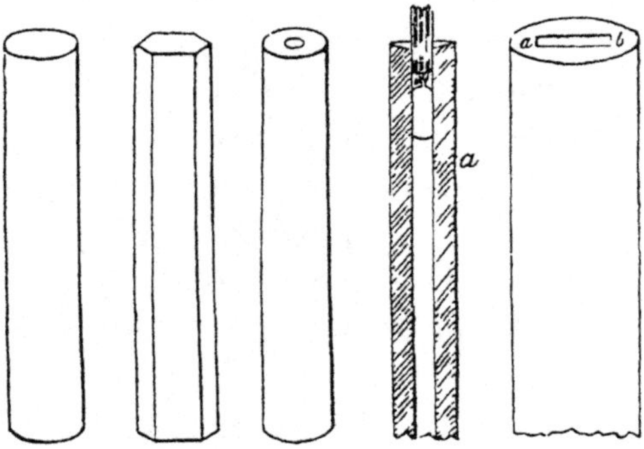

not being introduced until it is about to be lit. Another form given to the candle, (Fig. 132,) is a truncated pyramid with hexagonal base. This exterior angular shape is independent of the internal form of the candle. The third form is a perfect cylinder (Fig. 133) perforated in its centre as seen at Fig. 134, the diameter of this hole being regulated by the size of the wick. For general use a round hole is preferable, but the elliptic candles, (Fig. 135,) have great advantages. The opening, *a b*, should be in the form of a parallelopiped for the reception of a wide and slender wick, in order to the emission of a large flame of light.

The fourth and last form that we shall describe is that shown by Fig. 136. It is composed of two parts, an interior cylinder *a*, and a hollow external cylinder *b*, the opening of which, *c*, is of greater diameter by four millimetres than the interior cylinder. This last is enclosed by a ring or moulding around its base, of about twenty-five millimetres breadth and of sufficient thickness to exactly fill the hollow centre *a*, and thus preserve a concentric position when it is placed there, as show by *d* in Fig. 137. This concentric position, however, is only applicable to the lower part; indeed the remaining length of the cylinder, the internal *c*, should be filled by the

Figs. 136, 137, 138, 139.

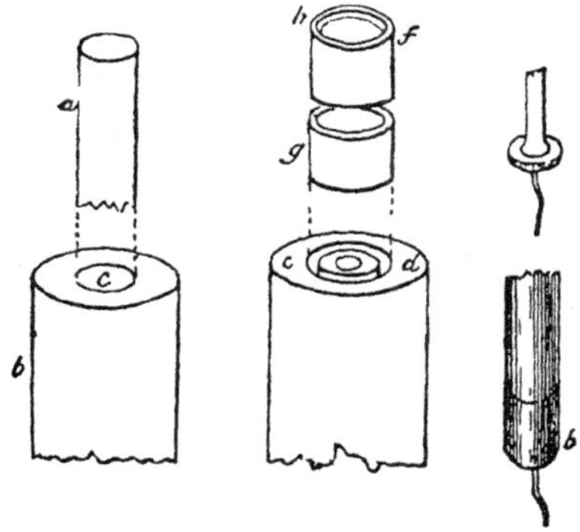

wick *f*, the annular form of which produces a slender circular flame, fed simultaneously both by the interior and exterior cylinders. The weight *g* is suspended by the wick *f*. This arrangement of the wick produces much more light than if it is imbedded as usual in the centre of the candle. This last form possesses the advantage of facilitating the excess of a double current of air to the circular flame; and if the interior cylinder *a*, Fig. 136, is made with an opening in its centre, as at *h*, Fig. 137, there will evidently be produced a current of air at each side of the flame, and the candles thus possess all the illuminating advantages of an Argand lamp. Fig. 138 shows a very profitable form for the wicks. It can be used either by hooking on a small weight beneath, as at *a*, Fig. 134, the size of which should nearly fill the circular hole in the centre of the candle, or by the use of an elastic spring placed in the bottom of the candlestick. The better and most usual form of movable wick is that shown at Fig. 139. Its whole length is from twenty-five to one hundred millimetres, according to that of the candle which it is to consume. It is previously impregnated with the same material of which the candle is made, in order to facilitate the first ignition. There is suspended beneath, a small weight *a*, Fig. 134, sufficiently heavy to cause the descent of the wick proportionally to the combustion. This wick is retained, without friction, by its collet *b*, Fig. 139, in the hollow centering the candle or bougie. In every case, the wick, by a little attention and care, is so thoroughly fed by the liquid fat as to require no snuffing, it being only necessary to remove the small quantity of carbon which oocasionally collects, after a long burning, at the top of the wick. It has been said above that the wicks should be previously saturated with the fat of which the candle is made; the better way is to impregnate but twenty-five millimetres of its length, leaving the rest dry, so that it may more readily suck up the liquefied fat in the surrounding basin, and which in ordinary candles is difficult of absorption by reason of the wick being already surcharged; hence an overflow of the excess causing the candles to run. Another defect in the manufacture of ordinary candles is the too great torsion of the wick, by which the easy penetration of the fluid fat is hindered. In those above recommended, the threads are placed longitudinally, and thus, independent of the capillary tube, which each thread is in fact, are formed, by the interstices between the threads, numerous other capillary tubes, which facilitate the ascension of the fluid fat towards the point of combustion, as well as the access of the air by the perforated centre; that which considerably augments the brilliancy of the flame. The application of these wicks to the candles after they have been formed, presents two other advantages, first the ability of properly proportioning the wick to the candle, so that during its combustion only so much of the tallow or wax is melted as is requisite for the nourishment of the wick. A few essays with wicks of different sizes speedily determine this important point, and the manufacture of the candles on a large scale can be regulated accordingly; secondly, the convenience of preparing the wicks previous to their introduction into the candles, in such manner as will tend to an increased brilliancy of flame or an emission of agreeable aroma.

There is a variety of methods for making candles upon this new principle; they can be formed, as usual, in the ordinary moulds, with some additional apparatus for the perforation of their centre. It is much better, however, to cast them in an assemblage of moulds, Fig. 140, with-

Fig. 140.

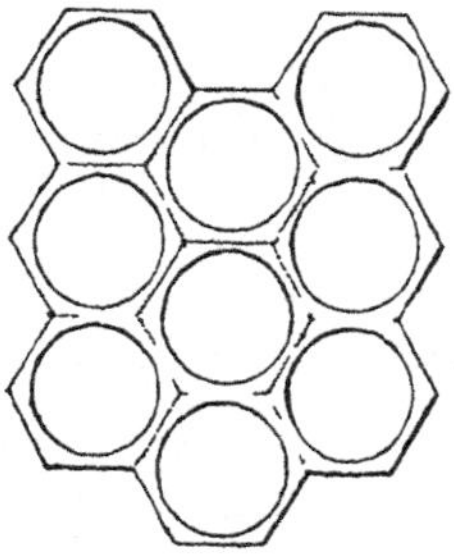

Fig. 141.

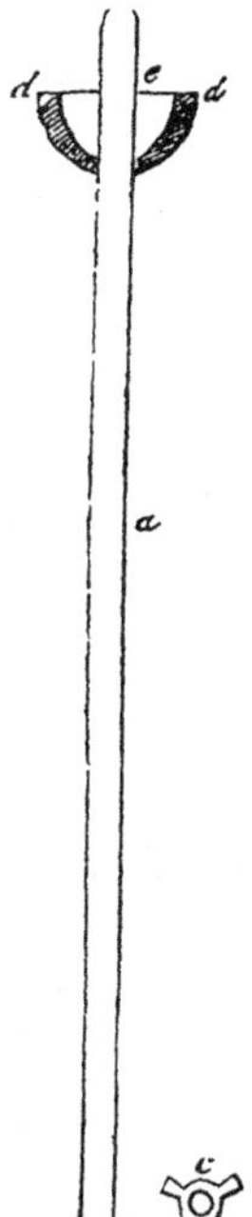

out intervals, and arranged with the least possible loss of space. These moulds are open at each end, and when the candles are being moulded, in order that their centres may be hollow, there is introduced an iron piercer *a*, Fig. 141, equal sized with the hole, and having screwed at its end a piston or plug *b*. This piston is not round but triangular, as is seen at *c*, and exactly fills the end of the mould, the three notches allowing a facility to the liquid fat of making its way into the mould, as will be seen. The other end of the piercer has a regulator *d d*, which is lowered until it reaches the top of the mould, in order to insure to the piercer a truly central position. The moulds, regulated and prepared as directed and shown by Fig. 140, are introduced into a chest just large enough for their admission, and but one inch higher. The melted tallow, at the proper degree of temperature, is then poured in until the moulds are completely submerged. If the candles are to be of ordinary quality, when cool, draw them by grasping the piercer by the handle part *e*, and lifting it upwards. The excess of tallow resting in the interstice between the moulds and sides of the box is left undisturbed; it serves to hold them steady for a subsequent operation.

When it is desired to subject the fluid matter to the action of a strong pressure, and by means of a vacuum, the trough, as soon as filled, must be slipped into a cast iron chest, to which is adapted a pneumatic apparatus, with a gauge and safety valve.

This, however, is an impracticable arrangement for manufacturing on a large scale, for, otherwise, if the pressure is so useful, it will be much more simple to employ that of a column of water aided by a suitably constructed apparatus.

The moulds may be of the usual material, and the most expeditious manner of making these candles seems to be to inject them with a syringe, through a tube, in the following manner.

When the wax or tallow is melted, it is poured into a double bottomed vessel, Figs. 142, 143, at the base of which is fixed a double acting force pump, A A A, and adapted to which is a hollow tube B, Fig. 143, of the shape of an ordinary mould. This mould tube is secured at the point *b* that it may be easily taken off and cleansed or changed; and when it is operated with for the facture of the hollow candles, it carries in its centre a well polished steel spike *e e*, of size according with that of the hollow. This spike or pin is fixed to *b*, and by the open

Fig. 142.

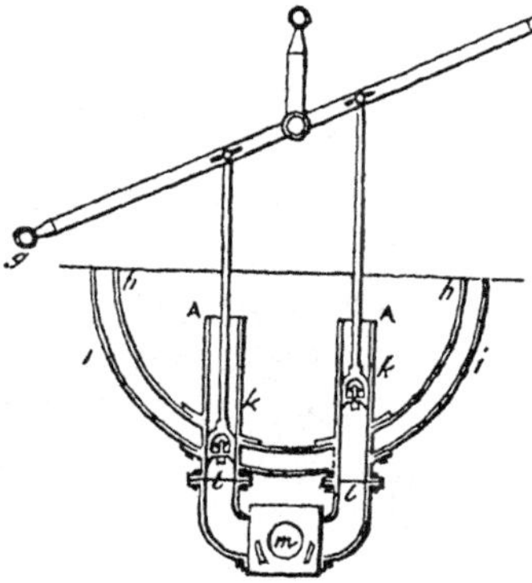

circular piece *d*, it is fastened by means of cross pieces, and a screw which insures its central position without disturbing the circulation of the melted substances. This spike or pin extends thirty centimetres beyond the end of the mould tube to the point *e*.

Fig. 143.

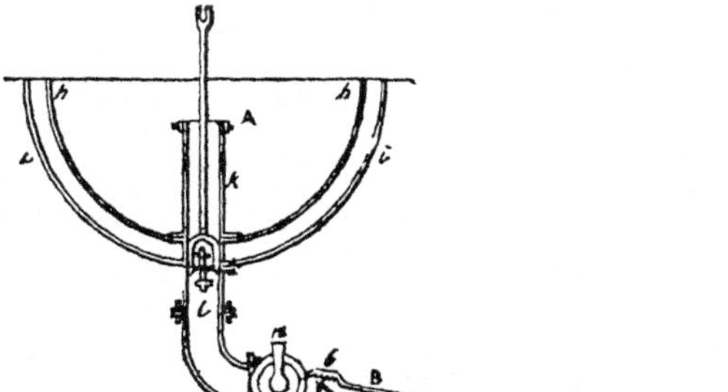

The temperature of the fused material should, for this operation, be regulated by the thermometer, and should be held midway between the points of congelation and fusion, by means of hot water placed in the double bottom before mentioned. The tubes leading from the base are also enveloped with an outer jacket communicating with the preceding, and consequently equally heated by hot water.

In this position the melted matter is driven by the pistons of the pump into a tube which forms a continual thread of tallow or stearin, traversing the mould tube B, under the form of a candle, perforated by the end of the rod *e*, extending into a tub of water, the heat of which is kept at about 10° F. lower than that requisite to produce a prompt cooling. The candle breaks off and falls into this tub.

The resistance of the water is counterbalanced and the straightness of the candles maintained, by a semi-cylindrical tube placed under the elongated rod *e*, upon the breadth of the tub, and which receives and conducts the ejected tube. As soon as the full length of a candle is out a workman cuts it at the point *e*, and lifts it up successively, to make place for the consecutive candle, and so in this manner the process is continued until all the melted tallow is used up. The tube candles, after having cooled, are cut off at a proper length, and the wicks, either dry or impregnated, are then applied.

The pumps are worked by a double lever *g g*. The bodies of the pumps *k k k*, are pierced throughout their length with holes for the admission of the melted suet; the pistons or suckers carry their valves reversed.

In the box *m* the valves of the two pumps are united.

Finally, at *n*, Fig. 143, is seen the safety cock, which is opened only at the moment of injection. The interior of the chest is designated by the letters *h h*; the exterior, or jacket, by the letters *i i*. This manipulation also enables the manufacture of a tube or casing of wax, which can afterwards be filled with tallow so as to form a low priced candle of handsome bougie like appearance.

Goddard's candles.—In 1845 Mr. Samuel Goddard, of

the city of London, proposed an improvement preventive of the guttering, or running of candles, whilst burning; consisting, first, in the peculiar modes of constructing the moulds in which the candles are to be cast; secondly, in the novel forms or shapes for the shafts of candles, which may be hollow for the introduction of dry wicks, or cast solid without wicks, or with wicks in the ordinary way; thirdly, in crystalizing the materials of which the candles are made; fourthly, in a novel method of covering or coating the surfaces of candles, if required, with a material for ornamental purposes; and. fifthly, in peculiar constructions of metallic caps to be applied to candles, in order to prevent their guttering whilst burning, or to conduct a descending wick.

In reference to the construction of moulds for the casting of candles, the inventor makes them of thin malleable metal, by first bending a suitable strip of the metal into a tubular shape, then soldering the edges together, so as to produce the perfect tube, and afterwards drawing the same to the required figure by the ordinary drawing process, upon a polished steel mandril. For the purpose of more readily discharging the cast candle from the moulds, the mandril may be made slightly tapering. These moulds may be made of a round, elliptical, or any other desired form, the main feature being to construct them of very thin malleable metal, drawn upon a mandril, for the purpose of operating upon the candle in the mould in the way hereafter described. In the construction of these moulds, zinc plates, tinned on their surfaces, and of about from the thirtieth to the fiftieth part of an inch in thickness, before drawing, have been found to answer the purpose. These moulds may be arranged in boxes, and made with tips or ribs, in the ordinary way; or they may be made with sliding tips or ribs, which will be necessary for discharging such candles as are made without wicks, from the moulds as above mentioned.

Fig. 144.

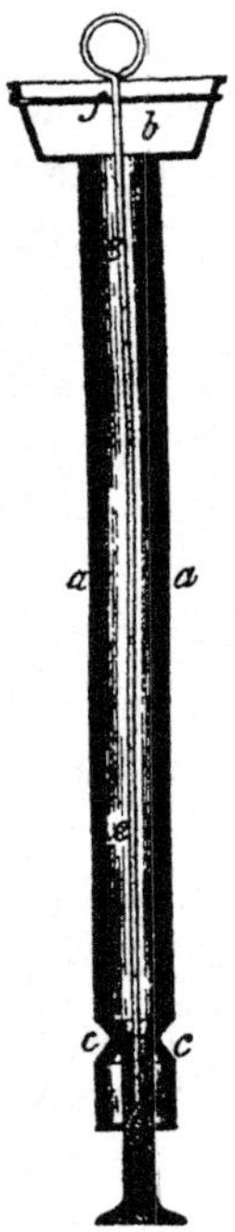

Fig. 144 represents in longitudinal section, one of the improved moulds. *a a* is the tube or shaft of the mould; *b*, the box or trough, into which the melted material is poured for casting the candle. The lower part of the tube is compressed at *c c*, to form a ledge or resting place for the sliding tip or nib *d*; which tip has a button at the outer end, by which it may be forced up, in order to relieve the candle previously to its being discharged from the mould, and to replace the rib again in its proper situation.

This rib *d*, slides through an aperture in the end of the mould, which aperture forms a guide. The mould here represented, is designed for casting hollow candles, through which a dry wick is to be drawn after the candle is made; and for this purpose, a long rod *e*, is inserted, its lower end resting in the centre of the rib, and the upper end being held by an adjustable guide *f*. By another form of rib, the end of which may be flat, or of any other required shape, solid shafts for candles, without wicks, may be cast in these moulds.

The novel forms or shapes for the shafts of candles, the inventor proposes to be, in some cases, elliptical, with flat wicks inserted, and manufactured in the ordinary way. In other cases, he proposes to make these elliptical shafts and also cylindrical as well as other shaped candles with an aperture through the centre, (for the reception of a dry wick to be hereafter inserted;) which aperture is formed by means of the central rod *e*, before mentioned. The patentee also makes elliptical and other shaped shafts for candles, solid, and without wicks or apertures for wicks; these candles are to be burned by wicks mounted in apparatus hereafter described.

The mode of crystalizing the material, such as spermaceti or other crystalizable substance, of which candles may be made, consists in employing, by preference, the thin metallic moulds such as have been above described, and immersing the moulds in hot water, immediately on their being filled with the melted fatty material; the moulds must be left immersed in the hot water, and allowed to cool as the temperature of the water becomes naturally reduced. By this means the material, at an uniform temperature throughout its mass, forms itself into crystals, which will produce upon the surfaces of the candles so operated upon, beautifully variegated appearances, and by the introduction of coloring matters, in small quantities, in the way well known, into the material, whilst melting, the variegated and beautiful appearance of the candle will be improved.

The novel mode of covering or coating the surfaces of candles with a material for ornamental purposes, is effected by the use of the above described thin metallic moulds. The moulds being filled with the melted material of which the coating of the candle is to be formed, are, immediately on being filled, to be immersed in cold water for one or two minutes, in order that a thin stratum of the material may become chilled, and incrusted on the internal surface of the mould. The thickness of this interior incrustation, will of course depend upon the length of time that the mould is immersed, and will be regulated according to the judgment of the workman. When the incrustation has been thus formed, the fluid part of the material must be poured out of the mould, and the mould then filled with such fatty matter to form the substance of the candle as may be desired; the moulds must then be immediately immersed in cold water again, in order to prevent the incrustation from melting and running into the other material; and having remained immersed in this cooling liquid for two or three minutes, according to the temperature of the water, the moulds are to be withdrawn, and placed in a cooling medium, surrounded by an atmosphere not lower than 70° F. This mode of treating the candles will prevent their cracking, and produce, particularly in spermaceti candles, a smooth face without the aid of wax. It is scarcely necessary to say, that the covering material may be colored in any way to suit the desire of the manufacturer. The correct performance of this process of covering candles with a very thin coating, depends upon the thinness of the metal of which the mould is constituted, and the sudden action of the cooling medium on the immersion of the mould; whereas it must be obvious that the same cannot be so delicately and rapidly effected by the employment of thick moulds, made in the ordinary way.

The peculiar constructions of metallic caps to be applied to candles for the purpose of preventing them from guttering whilst burning, and also for conducting a detached wick, will be seen by reference to Fig. 145, which represents in several views, a simple cap of thin metal, to be placed on the top of any candle, which, as it burns

Figs. 145, 146, 147.

will be prevented from guttering by the cap descending, the lip or rim around its edge guiding it, and preserving it in its proper position. This cap may be of any form suitable for the candle to which it is to be applied. In some cases the inventor proposes to attach to this cap a circular or elliptical tube of glass or mica, to act as a chimney, which may be supported by small arms extending from the cap, as shown in Fig. 146. This cap may also have a small metal tube standing in its centre, supported by small arms, for the purpose of holding a short piece of wick, which is to be saturated with some oleaginous material. The cap with its wick, being placed upon the top of a candle, formed with a hollow channel down its centre, as above described, or upon a solid candle as shown at Fig. 147, will, on being lighted, heat the fatty material below, and gradually descend as the candle burns; the tube and wick continually conducting up the fatty matter which will support the combustion as long as the candle lasts, and it will not require snuffing. The wick is also capable of being elevated or depressed in the tube, for the purpose of increasing or diminishing the light at pleasure.

We present the specification of the patent without comment, as the reader. after having progressed this far in the perusal of our work, will be as competent as the author, to decide upon its merits or demerits

Tuck's candles on continuous wicks.—In 1842, Mr. J. H. Tuck, of Middlesex county, England, obtained a patent for a peculiarly constructed apparatus, by which to make candles on continuous wicks. Fig. 148 represents a front elevation of the machine, all the several parts being in the positions they would occupy previously to filling the moulds. Fig. 149 is an end view of the same; and Fig. 150 a transverse section. The other figures given further on exhibit different views of some of the working

Figs. 148, 149, 150.

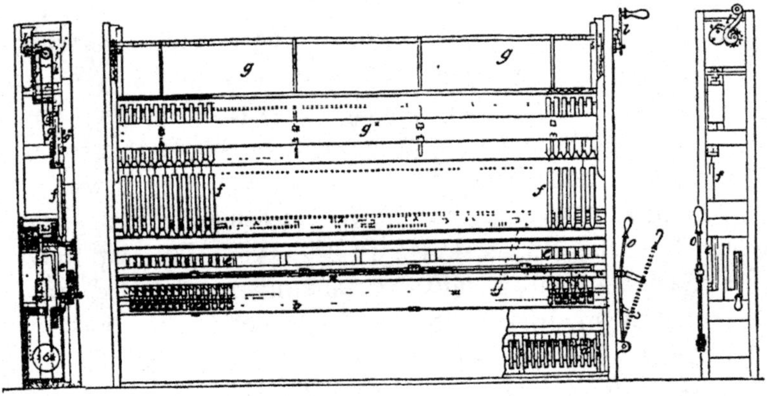

and internal parts of the machine, drawn to an enlarged scale. *a, a, a*, represent the bobbins or spools, upon which the continuous wick of cotton is wound; or balls of cotton may be used if preferred. These bobbins are mounted loosely upon axles, which hang in suitable bearings, attached to the frame-work. The continuous wicks from the bobbins *a, a, a*, pass upwards over a breast beam or board *b, b, b*, and through small holes *c, c, c*, made in the lower part of the stop cocks *d, d, d*, (see the detached views of this part of the apparatus, Figs. 151, and 152.) The wicks then proceed upwards through the cocks *d, d, d*, and through the centre of the moulds *e, e, e*, to the tips of which the cocks *d, d, d*, are connected as seen in the side views, Fig. 153, and in front view, Fig. 154.

Figs. 151, 152, 153.

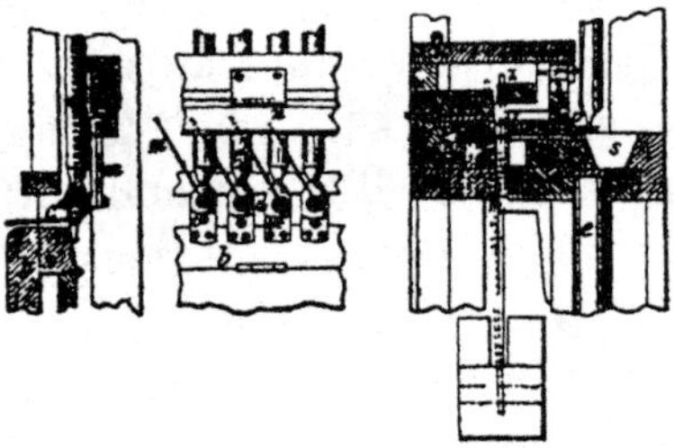

These stop-cocks may be constructed in any ordinary manner, but having a small hole through the plug, just large enough to allow the wick to pass easily into the cylindric tube or mould. They are furnished with spring levers *m*, *m*, *m*, one end of which is connected to the plug of the cock, and the other enters a hole or slat, made in a strip of thin iron, which is screwed or fixed on to the lower side of an horizontal sliding rail, *n n*. The end of this horizontal sliding rail, *n*, *n*, is connected by a joint, to the lower extremity of the hand lever *o*, *o*, (see Fig. 148;) and when this lever *o*, is brought from the position represented by dots in Fig. 148, into that shown by lines; the rail *n*, *n*, will be moved along, which movement will force the spring levers *m*, into the diagonal position represented in Figs. 148, and 152, and thereby close the several stop-cocks. It then becomes necessary to place the wicks above, in their proper position, in the centre of the cylindric moulds, and hold them up in this position, which is done by tying a knot in each, and attaching it to a hook in the horizontal rail *q*. Then a sliding rail *r*, is brought forward, covering about half of the apertures of the moulds, as shown in Fig. 153; and the rail *q*, at the same time receding, draws the wicks, in tension, into the vertical grooves, in the plate, on the face of the rail *r*.

The melted material may now be poured into the trough, extending along the part of the machine, as shown at *s*, *s*, *s*, when the candles will be formed by casting in the moulds. The materials having been allowed to remain in the moulds or cylindric tubes, until cold, the surplus matter, occupying the trough above the tubes or moulds, must be removed by a knife or instrument, suitably constructed, previous to "drawing the candles." The first row of candles having become cold and hard, and the positions of the various parts of the machine reversed,—that is, the wicks unhooked and the bar *r*, drawn back,—the candles, after being tapped or punched in the usual manner, are to be severally drawn out of the moulds, by the hand of the workman, who takes hold of the knot, and lifting it perpendicularly, the candle slips readily out of its mould, drawing the length of the cotton wick up through the mould, for the next row of candles. The knot of the wick, before mentioned, is then placed in a conical notch, cut in the rail above, at the bottom of the frame *g*, and, in that situation, the first row of candles hang, whilst a second row is cast in the moulds.

In order to carry out the manufacture of successive rows of candles in this apparatus, certain motions of the machine are required, which may be now explained, it being remembered that the stop-cocks have been opened by the movements of the machine, hereafter described, for the purpose of allowing a sufficient quantity of wick to be drawn up through the moulds, by drawing out the candles previously made, as described.

In order to open or loosen the stop-cock, it is necessary to reverse the lever *o*, *o*, and bring it into the position shown by dots in Fig. 148; and by this means, the spring levers *m*, *m*, will be drawn over into the opposite inclined position, and the wicks allowed to pass freely through the cocks.

The first row of candles, as at *f f f*, is suspended, in the manner above mentioned, by means of the knots from the lower side of the frame *g*, *g*, *g*, which frame is connected to endless chains *h*, *h*, (see Figs. 148, 149, and 150.) These chains pass over toothed pulleys *i*, *i*, and a similar frame *g*, *g*, is connected to the endless chains, *h*, *h*, one frame hanging in front of the machine, and the other a little distance behind, as is seen in Fig. 150. These frames are connected to the chains by swing pivots, so that when the front board is raised, the pivots pass over the pulleys *i*, *i*, and go into the back position, at the same time bringing the hinder board forward. On the same axle as the pulley *i*, a small cog-wheel *j*, is mounted; this wheel gears into and is turned by a small pinion *k*, to the axle of which power is applied, by means of the winch *l*, Figs. 148, and 149. Now, as the winch *l*, is turned, it causes the wheels and pinions, *i*, *j*, and *k*, to revolve, and consequently to raise to the highest point the frame *g*, and at the same time to lower, into its place, the other frame which was behind. This frame, by exchanging positions with the former, and being, as seen, advanced, causes all the conical notches to take hold of the taper ends of the candles of the upper row, and a straight board *g**, is then placed against the front of the row of candles, which holds them securely. These frames are furnished with a top and bottom row of notches, as seen in Fig. 1, formed so as to prevent the candles from falling sideways or dropping down.

The knots, by which this first row of candles were suspended from the conical notches of the upper frame *g*, are then cut off, by passing a sharp knife or other instrument under the lower edge of the top frame *g*, which is then passed into the back position. This operation having been performed, a fresh supply of wick will be found in the moulds *e*, *e*, *e*; and it now becomes necessary to tighten or close all the cocks *d*, *d*, *d*, which is effected as before mentioned, by reversing the position of the lever *o*, and bringing it into the position shown in Fig. 148. The next thing to be done is to place the wicks in the centre of the moulds, and to effect this readily, they are taken hold of, and held by the hooks *p*, *p*, *p*, on the horizontal bar or rail *q*. In order to take hold of the wicks, these hooks have to perform a forward, lateral, and a retrograde motion, at the same time the metal slide *r*, *r*, *r*, which is furnished with vertical grooves, (see Fig. 156,) is moved outwards; and when the hook bar *q*, *q*, retrogrades, it brings the wicks into the grooves of the metal slide *r*, *r*, *r*, and securely holds them there, the metal slide having, as before mentioned, been moved outwards, until it covers about one half of the aperture of the moulds, as seen in Fig. 153. These movements are effected in the following manner.

To the back part of the hook-bar *q*, a flat metal plate *t*, is screwed in an horizontal position (see Fig. 156); which plate is represented detached at Fig. 154. In this plate there is cut a peculiarly shaped slat *u*. The plate ex-

Figs. 154, 155, 156, 157.

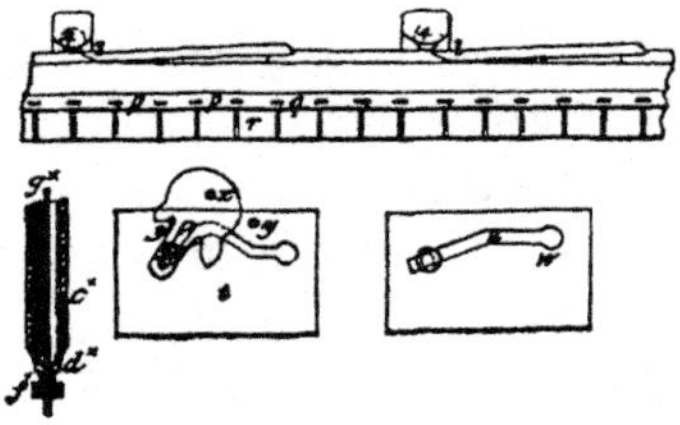

tends backwards through a groove or space cut in the sliding bar *v*, *v*; this bar supports an upright axle, which carries two horizontal brass pulleys *w*, *w*, that work in the slats cut in the plate *t*, and also in the one below it, hereafter to be described. A peculiarly shaped cone *x*, is mounted on a fixed centre, in the back framing of the machine; and the axle of the brass pulley *w*, works in a long slat made in the cam *x*, as seen in Fig. 154.

The metal plate *t*, is also furnished with pins or stops, *y*, *y*, which, by abutting against the cams, as aforesaid, prevent the plate from having a lateral motion, until the stop *y*, escapes from the ear of the cam. The sliding bar *v*, has a lateral motion, communicated to it by means of the hand lever *x*. This lateral motion causes the brass pulleys *w*, to advance along the slat *u*; and as the plate *t*, and hook bar *q*, to which it is attached, are prevented by the pin stops or studs *y*, from moving laterally, the said plate or hook-bar *q*, are forced outwards, until one of the studs *y*, escapes from the ear of the cam *x*, which it will do the moment the brass pulley has arrived at the apex of the angle of the slat *u*; then, as there is nothing to prevent it, the plate *t*, and hook bar *q*, move, or slide laterally with the assistance of the other stud *y*, until the end of the hook bar abuts against the end framing of the machine, when the continued forward motion of the rod *v*, and pulley *w*, pressing against the opposite inclined side of the slat *u*, causes the hook bar to retrograde; this motion of the pulley *w*, causes the hooks to traverse three sides of a square, viz., outwards, lateral and retrograde, and thereby to take hold of the wicks; which movement is completed by the advance of the metal slide *r*, *r*, simultaneously with the movement of the hook-bars. The metal slide *r*, *r*, is affixed to a wooden bar, to which is also affixed a horizontal metal plate 1, 1, similar in size and shape to the plate *t*, immediately beneath which it is situated as seen in Fig. 153. This plate 1, 1, shown detached at Fig. 155, has also a slat of peculiar form, one-half being inclined, and the other half, straight. The under brass pulley *w*, works in this slat in a way similar to the other pulley *w*, in the plate *t*. The ends of the metal slide *r*, *r*, fit accurately to the end framing, and thereby form a trough for receiving the melted material; and therefore, when by moving the sliding rod *v*, laterally, the brass pulley *w*, advances along the inclined slat, the plate 1, 1, and the metal slide *r*, *r*, to which it is connected, are obliged to move outwards, until the pulley *w*, arrives at the end of the inclined slat, when it will advance along the straight part of the slide, without acting upon the plate. The parts of the machine being thus situated, it only remains to make everything tight previous to pouring in the melted material; this is effected by shogging the lever *z*, *z*, to the position shown by dots in Figs. 150 and 153, when it will enter a notch cut in the back part of the slide 2, 2; on the front part of which several inclined planes 3,.3, are placed immediately beneath a corresponding number of rollers 4, 4, mounted in suitable bearings as seen at Fig. 156. Having shogged the lever *z*, it is now to be raised at one end, by which it will move the slide 2, 2, laterally, thereby forcing the inclined planes under the rollers 4, 4; the bearings of which being fixed, cause the metal slide *r*, *r*, to press tightly against the bed, and prevent any of the melted matenal from getting under it.

The apparatus is now quite ready for a fresh supply of the melted material, which is, as before mentioned, poured into the trough *s*, *s*; and when cold and ready to draw, the cocks *d*, *d*, are loosened by means of the lever *o*, *o*, and the wicks are disengaged from the hooks at the same time that the metal slide *r*, *r*, and the inclined planes 3, 3, are drawn back. After the candles have been tapped or punched, as before mentioned, the winch handle *l* is turned, and the empty frame *g*, above, is brought down to receive the second row of candles, and the full frame is at the same time raised to the highest point, thereby drawing the new candles out of the moulds. A front board or guard *g** is now put on the hooks in the frame *g*, as before, to prevent the candles from falling out. The top row is then disengaged from the wick, by simply passing a sharp knife below the frame *g*, which will thereby leave the candle a sufficient length of wick for lighting. The full frame may then be passed over to the back part of the machine and the candle removed therefrom at leisure.

In cases where considerable heat is not required to melt the tallow or candle material, the patentee substitutes a peculiarly formed tip to the moulds, in place of the stop-cocks before described. This construction is shown at Fig. 157, in which *c c* represent a portion of the candle mould, in vertical section. The tip, constituting the improved form, consists of a narrow tube and flange *d*, soldered or otherwise attached to the bottom of the mould. At the reverse end of this tube *d*, is formed a block *f*, through which the aperture of the tube is continued for the passage of the wick *g* (as above described with reference to the stop-cocks), the aperture being of such a diameter as to compress the wick rather tightly as it passes through it. By this arrangement the melted tallow or candle material will be prevented from passing beyond the tube *d*; for, on its arrival at the bottom thereof, the enlarged portion *f* will abstract the heat therefrom, and, consequently, reduce it to a solid state. The improved tip may be made of any suitable material, and the enlarged portion thereof, of any required shape or size, provided the bulk is sufficient to solidify the fluid material as it descends.

The apparatus employed for trimming or finishing the butts or bottoms and also for forming tops to such candles as may require it, is shown at Figs. 158 and 159. Fig.

Fig. 158.

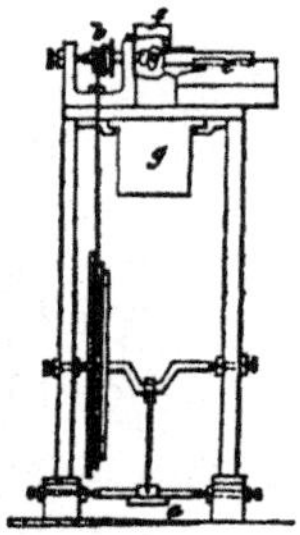

Fig. 159.

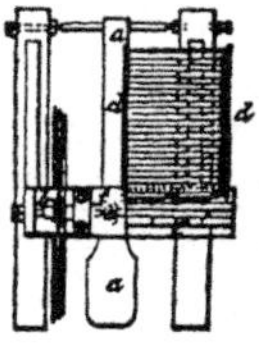

158 represents a front elevation of the machine, some of the parts being shown in section in order that they may be more clearly seen. Fig. 159 exhibits a plan view of

the same. The apparatus in most parts resembles a common lathe moved by a treadle *a a* in the ordinary manner. Rotary motion is communicated from the large pulley to the small pulley *b*, above, by means of a band. A knife or cutting instrument *c*, of peculiar construction, hereafter described, is screwed, or otherwise fastened to the axle of the small pulley *b*. The candles to be operated upon are placed in a series upon an inclined board *d d*, attached to the side of the machine, and they are severally brought from thence to a block *e*, having a curved groove at top to receive a candle. The candle is firmly held in this groove and its end pressed against the edge of the knife *c* by the hand of the attendant, or in any other convenient manner; and rotary motion being given at the same time to the knife by means of the treadle, the butt or bottom is thereby made perfectly smooth. The cuttings or shavings from the candle are confined to the cover *f f*, and made to fall into a box or receiver *g* below. The peculiarly shaped knife or cutting instrument *c*, is shown in plan at Fig. 160. The knife for forming the tips is shown detached from the holder at Fig. 161. Figs. 162 and 163 are two views of a knife

Figs. 160, 161, 162, 163.

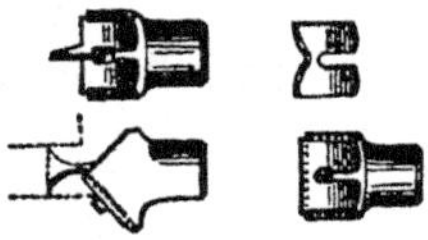

of somewhat different shape from the former, which may also be employed for the same purpose as Fig. 161.

Bougies which burn in water.—These are mostly used in nurseries or sick chambers, where it is requisite to keep a constant light during the night. They are made of wax and in manner the same as the dipped or ladled bougies; and that a vertical position may be

Fig. 164.

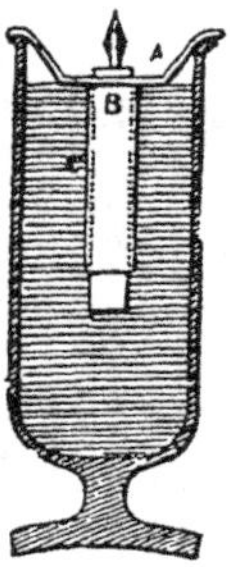

maintained in the water, they are fitted with a guide, which is a tin tube C, Fig. 164, of diameter slightly larger than the bougie, in order to a free motion, and soldered to a tinned iron plate A pierced in its centre with a hole B. This plate being placed upon a glass full of water, the bougie burns at the surface and always rises or mounts proportionably as it consumes. A slight improvement upon this arrangement renders the apparatus a time-teller, and, as such, a very convenient assistant to the nurse in cases of sickness, and as a substitute for a clock where medicine is to be administered at certain hours.

The modified candle burner is seen at Fig. 165. In

Fig. 165.

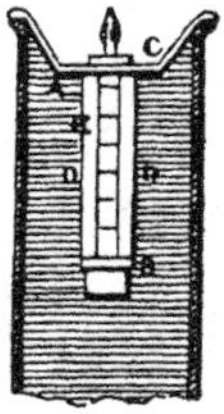

place of the entire tube there are two rings, A and B; the first, soldered upon the plate C, within it being fastened by one of their ends, four iron wires D D, equi-distant apart; the other ends are fastened on the other side of the ring B, so as to form by this means an open tube After having ascertained the amount of wax consumed in a bougie of determined and uniform size, a scale E is then made, the divisions of which are equal, and extended throughout a length pre-ascertained by experiment, and are each represented by figures 1, 2, 3, &c., commencing at the base. This scale must be managed so that it can slide easily upon the length of one of the iron wires that form the tube. When the bougie is lit, let it be at a moment simultaneous with the commencement of an hour marked by the time-keeper; the division 0 being coincident with the lower edge of the bougie. Proportionally as the candle rises, the scale indicates how much has been its consumption since it was first lit; it being only necessary to remember the hour at which you inflamed it to know very nearly the hour at the time of examining it during the night.

These kind of bougies can be made of length proportional to that of the night at different seasons of the year. They can be so that twenty to a pound shall last for ten or twelve hours; the thirty-two's nine hours; the forties eight hours, the sixties, four to five hours.

Perfumed candles.—These are of the ordinary candle material, aromatized by the addition of any essential oil that is grateful to the smell. When made of the pressed tallow, then, in order that they may not be greasy to the touch, a coating of varnish is applied to each. In this way an appearance is imparted equal to wax, whilst the candle is not necessarily of price exceeding that of the ordinary vaxeme light. The following are the ingredients, in proper proportions, for the varnish. Three parts, by weight, of gum damar (pinus dammara), a colorless resin, and two parts of white wax, thoroughly melted and mixed together.

The temperature requisite for the fusion of these ingredients forming the varnish being higher than that sufficient to melt the material of which the candle is made, they consequently burn without impediment, and while equal in gloss, feeling, and apparent hardness, to those of wax, have the advantage of costing but one-third their price.

CHAPTER LIV.

CONCLUDING REMARKS.

In closing our work we feel the conviction of having omitted nothing which might either directly or indirectly assist in furtherance of the art upon which we have written, and in the acquisition of a perfect knowledge of its principles and practice. The language and arrangement, too, of the recorded matter are such, we think and trust, as will be intelligible to the most ordinary intellectual capacity. It has been the author's desire to make it emphatically an useful and practical work. He has accordingly been guided throughout by that experience obtained by actual employment in the laboratory, which has taught him the inefficiency of books prepared by those who, though adepts in theory, are unskilful in the modes of imparting information to the operative. If after an attentive review of this book, the studious and ambitious savonnier should derive no benefit from its contents, the design of the author will then have failed in being realized, and he can only tender his humble regrets that the time consumed in writing and reading the work, has been productive of so little good to either party.

GLOSSARY.

Abreuvage.—A technical expression used by the French savonniers, for instance, when the boiling paste needs more ley, it is said to be "thirsty," or "dry," and the workman then drenches it with a fresh quantity or, in a word, gives it an abreuvage.

Achever.—The finishing dip.

Acrolein.—Formula $C_6H_4O_2$. A pungent oily liquid, resulting from the decomposition of glycerin in the destructive distillation of glycerated fats.

Accendible point.—That degree of temperature at which a combustible takes fire and inflames.

Acid Salts.—See Salts.

Adamantine.—As hard as adamant. A figurative title applied by Hancock & Mann to stearic candles on account of their great hardness and dryness.

Alcohol.—See next page.

Alembic.—Properly speaking, it is a glass vessel used in distillation, but the term is indiscriminately applied to stills of all sorts.

Alumino-ferruginous.—The colored portion which separates from white curd paste upon the addition of water. When thoroughly incorporated with the white stratum, it gives the cooled soap a mottled appearance.

Amygdalin.—In soap making, this word refers to substances, of which almonds or their oil are an ingredient.

Anhydrous.—A term used in contradistinction to hydrated, as expressive of the absence of water from substances. Those bodies which unite in definite proportions with water to form hydrates are true chemical compounds, and when perfectly freed of this water by any means become anhydrous. It is not synonymous with dry, for a body may contain chemically combined water, and yet be dry to the touch.

Azotized.—Bodies which comprise nitrogen among their components are termed azotized or nitrogenous.

Alcohol.—The spirit obtained by distillation of liquids that have undergone vinous fermentation. Its solvent and other useful properties make it a useful menstruum. The variation in the strength of commercial alcohol frequently renders it necessary to test its proof, which is generally done by the hydrometer. As a convenience we insert below

Lowitz's Table showing the quantity of Absolute Alcohol, by weight, in Spirits of different Specific Gravities.

100 parts.		Specific gravity.	100 parts.		Specific gravity.	100 parts.		Specific gravity.
Alcohol.	Water.	At 60° F.	Alcohol.	Water.	At 60° F.	Alcohol.	Water.	At 60° F.
100	0	0.796	66	34	0.881	32	68	0.955
99	1	0.798	65	35	0.883	31	69	.957
98	2	0.801	64	36	0.886	30	70	.958
97	3	0.804	63	37	0.889	29	71	.960
96	4	0.807	62	38	0.891	28	72	.962
95	5	0.809	61	39	0.893	27	73	.963
94	6	0.812	60	40	0.896	26	74	.965
93	7	0.815	59	41	0.898	25	75	.967
92	8	0.817	58	42	0.900	24	76	.968
91	9	0.820	57	43	0.902	23	77	.970
90	10	0.822	56	44	0.904	22	78	.972
89	11	0.825	55	45	0.906	21	79	.973
88	12	0.827	54	46	0.908	20	80	.974
87	13	0.830	53	47	0.910	19	81	.975
86	14	0.832	52	48	0.912	18	82	.977
85	15	0.835	51	49	0.915	17	83	.978
84	16	0.838	50	50	0.917	16	84	.979
83	17	0.840	49	51	0.920	15	85	.981
82	18	0.843	48	52	0.922	14	86	.982
81	19	0.846	47	53	0.924	13	87	.984
80	20	0.848	46	54	0.926	12	88	.986
79	21	0.851	45	55	0.928	11	89	.987
78	22	0.853	44	56	0.930	10	90	.988
77	23	0.855	43	57	0.933	9	91	.989
76	24	0.857	42	58	0.935	8	92	.990
75	25	0.860	41	59	0.937	7	93	.991
74	26	0.863	40	60	0.939	6	94	.992
73	27	0.865	39	61	0.941	5	95	
72	28	0.867	38	62	0.943	4	96	
71	29	0.870	37	63	0.945	3	97	
70	30	0.872	36	64	0.947	2	98	
69	31	0.874	35	65	0.949	1	99	
68	32	0.876	34	66	0.951	0	100	
67	33	0.879	33	67	0.953			

Barlongue.—This term applies to wicks whose threads or strands are uneven at their ends.

Ben, oil of.—An inodorous saponifiable oil obtained from the *mohringa optera.* It yields stearic, margaric, benic and moringic acids, and is much used in perfumery as a menstruum for the aroma of delicate flowers.

Bibasic.—See *Salts.*

Bougie.—Strictly speaking, a wax candle, but the term is used in a general sense, for instance, sperm bougie, stearic bougie, &c.

Carnivorous.—Flesh eating. Carnivorous animals are those which feed and live upon animal food.

Cetaceii.—The distinctive appellative of a family of sea-fish.

Compound radical.—A compound body assumed or known to act the part of an element, forming acids, bases, and their compounds, especially in organic chemistry. Thus cyanogen NC_2, resembles the halogen bodies (chlorine, &c.) and forms a large series of corresponding compounds. Ammonia, NH_3, when one equivalent of water, HO, is present, acts precisely like the oxide of an alkaline metal, which would be written $NH_3, HO{=}NH_4O$, and although the metal NH_4 has not been isolated, yet it is assumed to exist. In like manner glycerin acting the part of an oxy-base to the fatty and other acids, and its formula, without water, being $C_6H_7O_5$, it is assumed to consist of the radical C_6H_7 united to O_5, and in conformity with modern nomenclature, the hypothetic radical is termed glyceryl, and the glycerin itself, the oxide of glyceryl.

Cotyledon.—In botany, the seminal leaf of a plant forming part of the embryo. Some plants have two, and others only one cotyledon.

Cruciform.—A natural order of plants, the petals of whose flowers have the form of a Maltese cross, are thus called.

Culot.—The culot of a candle mould, is the cup which surrounds its base, and by which it is supported upon the bed plate of the stand. In the French moulds the culot is separate, as seen in Fig. 70. The moulds in this country are made without culots, and fastened in the stands by a narrow flange, which is sometimes cut with a screw thread, so that they may be more firmly adjusted.

Detergency.—The property of cleansing substances.

Diaphaneity.—Transparency. The diaphaneity of a candle means its wax-like transparency.

Egouttoir.—A dripping frame on which the baguettes of candles are hung, so that the redundant liquid tallow may drain off into a recipient.

Empatage.—This is the technical term of one of the operations in the process of soap making. Literally, it means, the transformation of the ingredients into a pasty magma.

Epine.—The movable valved plug with which the exit pipe of a vessel is closed. The pipe itself is called the epine tube, but instead of being opened and closed by a plug, is furnished with a stop-cock as a substitute.

Extempore.—Extemporaneous soaps are those made in the cold way, and without being boiled, and hence without much delay.

Ferret.—A small tin tube used to protect the lumignon of the wick in the process of ladling the bougies *à la cuiller*.

Filet.—The workman's name for the dose of water given to suet previous to its fusion in order to prevent its being scorched during liquefaction.

Flaques.—The soap paste is said to flaque when having consumed all the strong ley, it begins to sink or subside.

Guano.—The excrement of birds found upon certain small islands in the South Sea. Its nitrogenous and phosphated components render it valuable as a manure.

Herbaceous.—That order of perennial plants whose stem perishes annually.

Hydracids.—The non-oxygenated acids in which hydrogen is the electro-positive element. For example, the hydrochloric, hydroidic, and hydrosulphuric (sulphuretted hydrogen) acids.

Hydrate.—See *Anhydrous.*

Hypothetic radical.—See *Compound radical.*

Inorganic.—This term applies to substances from the mineral kingdom, and is in contradistinction to organic, which refers to such as exist in organized matter, and are destructible or capable of resolution into other organic bodies.

Isomeric.—Consisting of equal parts. Organic chemistry shows the existence of bodies having precisely the same composition, but possessing different properties. Such bodies are said to be isomeric.

Isomorphic.—Having the same form. Certain bodies are found to crystalize in the same form when similarly combined. Thus the sulphate and the chromate of potassa assuming the same shape of crystal, the sulphuric and chromic acids are isomorphic; magnesia, protoxides of iron, and manganese, &c., all assume the crystaline form when united with sulphuric acid and the same amount of water, and hence those bases are termed isomorphic.

Labiatæ.—An order of plants, some of which yield fragrant flowers.

Leyed.—When the combination of the oil and alkali is perfect, the paste is said to be leyed.

Litmus.—A blue pigment obtained from the lichen *Rocella.* Paper tinged by it is used as a test for the presence of acids which turn it red.

Lumignon.—The end of the wick protruding from the tip of the candle, and left free from tallow or wax, in order to facilitate ignition.

Marc.—The residual impure portions remaining after the expression of the liquid fat. It is also a general term applicable to the dross left as refuse, after pressure has separated the more valuable fluid portion of matter.

Marmite.—A closed metallic vessel fitted with a valve and heated under pressure.

Mesentery.—The membrane by which the intestines are attached to the vertebræ.

Mettre-pres.—The dip preceding the achever of dipped candles.

Monobasic.—See *Salts.*

Omentum.—The membrane attached to the stomach, and lying on the anterior surface of the intestines.

Plumula.—The growing point of the embryo and rudiment of the future stem of a plant.

Pollen.—The pulverulent matter contained in the anthers of flowers.

Rationale.—The theory of an operation or process, or, in other words, an explanation of the nature of the chemical changes which occur.

Regaler.—To regaler the ribbons of wax is to stir them about so as to present new surfaces to the influence of the light.

Ruminants.—That order of animals which chew the cud.

Salts.—Although there is a great variety of salts, for all practical purposes in the present arts, we distinguish two classes, haloid and oxy-salts. The former consist of two elements, a halogen body and a metal, such as chlorine and sodium, which form chloride of sodium or common salt. The latter consist of more than two elements, first united as an oxy-acid and an oxy-base, which again unite to form the salt. Thus, sulphur and oxygen form sulphuric acid, sodium and oxygen form soda, and sulphuric acid unites with soda to form Glauber's salt. In general, when the properties of neither the acid nor the base are perceptible, the salt is neutral; if acid predominate, it is termed an *acid salt;* if base predominate, a basic or *sub-salt*. According to one theory, an acid is mono or bibasic, according as it unites with one or two equivalents of base. Thus, stearic is termed a bibasic acid, and in the acid stearate of soda, there is a stearate of soda and stearate of water, which water in the neutral salt is replaced by soda.

Savonnier.—The French for soap-manufacturer, and being more expressive and shorter, is a far better title than the vulgar title of "soap boiler."

Sebacic acid.—Formula $C_{10}H_8O_3+HO$. A concrete crystalizable fat obtained by the distillation of oleic acid or fats containing olein. It is soluble in water, alcohol and ether, fuses at 261°, and forms salts with the bases.

Tavelée.—The workman's term to designate the speckled and undulated appearance of a badly dipped candle.

Turmeric.—The root of *Curcuma longa*, which yields a yellow dye. Paper stained therewith is used as a test for caustic and carbonated alkalies, which turn it brown.

INDEX.

T.

V.

W.

THE END.

Printed in Dunstable, United Kingdom

91219640R00308